Methods in Molecular Biology

Series Editor
John M. Walker
School of Life and Medical Sciences
University of Hertfordshire
Hatfield, Hertfordshire, AL10 9AB, UK

For further volumes:
http://www.springer.com/series/7651

ErbB Receptor Signaling

Methods and Protocols

Edited by

Zhixiang Wang

Department of Medical Genetics,
University of Alberta, Edmonton, AB, Canada

Humana Press

Editor
Zhixiang Wang
Department of Medical Genetics
University of Alberta
Edmonton, AB, Canada

ISSN 1064-3745 ISSN 1940-6029 (electronic)
Methods in Molecular Biology
ISBN 978-1-4939-7218-0 ISBN 978-1-4939-7219-7 (eBook)
DOI 10.1007/978-1-4939-7219-7

Library of Congress Control Number: 2017946170

Printed on acid-free paper

This Humana Press imprint is published by Springer Nature
The registered company is Springer Science+Business Media LLC
The registered company address is: 233 Spring Street, New York, NY 10013, U.S.A.

Preface

There are more than 90 known protein tyrosine kinase genes in the human genome; 58 encode transmembrane protein receptor tyrosine kinases (RTKs) distributed into 20 subfamilies. Among them, the ErbB receptor family, also known as the EGF receptor family or type I receptor family, includes the epidermal growth factor (EGF) receptor (EGFR) or ErbB1/Her1, ErbB2/Her2, ErbB3/Her3, and ErbB4/Her4. Among all RTKs, EGFR was the first RTK identified and the first one linked to cancer. Thus, EGFR has also been the most intensively studied among all RTKs.

ErbB receptors were first implicated in human cancer approximately three decades ago, when the avian erythroblastosis tumor virus was found to encode an aberrant form of the human epidermal growth factor (EGF) receptor (EGFR). Scientific communities have since developed a substantial understanding of the cell signaling mediated by ErbB receptors, and the biology underlying the dependence of cancers on aberrant ErbB receptor signaling. ErbB receptors are activated after homo- or heterodimerization. The ErbB family is unique among various groups of RTKs in that ErbB3 has impaired kinase activity, while ErbB2 does not have a direct ligand. Therefore, heterodimerization is an important mechanism that allows the activation of all ErbB receptors in response to ligand stimulation. The activated ErbB receptors bind to many signaling proteins and stimulate the activation of many signaling pathways, including the Ras-Raf-Mek-ERK, PI3K-Akt-Tor, PLC-γ1, STAT, and Src pathways. The specificity and potency of intracellular signaling pathways are determined by positive and negative regulators, the specific composition of activating ligand(s), receptor dimer components, and the diverse range of proteins that associate with the tyrosine phosphorylated C-terminal domain of the ErbB receptors. Through the control of these diverse signaling networks, ErbB receptors regulate many critical cellular processes, such as cell proliferation, cell differentiation, cell survival, cell metabolism, cell migration, and cell cycle.

Most of the research protocols have been developed to study the activation, dimerization, phosphorylation, interaction with other proteins, and the functions of RTKs. These protocols have been primarily developed in studying ErbB receptors, especially EGFR, as a model system. The protocols used to study the signaling of ErbB receptors may be easily adapted to study the signaling of all other RTKs and many non-receptor protein tyrosine kinases.

This volume contains protocols specifically designed for studying cell signaling mediated by ErbB receptors. These protocols apply to the study of a broad range of ErbB receptor-mediated signaling from basic research to clinic applications, from cultured cells to various animal models and primary cancer cells from patients. This book provides the most comprehensive protocols, not only for the study of cell signaling mediated by ErbB receptors but also for cell signaling which is mediated by other RTKs and beyond. This book includes five parts. Part I includes several reviews that provide a general overview of the field and updated knowledge regarding ErbB receptor signaling and its relevance to cancer. Part II provides the most common protocols for studying various aspects of ErbB receptor-mediated cell signaling. Part III includes newly developed methods in biomedical research that are also widely used in the study of ErbB receptor signaling. Part IV provides a protocol for studying

EGFR signaling in Drosophila. Finally, Part V provides important protocols for studying ErbB receptor signaling in various animal model systems.

This volume includes the most commonly used protocols for studying cell signaling that is mediated by ErbB receptors. All of the protocols have been obtained from researchers who either originally developed these protocols or modified and used these protocols. The protocols are very detailed and easy to follow. Thus, this volume may serve as a handbook for any researcher who is studying the cell signaling mediated by ErbB receptors and other RTKs. In addition, several reviews included in this volume provide the reader with up-to-date information in this continuously evolving field.

Edmonton, AB, Canada *Zhixiang Wang*

Contents

Contributors

COSTIN N. ANTONESCU • *Department of Chemistry and Biology, Ryerson University, Toronto, ON, Canada; Graduate Program in Molecular Science, Ryerson University, Toronto, ON, Canada; Keenan Research Centre for Biomedical Science of St. Michael's Hospital, Toronto, ON, Canada*

JING-WEN BAI • *The Breast Center, ChangJiang Scholar's Lab, Cancer Hospital of Shantou University Medical College, Shantou, China*

SUKHMANI BILLING • *Department of Biochemistry, McMaster University, Hamilton, ON, Canada*

BRIAN P. CERESA • *Department of Pharmacology and Toxicology, University of Louisville, Louisville, KY, USA*

SHYAMBABU CHAURASIYA • *Department of Oncology, University of Alberta, Edmonton, AB, Canada; City of Hope National Cancer Centre, Duarte, CA, USA*

CHAO CHEN • *Department of Medical Microbiology and Immunology, Li Ka Shing Institute of Virology, University of Alberta, Edmonton, AB, Canada*

ALEXIA CONTE • *IFOM, The FIRC Institute for Molecular Oncology Foundation, Milan, Italy*

KIMBERLEY GAUTHIER • *Department of Anatomy and Cell Biology, McGill University, Montreal, QC, Canada; Centre for Translational Biology, Research Institute of the McGill University Health Centre, Montreal, Canada*

KE GONG • *Department of Neurology and Neurotherapeutics, University of Texas Southwestern Medical Center, Dallas, TX, USA; VA North Texas Health Care System, Dallas, TX, USA*

JULIE A. GOSNEY • *Department of Pharmacology and Toxicology, University of Louisville, Louisville, KY, USA*

GAO GUO • *Department of Neurology and Neurotherapeutics, University of Texas Southwestern Medical Center, Dallas, TX, USA; VA North Texas Health Care System, Dallas, TX, USA*

AMYN A. HABIB • *Department of Neurology and Neurotherapeutics, University of Texas Southwestern Medical Center, Dallas, TX, USA; VA North Texas Health Care System, Dallas, TX, USA*

NICHOLAS HARDEN • *Department of Molecular Biology and Biochemistry, Simon Fraser University, Burnaby, BC, Canada*

MARY M. HITT • *Department of Oncology, University of Alberta, Edmonton, AB, Canada*

HONGXING JIANG • *Department of Surgery, University of Alberta, Edmonton, AB, Canada*

ANIL ABRAHAM JOY • *Division of Medical Oncology, Department of Oncology, Faculty of Medicine and Dentistry, University of Alberta and Cross Cancer Institute, Edmonton, AB, Canada*

YAO-CHEN LI • *The Breast Center, ChangJiang Scholar's Lab, Cancer Hospital of Shantou University Medical College, Shantou, China*

STEFANIE LUCARELLI • *Department of Chemistry and Biology, Ryerson University, Toronto, ON, Canada; Graduate Program in Molecular Science, Ryerson University, Toronto, ON, Canada*

SHEENA H. MACLEOD • *Department of Oncology, University of Alberta, Edmonton, AB, Canada*

HAMID MAADI • *Department of Medical Genetics, University of Alberta, Edmonton, AB, Canada*

BABAK N. MOLLALOU • *Department of Medical Genetics, University of Alberta, Edmonton, AB, Canada*

BABAK NAMI • *Department of Medical Genetics, University of Alberta, Edmonton, AB, Canada*

STEVEN PENNOCK • *Department of Medical Genetics, University of Alberta, Edmonton, AB, Canada; Regeneron Pharmaceuticals Inc., Tarrytown, NY, USA*

KYLE G. POTTS • *Department of Oncology, University of Alberta, Edmonton, AB, Canada*

BRUCE REED • *Department of Biology, University of Waterloo, Waterloo, ON, Canada*

CHRISTIAN E. ROCHELEAU • *Department of Anatomy and Cell Biology, McGill University, Montreal, QC, Canada; Centre for Translational Biology, Research Institute of the McGill University Health Centre, Montreal, Canada; Division of Endocrinology and Metabolism, Department of Medicine, McGill University, Montreal, Canada*

RALPH CHRISTIAN DELOS SANTOS • *Department of Chemistry and Biology, Ryerson University, Toronto, ON, Canada; Graduate Program in Molecular Science, Ryerson University, Toronto, ON, Canada*

SARA SIGISMUND • *IFOM, The FIRC Institute for Molecular Oncology Foundation, Milan, Italy*

JUNFENG TONG • *Department of Medical Genetics, University of Alberta, Edmonton, AB, Canada*

YI WANG • *Canadian Nuclear Laboratories, Chalk River, ON, Canada; Signal Transduction Research Group, Department of Medical Genetics, Faculty of Medicine and Dentistry, University of Alberta, Edmonton, AB, Canada*

QIAN WANG • *Department of Pharmacology, Faculty of Medicine and Dentistry, University of Alberta, Edmonton, AB, Canada*

ZHIXIANG WANG • *Department of Medical Genetics, University of Alberta, Edmonton, AB, Canada*

PING WEE • *Department of Medical Genetics, University of Alberta, Edmonton, AB, Canada*

GUO-JUN ZHANG • *The Breast Center, ChangJiang Scholar's Lab, Cancer Hospital of Shantou University Medical College, Shantou, China*

YONG-QU ZHANG • *The Breast Center, ChangJiang Scholar's Lab, Cancer Hospital of Shantou University Medical College, Shantou, China*

XIAOFU ZHU • *Division of Medical Oncology, Department of Oncology, Faculty of Medicine and Dentistry, University of Alberta and Cross Cancer Institute, Edmonton, AB, Canada*

Part I

Introduction and Review

Part I

[illegible]

Chapter 1

ErbB Receptors and Cancer

Zhixiang Wang

Abstract

The ErbB receptor family, also known as the EGF receptor family or type I receptor family, includes the epidermal growth factor (EGF) receptor (EGFR) or ErbB1/Her1, ErbB2/Her2, ErbB3/Her3, and ErbB4/Her4. Among all RTKs, EGFR was the first RTK identified and the first one linked to cancer. Thus, EGFR has also been the most intensively studied among all RTKs. ErbB receptors are activated after homodimerization or heterodimerization. The ErbB family is unique among the various groups of receptor tyrosine kinases (RTKs) in that ErbB3 has impaired kinase activity, while ErbB2 does not have a direct ligand. Therefore, heterodimerization is an important mechanism that allows the activation of all ErbB receptors in response to ligand stimulation. The activated ErbB receptors bind to many signaling proteins and stimulate the activation of many signaling pathways. The specificity and potency of intracellular signaling pathways are determined by positive and negative regulators, the specific composition of activating ligand(s), receptor dimer components, and the diverse range of proteins that associate with the tyrosine phosphorylated C-terminal domain of the ErbB receptors. ErbB receptors are overexpressed or mutated in many cancers, especially in breast cancer, ovarian cancer, and non-small cell lung cancer. The overexpression and overactivation of ErbB receptors are correlated with poor prognosis, drug resistance, cancer metastasis, and lower survival rate. ErbB receptors, especially EGFR and ErbB2 have been the primary choices as targets for developing cancer therapies.

Key words ErbB family receptors, EGFR, ErbB2, ErbB3, ErbB4, Cell signaling, Cancer

1 Introduction

ErbB receptors were first implicated in human cancer approximately three decades ago, when the avian erythroblastosis tumor virus was found to encode an aberrant form of the human epidermal growth factor (EGF) receptor (EGFR). Scientific communities have since developed a substantial understanding of ErbB receptor-mediated cell signalling, and the biology underlying the dependence of cancers on aberrant ErbB receptor signaling. Now, the ErbB receptor family consists of four members, EGFR/HER1/ErbB1, HER2/ErbB2, HER3/ErbB3, and HER4/ErbB4 [1, 2]. ErbB receptors are activated after homodimerization or heterodimerization [2–4]. The ErbB family is unique among the various

Zhixiang Wang (ed.), *ErbB Receptor Signaling: Methods and Protocols*, Methods in Molecular Biology, vol. 1652, DOI 10.1007/978-1-4939-7219-7_1,

groups of receptor tyrosine kinases (RTKs) in that ErbB3 has impaired kinase activity, while ErbB2 does not have a direct ligand. Therefore, heterodimerization is an important mechanism that allows the activation of all ErbB receptors in response to ligand stimulation [2, 5–7]. The activated ErbB receptors bind to many signaling proteins and stimulate the activation of many signaling pathways, including the Ras-Raf-Mek-ERK, PI3K-Akt-Tor, PLC-γ1, signal transducer and activator of transcription (STAT), and Src pathways (reviewed by [8]). The specificity and potency of intracellular signaling pathways are determined by positive and negative regulators, the specific composition of activating ligand(s), receptor dimer components, and the diverse range of proteins that associate with the tyrosine phosphorylated C-terminal domain of the ErbB receptors [9].

Except for ErbB4, the aberrant activation of ErbB receptor kinase activity contributes to the tumorigenesis and progression of many cancers, including non-small cell lung cancer (NSCLC, squamous), head and neck, glioma, breast, esophageal, colon, colorectal, anal, lung, gastric, bladder, endometrial, melanoma, medulloblastoma, prostate, pancreas, and ovary cancers (reviewed by [10]).

2 ErbB Receptors

The four members of the ErbB receptor family are among the approximately 60 receptor tyrosine kinases (RTKs) that have been found in the human genome [11]. Like other RTKs, ErbB receptors are composed of an N-terminal extracellular ligand-binding domain, a single transmembrane helix, and a cytoplasmic domain containing a tyrosine kinase domain followed by a C-terminal regulatory domain [12]. The extracellular domain can be divided into four subdomains, including domain I (L1), II (CR1), III (L2), and IV (CR2). L1 and L2 are believed to represent ligand-binding regions. CR1 and CR2 are regions that are very rich in cysteine residues and responsible for receptor dimerization (Fig. 1a, b) [13].

2.1 EGFR

EGFR was first identified in the 1970s [14] and subsequently, its intrinsic kinase activity was revealed. In 1984, the full-length receptor was cloned [15]. Shortly thereafter, it was discovered that the expression of EGFR could lead to cell transformation [16]. Also, it has been shown that many cancer cells are characterized by EGFR hyperactivation, overexpression, or mutants with dysregulated signaling (reviewed by [10]). These EGFR-dependent perturbations are also correlated with poor patient prognosis [17–19]. Consequently, EGFR and its signaling activity have been targets for developing novel therapeutic drugs to treat a variety of cancers.

EGFR is a 170 kDa transmembrane glycoprotein composed of a single polypeptide chain. The heavily glycosylated 622-amino acid

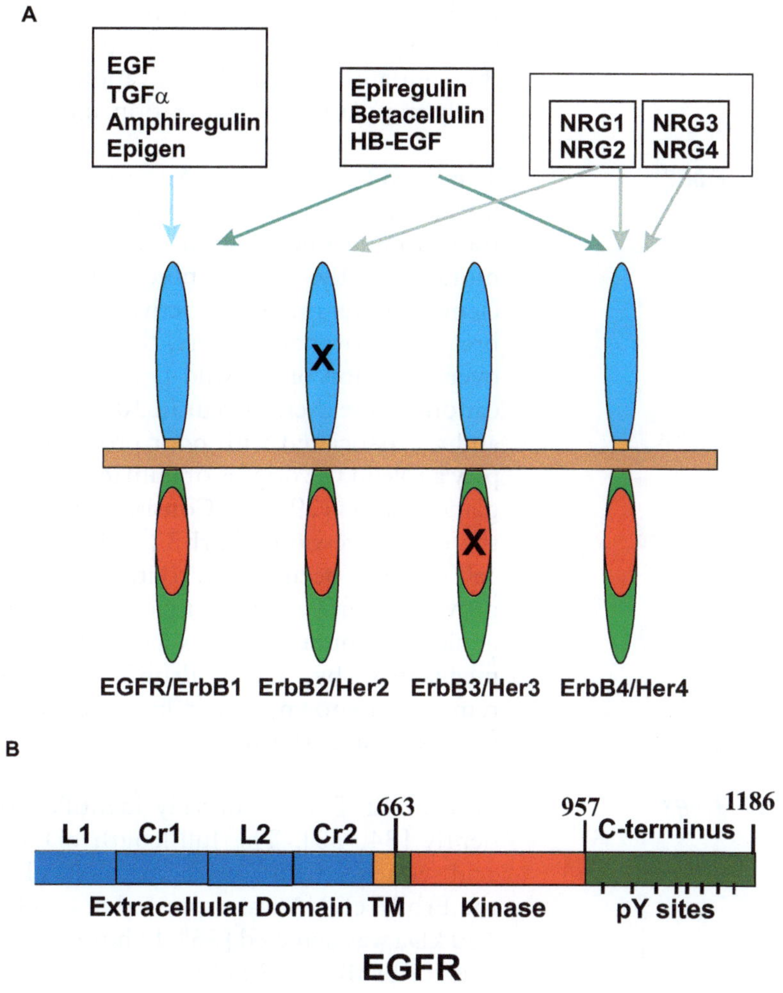

Fig. 1 ErbB receptor family and their ligands. (**a**) ErbB receptors and their ligands. The ErbB receptor family is composed of four members: EGFR/ErbB1/Her1, ErbB2/Her2/neu, ErbB3/Her3, and ErbB4/Her4. Eleven ligands have been identified for the ErbB receptor family. (**b**) Linear structure of EGFR. TM: transmembrane domain

extracellular domain contains two cysteine-rich regions and is responsible for ligand binding. The transmembrane domain is a single 23-residue α-helical transmembrane peptide. The 542-amino acid intracellular cytoplasmic domain contains a 250-amino acid conserved protein tyrosine kinase core, followed by a 229-residue C-terminal tail with regulatory tyrosine residues [20] (Fig. 1b). The binding of EGF at the cell surface induces the dimerization of EGFR, which results in the activation of EGFR tyrosine kinase activity and receptor trans-autophosphorylation [21, 22]. Sites of tyrosine autophosphorylation in the activated

EGFR interact with downstream signaling proteins to form large signaling complexes. The receptor-signaling protein complexes then initiate the activation of various signaling pathways, and eventually stimulate cell proliferation and survival.

2.2 ErbB2

The ErbB2 gene encodes a 185 kDa transmembrane receptor with intrinsic tyrosine kinase activity. ErbB2 was originally identified as a transforming oncogene in rat neuroglioblastomas, where a point mutation in the transmembrane domain of the receptor was sufficient to confer oncogenic activation [23, 24]. In contrast, in human breast and ovarian cancers, the receptor is activated through the overexpression of the wild-type gene, mostly due to gene amplification. These events occur in 20–25% of breast and ovarian cancers and are associated with poor prognosis [25–28]. ErbB2 may also play a role in the origins of human gastric, endometrial, and salivary gland cancers [29, 30]. Consistent with these clinical observations, the overexpression of ErbB2 in human breast and ovarian cancer cell lines has been shown to increase DNA synthesis, promote cell growth, improve soft agar cloning efficiency, and increase tumorigenicity in nude mouse xenograft models [31–33]. The precise mechanism by which ErbB2 overexpression transforms cells remains unknown, but it likely involves the activation of signal transduction pathways.

2.3 ErbB3

Human ErbB3 was initially identified by two groups, independently [34, 35]. The full-length cDNA encodes a 1342-amino acid sequence with a predicted molecular mass of 148 kDa. When the ErbB3 cDNA was expressed in COS7 or 293 cells, a protein of 160 kDa was detected [35]. In human cells, however, the molecular mass of ErbB3 is 180 kDa [36, 37], and the additional mass is due to posttranslational modifications by glycosylation [38] and phosphorylation [39]. The transmembrane domain contains a motif that is possibly involved in receptor packing during dimerization. The cytoplasmic domain contains a region that is broadly homologous to the catalytic domain of tyrosine kinases and 13 tyrosine residues that are possible sites of phosphorylation [40]. Unlike the other ErbB family members that are activated through autophosphorylation upon binding ligand, ErbB3 was shown to be kinase impaired, with only 1/1000 the autophosphorylation activity of EGFR, and incapable of phosphorylating other proteins [41]. Therefore, ErbB3 must act as an allosteric activator.

ErbB3 has been linked to cancer, mainly due to its mechanistic role in promoting oncogenic HER2 and EGFR signaling. Recently, however, somatic mutations have been found scattered throughout the ErbB3 gene in subsets of breast and gastric cancers [42]. Also, ErbB3 has been found to be overexpressed in some other cancers, including breast and bladder cancers [39].

2.4 ErbB4

ErbB4 is a 180 kDa glycoprotein. Unlike ErbB2, which cannot directly bind a ligand, and ErbB3, which does not have a functional kinase domain, ErbB4 is a fully functional RTK capable of signaling, both as a homodimer and heterodimer. ErbB4 can be activated by at least seven members of EGF-related peptide growth factor family. Activated ErbB4 forms a homodimer or heterodimerizes with another ErbB family member [43].

The ErbB4 receptor is unique among ErbB receptors in that it undergoes proteolytic processing after ligand binding. The cleaved 80 kDa intracellular domain enters the nucleus to participate in the regulation of gene transcription [151]. ErbB4 is necessary for the development of the heart, mammary gland, and the central nervous system. ErbB4 has also been implicated in diseases such as cancer, and cardiovascular and psychiatric disorders [43].

3 Ligands of ErbB Receptors

Members of the EGF family of peptide growth factors bind to ErbB receptors, serve as agonists for ErbB family receptors and are grouped into three categories based on their binding partners. One category, including EGF, transforming growth factor α (TGFα), amphiregulin (AR), and epigen (EPG), specifically binds to EGFR. A second category binds to both EGFR and ErbB4, and includes epiregulin (EPR), betacellulin (BTC), and HB-EGF. The third category includes all of the neuregulins (NRG1–NRG4). Among them, NRG1 and NRG2 bind to ErbB3 and ErbB4, whereas NRG3 and NRG4 only bind to ErbB4 (Fig. 1a) [10, 44–48].

The ErbB ligands are all expressed as single-pass integral membrane proteins [49]. These ligand precursors consist of an extracellular component, a transmembrane segment, and a small intracellular portion. The extracellular segment of the precursor is cleaved and released through proteolysis by members of the ADAMs (A disintegrin and metalloproteases) family [50]. The proteolytic processing and the release of membrane proteins function as a posttranslational switch that regulates the activity of the growth factor. This process is called protein ectodomain shedding, and the proteolytic enzymes are sometimes referred to as sheddases. In addition to growth factor processing, the ADAMs also process growth factor receptors, cytokines, cytokine receptors, and cell adhesion molecules. This family contributes to various physiological processes, including adipogenesis, fertilization, myogenesis, and neurogenesis [48, 51].

Together, these ligands regulate the activity of the ErbB receptors, each of which appears to contribute in a unique manner to a complex signaling network [2]. These ligands can stimulate different biological outcomes from the same receptor. One possible mechanism is by stimulating receptor phosphorylation on distinct sets of tyrosine residues. It has also been suggested that the conformation of the ligand-bound extracellular domain of the receptor is also important for the distinct patterns of ErbB receptor tyrosine phosphorylation and downstream signaling [52]. The expression of some EGF family members on tumor cells, most notably TGFα, AR, and HB-EGF, is associated with poorer patient prognosis or resistance to chemotherapeutics [53–57].

4 Receptor–Ligand Interaction and Receptor Dimerization

The extracellular modules of all four ErbB receptors have been crystallized and their structures determined with either bound ligand (for EGFR and Her4) or without ligand (for all four receptors), as well as in complex with antibodies or antibody mimics [12, 58, 59]. Together with the findings from other studies, a comprehensive picture regarding the interaction between ErbB receptors and their ligands has emerged. The four ErbB receptors can form ten different homodimers and heterodimers [1, 60].

The structural studies have revealed some key features of the ErbB receptor dimerization [12, 59]. First, for the dimerization of EGFR, while the ligand introduces the conformational change, it is the receptors themselves that form the dimer interface. Second, the receptors can only exist in two conformations: an extended form and a tethered form. The extended form represents the form of the receptor that is ready to dimerize and the tethered form represents the form in which the dimerization element is buried. Third, the mechanisms underlying the restriction to these two conformations are due to the rigid nature of the various extracellular domains, including domains I, II, III, and IV. As a result, the extracellular module appears to "click" into either the extended form priming for dimerization or the compact, tethered form that is not able to dimerize.

It is important to point out a unique feature of ErbB2 structure and its implication in receptor dimerization. Crystal structures of the orphan ErbB2 extracellular region have revealed that it has an extended configuration, where the arrangement of the four domains is similar to that seen in each molecule of the EGFR dimer. Therefore, ErbB2 resembles a constitutive, or ligand-independent, activated conformation. This is consistent with the fact that ErbB2 homodimers can spontaneously form in ErbB2-overexpressing cells, ErbB2 is the preferred dimerization partner for all of the other ErbB receptors [1, 48], ErbB2 (but not EGFR) overexpression transforms cells, and ErbB2 overexpression is associated with a high percentage of breast cancer and poor prognosis [60]. It appears that ErbB2 extracellular domains are already in extended form in the absence of ligands. Domains I and III of

ErbB2 interact directly to stabilize ErbB2 in the extended form. Since there is no room for a ligand between domains I and III, it is not surprising that ErbB2 is an orphan receptor [60].

On the other hand, the ErbB3 homodimer is apparently non-functional as it lacks kinase activity. However, since the kinase-impaired ErbB3 possesses 1/1000th of the phosphorylation activity of EGFR, it is not clear if the ErbB3 homodimer is functional [48].

5 Phosphorylation of C-Terminal Tyrosine Residues

Members of the EGF family differentially stimulate the coupling of receptors to signaling effectors by stimulating receptor phosphorylation on distinct sets of tyrosine residues, most of which are located in the C-terminal noncatalytic sequence [48, 52, 61]. Large-scale phosphoproteomic screening has revealed that ErbB receptors can potentially bind >100 proteins. Figure 2 illustrates the putative sites of EGFR, ErbB2, ErbB3, and ErbB4 tyrosine phosphorylation. It has been shown that all ErbB receptors have multiple tyrosine residues that are phosphorylated following their activation. Figure 2 also illustrates the intracellular signaling effectors that have been predicted or shown to bind to these phosphorylated sites [48, 52, 61–68]. The mapping of these tyrosine phosphorylation sites has revealed several interesting features. First, many individual phosphotyrosine residues of EGFR and ErbB4 are able to bind multiple downstream effector proteins. For example, EGFR phospho-Y974 (pY974) binds to SHC, STAT5, PTP-2c, Crk, and Src, and EGFR pY1173 binds to SHC, Grb2, PLC-γ1, and SHP1. Similarly, ErbB4 pY1150 binds to Abl, Syk, Crk, and RasA1, and ErbB4 pY1162 binds to Grb2, Abl, and Vav2. Second, for all four receptors, there are multiple phosphotyrosine residues that bind to the same effector. For example, EGFR pY1068, pY1086, pY1148, and pY1173 all bind to Grb2. ErbB2 pY1005, pY1196, pY1222, and pY1248 all bind to SHC. ErbB3 pY1054, pY1197, pY1222, pY1260, pY1276, and pY1289 all bind to the p85 subunit of PI3K. ErbB4 pY1162, pY1188, pY1202, pY1208, pY1221, pY1242, and pY1268 all bind to Grb2. Third, different ErbB receptors have different potentials in activating distinct downstream signaling pathways. For example, it has been shown that EGFR, ErbB2 and ErbB4 strongly activate the Ras/ERK pathway due to the multiple binding sites for Grb2 and SHC. ErbB3 strongly activates the PI3K/Akt pathway due to the multiple binding sites for the p85 subunit of PI3K. In addition to the Ras/ERK pathway, ErbB4 also strongly activates the Abl signaling pathway due to the multiple binding sites for Abl.

The activation/phosphorylation of ErbB receptors and the recruitment of signaling effectors are also significantly affected by

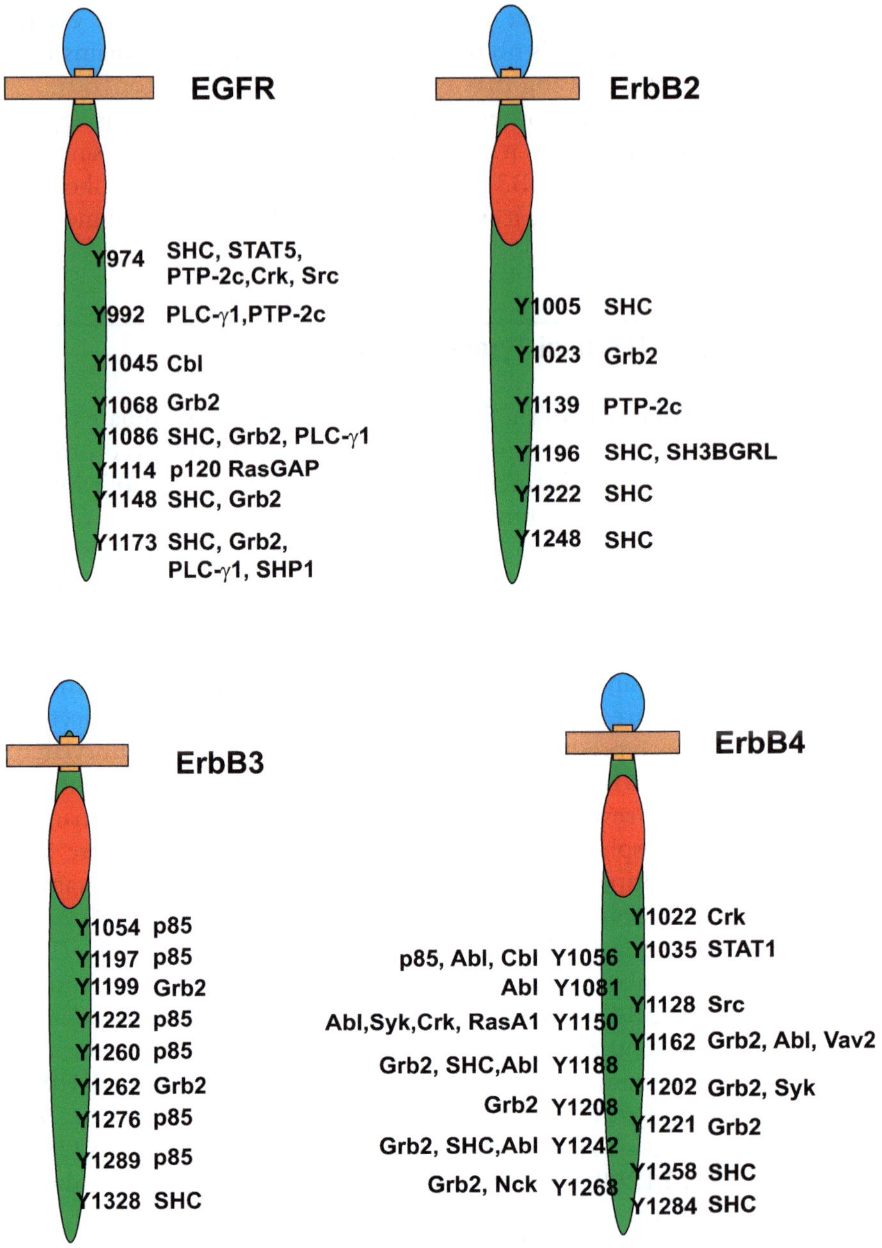

Fig. 2 Phosphorylation of the EGFR C-terminus and its association with signaling proteins. EGFR is homodimerized or heterodimerized with other ErbB proteins in response to ligand. The putative sites of tyrosine phosphorylation are indicated along with the intracellular signaling effectors that have been predicted or shown to bind to these phosphorylated sites

the type of the ligands that bind to the ErbB receptors. It has been shown that the complement of signaling effectors recruited to and activated by ligand-stimulated ErbB receptors is specified by the receptor and ligand, and these factors account for the differences in ligand intrinsic activity. The differences in the sites of EGFR tyrosine phosphorylation induced by ligands may account for the ligand-induced EGFR coupling to signaling effectors and biological responses.

6 Activation of Downstream Signaling Proteins and the Control of Cell Functions

ErbB signaling networks consist of several modules that are interconnected and overlapping. By regulating these signaling networks, ErbB receptors govern fundamental cellular processes, including proliferation, cell migration, metabolism, and survival. [2, 69]. The activated ErbB receptors (homodimers and heterodimers) form signaling complexes with many signaling proteins, as described above [1, 22, 70]. The formation of the receptor-signaling protein complexes then initiates the activation of various signaling pathways, including Ras/ERK, PI3K/Akt, PLC-γ1, Src, and STAT (Fig. 3).

6.1 Ras and ERK

As shown in Fig. 3b, the interaction between EGFR and SHC/Grb2 results in the recruitment of Sos to the plasma membrane to activate Ras. Activated Ras mediates Raf activation, which then phosphorylates and activates mitogen-activated protein kinase/ERK kinase (MEK), leading to the activation of extracellular signal-regulated kinases (ERK). Activated ERK phosphorylates RSK, which then translocates into the nucleus to activate transcription factors such as c-fos. Activated ERK may also translocate into the nucleus to activate transcription factors such as Elk1 and c-fos [1, 22, 71–75].

6.2 PLC-γ1

The activation of PLC-γ1 by EGFR regulates multiple cellular functions, including cell proliferation, differentiation, receptor endocytosis, cell motility, membrane ruffle formation, and branching tubulogenesis [76–83]. The overexpression and hyperactivation of PLC-γ1 have been implicated in breast and prostate cancers. In particular, PLC-γ1 activity has been linked to cancer cell invasion [84, 85]. PLC-γ1, a 145 kDa protein, contains two SH2 domains, one SH3 domain, and two pleckstrin homology (PH) domains, and catalyzes the hydrolysis of phosphatidylinositol-4,5-bis-phosphate (PtdIns-4,5-P2), creating inositol-1,4,5-triphosphate (IP3) and diacylglycerol (DAG) (Fig. 3c). These second messengers (IP3 and DAG) are known to stimulate the release of Ca^{2+} from internal stores and activate protein kinase C (PKC), respectively [76]. PLC-γ1 forms a complex in vivo with EGFR through its SH2 domain interaction [86–90]. Complex formation leads to the phosphorylation of PLC-γ1 on tyrosine residues and an increase in its enzymatic activity [62, 91, 92]. Recent studies have shown that PLC-γ1 is involved in broad cell signaling. Interestingly, most of the recently identified interactions between PLC-γ1 and its binding proteins are

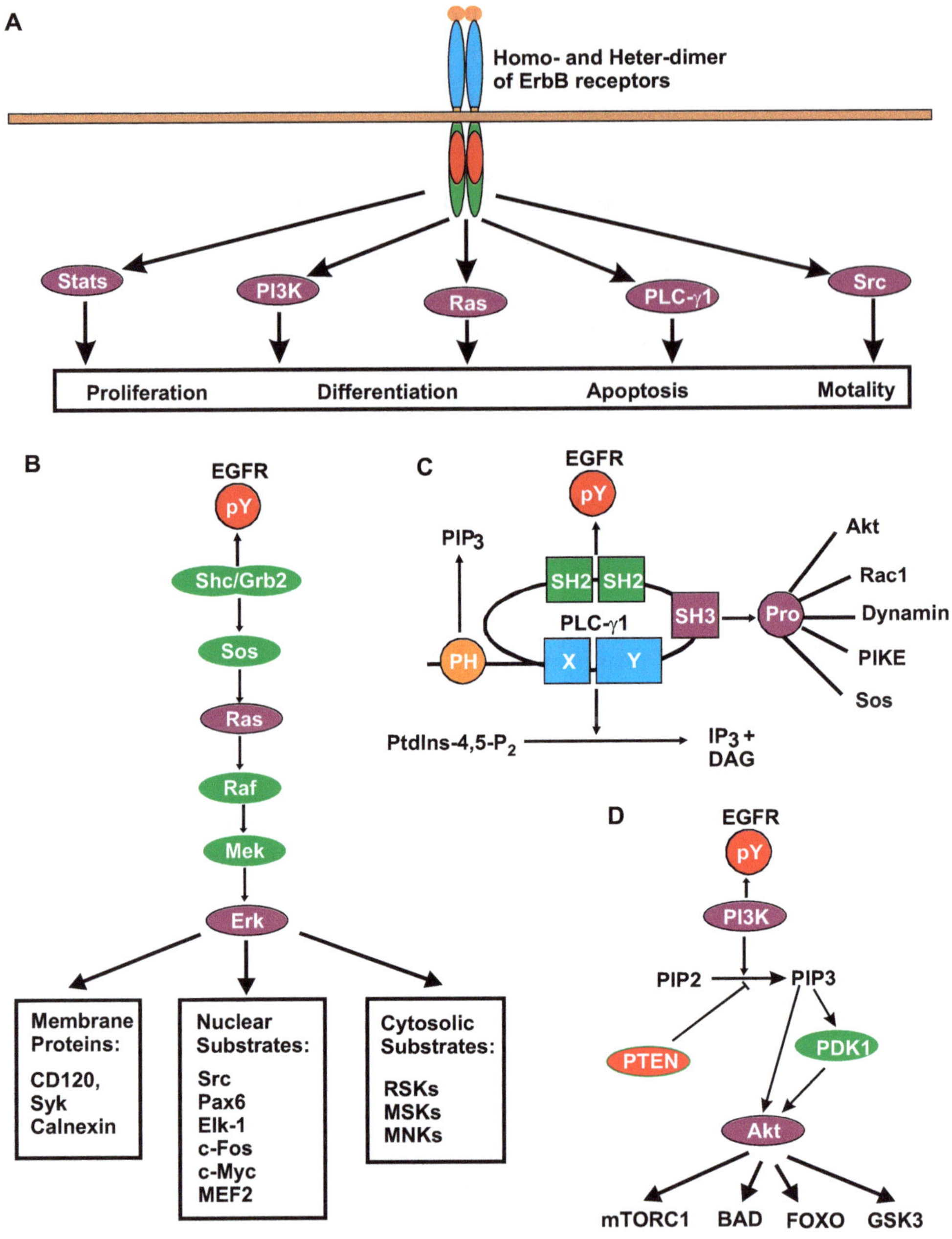

Fig. 3 Signaling pathways activated by ErbB receptors. (**a**) Binding of EGF to EGFR at the plasma membrane initiates the activation of various signaling pathways. Well-defined pathways include the Ras-ERK pathway, PI3K-Akt pathway, PLC-γ1 pathway, STATpathway, and Src pathway. (**b**) The signaling cascade of the Ras-ERK pathway. (**c**) The signaling cascade of the PLC-γ1 pathway. (**d**) The signaling cascade of the PI3K-Akt pathway

mediated by its SH3 domain. EGF stimulates the interaction between PLC-γ1 and PLD2, which is mediated by the PLC-γ1 SH3 domain [93]. PLC-γ1 binds directly to Akt in response to EGF through its SH3 domain [94]. The PLC-γ1 SH3 domain acts as a guanine nucleotide exchange factor (GEF) for PIKE [95], dynamin [83] and Rac1 [96] (Fig. 3c). The activated PLC-γ1 regulates cell mitogenesis and migration [94, 96–98].

6.3 PI3K and Akt

Activated EGFR also activates PI3K either through its direct interaction with the p85α subunit or through activated Ras [99, 100] (Fig. 3d). Activated PI3K then catalyzes the production of the second messenger phosphatidylinositol-3,4,5-trisphosphate (PIP3) by phosphorylating phosphatidylinositol-4,5-bisphosphate (PIP2). A direct antagonist of PI3K is the phosphatase and tensin homologue deleted on chromosome 10 (PTEN). PTEN dephosphorylates PIP3 into PIP2 to reverses the activity of PI3K, thereby functioning as an important negative control of incoming signals. PIP3 transduces activating signals by binding to the PH domains of proteins to recruit them to the cell membrane. One key downstream mediator of the PI3K signaling cascade is the serine/threonine (Ser/Thr) kinase Akt. The activation of PI3K by ErbB receptors stimulates Akt activity, which protects the cell from undergoing apoptosis [99, 101, 102]. As discussed above, among ErbB homodimers and heterodimers, the ErbB2/ErbB3 heterodimer stimulates the strongest activation of the PI3K/Akt pathway, which is implicated in many cancers. Akt is a central player in the signal transduction pathways activated by ErbB receptors and is thought to contribute to several cellular functions, including nutrient metabolism, cell growth and apoptosis. The alteration of Akt activity is associated with several human diseases, including cancer and diabetes [103]. Akt is composed of an amino terminal PH domain, a central kinase domain and a carboxyl terminal regulatory domain. Akt is recruited to the plasma membrane by its SH3 domain interaction with PIP3, which exposes Akt Thr308 for phosphorylation by 3-phosphoinositide-dependent kinase 1 (PDK1), which is already located at the membrane. Rapamycin complex 2 (mTORC2) phosphorylates Ser473 in the C-terminus, which leads to full Akt activation. Activated Akt then mediates signals promoting cellular growth and survival and suppresses pro-apoptotic signals. Akt phosphorylates several intracellular proteins, including forkhead box O transcription factors (FoxO), the BCL2-associated agonist of cell death (BAD), and the glycogen synthase kinase 3 (GSK3), to promote cell cycle entry and cell survival. The proteins TSC1 (hamartin) and TSC2 (tuberin) form a complex that inhibits the activity of the small G-protein Ras homologue enriched in the brain (Rheb), which is necessary for mTORC1 activation. The Akt-mediated phosphorylation of TSC2 releases Rheb from its inhibited state. Rheb then accumulates in a

GTP-bound state and can directly activate mTORC1, which phosphorylates the p70S6 kinase (S6 K1) and the eukaryotic translation initiation factor 4E binding protein 1 (4EBP1), leading to increased protein translation, which protects the cell from undergoing apoptosis [99, 101, 102].

6.4 STAT and Src

The activated EGFR also activates STATs directly, by binding to and phosphorylating STATs, or indirectly, by activating c-Src. Activation can occur as shown via cytokine signaling (IL-6), growth factor receptor signaling (EGFR), or non-receptor tyrosine kinase signaling (Src). JAK is not required when STATs bind directly to EGFR for activation, but JAK allows the maximal activation of STATs via phosphorylation by EGFR-activated Src. Grb2 can inhibit STAT-mediated EGFR signaling by binding to the STAT activation site on EGFR, whereas SOCS inhibits STATs by binding to JAK which suppresses the activation of STATs by Src. Once activated, STATs dimerize and translocate to the nucleus where they activate the transcription of genes involved in proliferation, differentiation, and survival [48, 104].

The Src kinase family is composed of ten proteins: Src, Frk, Lck, Lyn, Blk, Hck, Fyn, Yrk, Fgr, and Yes. Src family kinases are membrane-associated, non-receptor tyrosine kinases that act important signaling intermediaries regulating a variety of outputs, such as cell proliferation, differentiation, apoptosis, migration, and metabolism [105–107]. Src signaling is cross-connected with many signaling pathways, such as the PI3K and STAT pathways [105, 106]. Importantly, Src kinases, which are activated in many cancers with high EGFR levels, have been shown to potentiate EGFR signaling [108–110]. The c-Src potentiation of EGFR has been associated with the c-Src-dependent phosphorylation of the EGFR receptor and the complex formation between c-Src and EGFR [109, 110]. In addition to focal adhesion kinase (FAK), which is involved in the regulation of adhesion and migration, PI3K and STAT3 are also substrates for c-Src [111]. Although the Src kinase has been linked to the development and progression of cancer for many years, we still do not completely understand its role in cancer [112]. The tyrosine kinase activity of Src is independent of RTK signaling, but Src may interact with RTKs such as EGFR. As such, the Src-EGFR interaction may enhance EGFR signaling, but on the other hand, it may be involved in resistance to EGFR-targeted therapy [112, 113].

7 Other Signaling Pathways Mediated by ErbB Receptors

7.1 Signaling from Endosomes

The concept of EGFR signaling from endosomes or "signaling endosomes" has developed gradually. The early evidence that supported signaling from endosomes was reported in the mid- to late

1980s. These studies showed that internalized EGFR was autophosphorylated and catalytically active [114–116]. Various signaling molecules that regulate Ras activity, including Grb2, SHC, Sos, and GAP, are cointernalized with EGFR into endosomes and remain associated with the receptor in endosomes [117–121]. Subsequently, additional results have confirmed the sustained interaction between the EGFR and various signaling proteins in endosomes [122–126].

The major evidence supporting endosomal EGFR signaling has come from endocytosis inhibition experiments. Since the mid-1990s, researchers have developed many ways to inhibit EGFR endocytosis and then examine the effects on cell signaling. These experiments have yielded mixed results regarding which signaling pathways are activated by the endosomal EGFR, and the physiological relevance of EGFR signaling from endosomes. The inhibition of EGFR endocytosis by a dominant-negative dynamin mutant enhanced the activation of PLC-γ1 and cell proliferation, but decreased ERK activation [127]. In a study of EGFR transactivation by G-protein coupled receptors, it was found that the inhibition of EGFR endocytosis by either mutant dynamin or β-arrestin abolished ERK activation [128, 129]. The inhibition of EGFR endocytosis by phospholipase D also blocked EGF stimulated ERK activation [130]. However, none of these studies have provided a mechanism to explain why the activated EGFR at the plasma membrane was unable to activate ERK. On the other hand, other research has shown that the inhibition of EGFR internalization enhanced ERK activation [131, 132]. EGFR efficiently activated ERK by overexpression of dominant negative dynamin in HeLa cells and Hep2 cells, which are conditionally defective in clathrin-dependent endocytosis [131]. Sprouty2 has been found to attenuate EGFR ubiquitination and endocytosis, and consequently, enhance Ras/ERK signaling [132]. Initially, in the few cases where biological end points were measured, the inhibition of endocytosis did not result in the attenuation of biological effects [127, 133]. These results argued against a physiological relevance for endosome-originated signals [134].

The controversy over endosomal signaling and its physiological relevance is due, in part, to the limitation of current approaches. For example, while the endocytosis-inhibition approach has made significant contributions and remains a powerful tool in studying endosomal signaling, it has its limitations. Although the inhibition of EGFR endocytosis eliminates endosomal signaling, the retention of EGFR at the cell surface also enhances signaling from the plasma membrane. Therefore, it is difficult to determine whether the observed effects are due to the lack of endosomal signaling or prolonged plasma membrane signaling. Blocking EGFR endocytosis by mutant dynamin or β-arrestin affects all endocytic events mediated by these factors. Consequently, it is difficult to determine whether the observed effects are due to the inhibition of EGFR endosomal signaling specifically, or a broad inhibition of endocytosis. Moreover, this approach is not suitable for studying the dynamics of endosomal signaling. None of these approaches have offered the means to allow activated receptors inside a cell without their initial activation at the cell surface [135].

In the early 2000s, a novel system was established to allow the specific activation of endosome-associated EGFR without the initial activation at the plasma membrane, and without disrupting the overall endocytosis pathway. To specifically activate endosomal EGFR, cells were treated with EGF in the presence of AG1478, a specific EGFR tyrosine kinase inhibitor, and monensin, which blocks the recycling of EGFR. This treatment led to the internalization of inactive EGF-EGFR complexes into endosomes. The endosome-associated EGFR was then activated by removing AG1478 and monensin, and under these conditions, no surface EGFR phosphorylation was detected [136, 137]. The specific activation of endosome-associated EGFR was also achieved without using monensin [137]. In this system, EGFR followed the same endocytic pathway as in the untreated control cells: the EGF receptor was first internalized into Rab5-positive endosomes and eventually trafficked to lysosomes for degradation. The only difference was that the EGF receptor was not activated during its internalization from the plasma membrane into endosomes, and was stopped at the endosomes until it was activated. Thus, this system not only allowed the generation of specific endosomal signaling of the EGFR, but also provided conditions very similar to the endosomal signaling of EGFR following its normal activation at the plasma membrane. By using this system, it was shown that (1) endosomes can serve as a nucleation site for the formation of signaling complexes, (2) endosomal EGFR signaling is sufficient to activate the major signaling pathways leading to cell proliferation and survival, and (3) endosomal EGFR signaling is sufficient to suppress apoptosis induced by serum withdrawal [137] and to stimulate cell proliferation [138] (Fig. 4).

In most cases, endosomal EGFR signaling is the continuation of EGFR signaling that originated at the plasma membrane, serving to maintain EGFR signaling and provide spatial–temporal regulation of EGFR signaling. However, in some cases, specific and novel signaling may be initiated only from endosomes, since these signaling events require factors to be brought together by endocytosis. While specific signaling complexes can be assembled through their recruitment to the early endosomal resident protein Rab5, there are no convincing examples that specific and novel signaling is initiated from endosomes in the context of EGFR signaling. However, it is well known in TGFβ signaling that specific and novel signaling may be initiated only from endosomes. TGFβ receptors (TGFβR) become phosphorylated at Ser

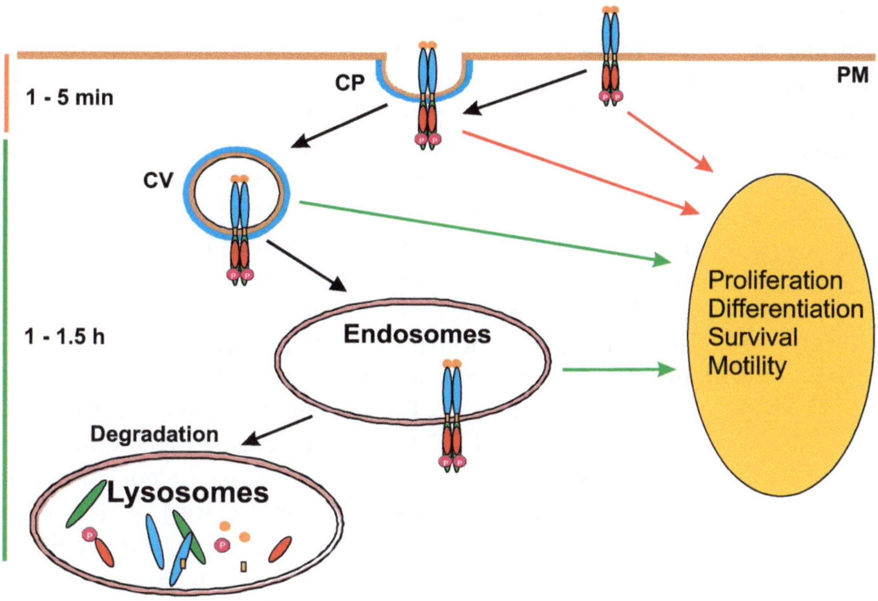

Fig. 4 A model to describe EGFR signaling along its endocytic route. Following its activation at the plasma membrane, EGFR continues to signal along its endocytic route until its degradation in lysosomes. EGFR signaling from both the plasma membrane and the intracellular endocytic compartment regulates major signaling pathways leading to cell proliferation and survival. The activated EGFR may remain 1–5 min at the plasma membrane and 1–1.5 h along its endocytic compartments. *PM* plasma membrane, *CP* coated pit, *CV* coated vesicle

residues and are internalized by endocytosis following ligand binding. Once localized into endosomes, TGFβR can bind to the SMAD anchor for receptor activation (SARA). The protein complex has been shown to induce the phosphorylation of transcription factors SMAD1 or SMAD2 by their Ser/Thr kinase receptors. Upon phosphorylation, SMADs are released into the cytoplasm, bind to a cofactor (SMAD4), enter the nucleus, and promote gene transcription [139, 140].

Together, it is clear that EGFR signals from both the plasma membrane and endosomes, and that the signals from both locations are able to activate major signaling pathways that stimulate cell proliferation and promote cell survival. However, following EGF stimulation, activated EGF receptors only stay at the plasma membrane briefly (5-10 min), but remain in endosomes much longer (1 h) (Fig. 4) [141]. This finding argues for a more physiologically important role for endosomal signaling. Plasma membrane EGFR signaling has been usually exaggerated by the studies with the inhibition EGFR endocytosis, since activated EGFRs that stay at the plasma membrane are artificially over extended.

7.2 Nuclear Translocation and the Role of ErbB Receptors in the Induction of Gene Expression

In addition to signaling from the plasma membrane, some ErbB receptors have been shown to translocate to the nucleus and are involved in the regulation of transcription within nucleus. ErbB4 is a prototype for receptors that can undergo regulated intramembrane proteolysis (RIP). In 2001 and 2002, two laboratories reported that ErbB4 was able to signal via RIP, in addition to activating signaling cascades indirectly [142, 143]. Previously, this mechanism has been best characterized for its role in the signaling of the amyloid precursor protein and Notch receptors [144]. RIP involves two proteolytic cleavage events that release ectocellular and intracellular domains (ECD and ICD, respectively) of type 1 integral membrane proteins. The RIP of ErbB4 is conducted by the sequential activities of the tumor necrosis factor-α converting enzyme (TACE) [145] and γ-secretase [143]. Once released, the soluble ICD of ErbB4 can translocate to the nucleus and operate as a transcriptional coactivator or corepressor [146]. Currently, transcriptional regulators that have been shown to coregulate transcription with the released ErbB4 ICD include the Yes-associated protein (YAP) [147], signal transducer and activator of transcription 5A (STAT5A) [148], ETO2 [149], estrogen receptor α [150], the TAB2-NCoR complex [151], Krab-associated protein 1 (Kap1) [152], and AP-2 [153].

It has long been known that many functions of EGFR, such as EGF-induced DNA synthesis and its mitogenic effect, require mechanisms other than the early, transient responses to receptor activation [47, 154, 155]. In addition, stabilized complexes of EGF-EGFR on the cell surface have not been found to induce DNA synthesis, although transient responses were activated [156]. It is evident that a critical activity of EGFR signaling is still missing from the current model. For years, EGFR and its ligands have repeatedly been observed in the nucleus of cells; for example, in cell lines, human placenta, regenerating liver [156] and in many different types of cancer [157–160]. Further, it has been revealed that the nuclear localization of EGFR is correlated with highly proliferative cells in various tissues. In addition, there is a potential role for nuclear EGFR as a transcription factor or coactivator, which may activate the genes required for its mitogenic effects, such as the gene that encodes cyclin D1 [161–163].

7.3 EGFR Signaling at Different Phases of the Cell Cycle

Most of the findings regarding ErbB receptor signaling as described above may only reflect receptor activation and function during interphase, especially G1 phase, as these studies were conducted with cells primarily in G1. The cell cycle is a series of events leading to cell replication. When plated at low cell densities in serum-containing medium, cultured cells start to proliferate, moving through the four phases of the cell cycle: G1, S, G2, and M. Growth factors, including EGF, regulate cell cycle progression, particularly the G1–S progression. Growth factors must be present until the restriction point (R point) in G1 phase to stimulate entry into the cell cycle and proliferation. After the R point, growth factors are not required to complete the other stages of the cell cycle [164, 165]. Furthermore, it has been shown that two pulses (30 min) of growth factors, including platelet-derived growth factor and EGF, separated by 8 h, are also able to drive the cell cycle [138, 166]. While it has been proposed that the G0 to S interval is the only portion of the cell cycle that is regulated by growth factors [164], there have been sporadic publications showing a minor role for growth factor-mediated cell signaling in S and G2 phases [167–173].

The role of growth factor-induced cell signaling in mitosis has been rarely and poorly studied. Mitosis (M phase) is the most dynamic period of the cell cycle, involving a major reorganization of virtually all cell components and structures. A hallmark of cancer is the ability of the cancer cell to sustain chronic proliferation [174]. A major difference between cancer cells and normal cells is that the cancer cells are much more mitogenic and show a higher frequency of mitosis. Therefore, most cancer drugs are designed to specifically target mitotic cells [175]. It has been reported that while there is no change in the number of surface-exposed EGFRs between interphase and M phase of the cell cycle, EGF-induced activation of EGFR and downstream signaling is tightly suppressed in M phase due to a decrease in ligand binding affinity and the inability of EGF to induce receptor dimerization [176]. It has also been shown that EGFR, ERK2, GTPase-activating protein (GAP), and PLC-γ1 are less phosphorylated in M phase-arrested cells than they are in interphase [177]. Further research has shown that Cdc2 inhibits EGFR-stimulated ERK activation during mitosis by primarily targeting signaling proteins that are upstream of MEK1, including EGFR [178]. These studies suggest that inhibition of EGFR activity and its downstream signaling pathways underlie the importance of keeping the cell sheltered from extracellular signals when it undergoes division. Moreover, this inhibition is beneficial for preventing gene expression so as to preserve the energy needs that are required for mitotic structural changes [176–178].

However, in a recent study of EGF-induced EGFR endocytosis during M phase, we have shown that in M phase, the EGFR is expressed at the same level as in interphase, and is also activated by EGF to the same degree as in interphase, as shown by the phosphorylation of Y992 [179]. In addition, we have shown that EGFR is strongly activated by EGF during mitosis, as all five major tyrosine residues (Y992, Y1045, Y1068, Y1086, and Y1173) are phosphorylated to a level similar to that in interphase. Moreover, we have shown that the activated EGFR is able to selectively activate some downstream signaling pathways, while avoiding affecting others (Fig. 5) [180]. The activation of EGFR results in the activation of Akt2, but not Akt1. Thus, Akt2 is likely to mediate the

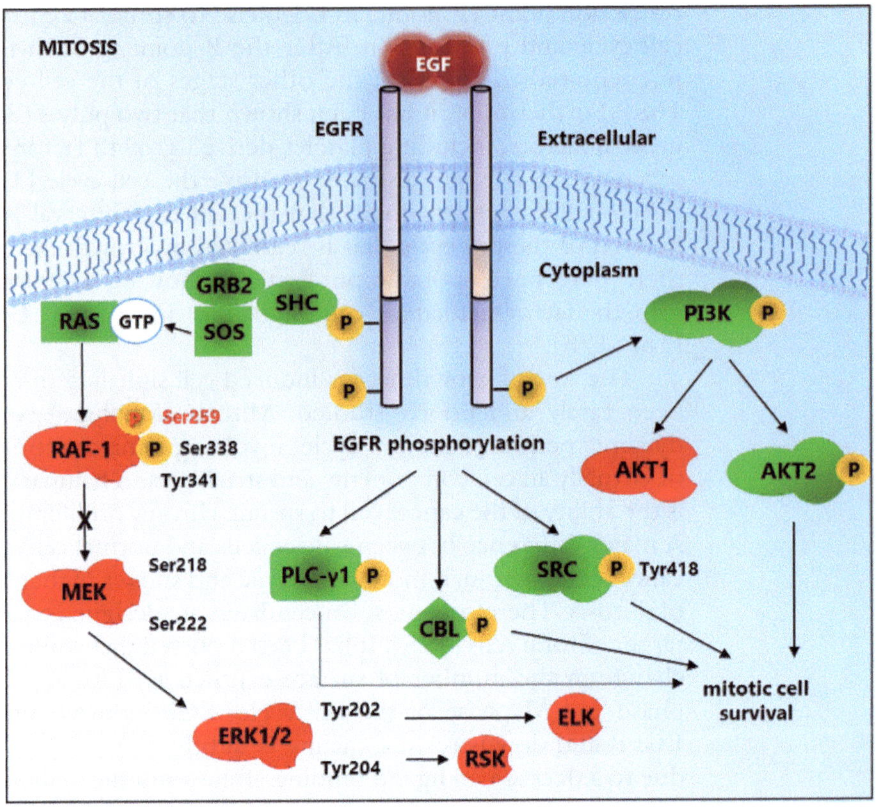

Fig. 5 Schematic illustration of EGFR signaling pathways during mitosis and interphase. The green components indicate the proteins/pathways activated in mitosis, and the red components indicate the proteins/pathways activated in interphase, but not activated in mitosis

antiapoptotic effects of EGF against nocodazole-induced cell death. Activated EGFR also activates c-Cbl and PLC-γ1, two signaling proteins with multiple cellular functions, during M phase. However, activated EGFR is unable to activate ERK1/2 and their downstream substrates RSK and Elk-1. Although it is able to activate Ras, EGFR fails to fully activate Raf-1 in mitosis due to the lack of phosphorylation at Y341 and the lack of dephosphorylation at pS259 [180]. These findings indicate that in contrast to interphase, EGFR-mediated cell signaling is specifically and differentially regulated in mitosis to serve the special needs of mitotic cells.

7.4 Transactivation of G Protein-Coupled Receptors (GPCRs)

Ullrich's group provided the first evidence supporting the transactivation of GPCRs by RTKs. They showed that EGFR and ErbB2 were rapidly phosphorylated in Rat-1 cells following the addition of various GPCR agonists, including lysophosphatidic acid, thrombin, and endothelin-1 [181]. The activation of EGFR and ErbB2 was blocked when EGFR kinase was inhibited either by the inhibitor

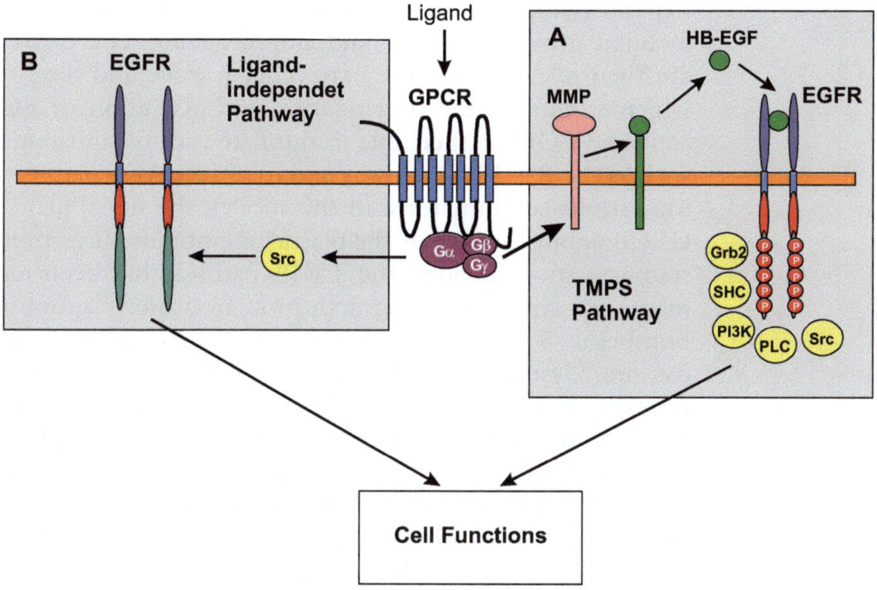

Fig. 6 Schematic illustration of EGFR transactivation by the GPCR. Two major mechanisms have been proposed and supported by the available data: TMPS and ligand-independent intracellular pathways. *Box A*: TMPS. In this model, GPCR-mediated EGFR transactivation is dependent on the activation of MMPs that are able to cleave EGFR ligands, including heparin-binding EGF-like factor (HB-EGF), and then stimulate ligand shedding into the extracellular space. The EGF ligands are then able to bind to EGFRs and stimulate the dimerization and activation of the receptors, which leads to the activation of signaling cascades. *Box B*: Ligand-independent pathway. This pathway involves the activation of intracellular protein tyrosine kinases (PTKs) such as the Src kinase family. Src mediates the transphosphorylation of tyrosines in the cytosolic region of EGFR, which then provide docking sites for the assembly and activation of signaling complexes. Also, Src could be activated by GPCR through different mechanisms

AG1478 or by the expression of dominant-negative EGFR [181]. Furthermore, they showed that EGFR transactivation occurred in diverse cell types and with different types of G proteins [182]. Subsequently, the transactivation of EGFR by GPCR has been reported for various receptor tyrosine kinases, including PDGFR [183, 184], Trk [142, 185], insulin-like growth factor receptor [186, 187], vascular endothelial growth factor receptors [188, 189], and fibroblast growth factor receptors [190]. Here, we focus on the EGFR since it is the most well studied.

Accumulated evidence has suggested that there are several mechanisms for the transactivation of the EGFR by GPCRs [191–193]. There are two major mechanisms (Fig. 6) [194]. The first mechanism is the "triple membrane passing signal" (TMPS) pathway. According to this model, EGFR transactivation by the GPCR is controlled by the activation of membrane-bound matrix metalloproteases (MMPs). The group of MMPs that is most implicated in this regulation is the ADAM (a disintegrin and metalloprotease) family. MMPs are able to cleave EGFR ligands such as heparin-binding EGF-like factor (HB-EGF), neuregulin, transforming growth factor-α, and amphiregulin. The cleaved ligands are then released into the extracellular space and bind to EGFR, which stimulates the dimerization and activation of EGFR. The activated EGFR is then able to simulate various signaling pathways such as the Ras-ERK pathway and the PI3K-AKT pathway, regulating various cell functions. In this model, the signal generated by a GPCR agonist will cross the plasma membrane three times. EGFR transactivation through the TMPS pathway has been reported in many cell types following activation by various agonists such as bombesin, 5-hydroxytryptamine, carbachole, angiotensin II, bradykinin, lysophosphatidic acid, endothelin 1, gonadotropin-releasing hormone, phenylephrine, leptin, thrombin, deoxycholic acid, and prostaglandin E2 [191–193, 195, 196]. EGFR transactivation through this mechanism has been implicated in the regulation of many normal cell functions, as well as the growth, development and progression of many diseases, such as cancers, kidney disease, and cardiovascular disease.

In the second major mechanism, EGFR is transactivated by GPCR without any detectable EGF-like ligands, which suggests that EGFR transactivation by the GPCR can also be through intracellular signaling pathways that are ligand independent. All of the ligand-independent mechanisms involve the activation of intracellular protein tyrosine kinases (PTKs), such as the members of the Src kinase family. The increased PTK activity mediates the phosphorylation of EGFR in its cytosolic domain. The phosphorylated EGFR, in turn, associates with various signaling proteins and initiates the activation of multiple signaling pathways. Several mechanisms have been suggested regarding the pathways leading to the enhanced activity of PTKs [191–193, 195, 197].

The second mechanism requires the production of reactive oxygen species (ROS). In this model, the activation of GPCRs by agonist stimulates the phosphorylation of p47phox and the activation of NADPH oxidase, which produces ROS through O2 by using NADPH as electron donor. ROS may cause an imbalance in the equilibrium of intracellular phosphorylation and enhance the activity of intracellular PTKs through two mechanisms: (1) the inactivation of protein tyrosine phosphatases (PTPs) by oxidation of the cysteine residues in their catalytic sites. This leads to the enhanced activation of PTKs, and (2) the stimulation of the proteolysis of regulatory proteins that block PTK activity, which also leads to the higher activity of PTKs [191–193]. The increased PKT activity stimulates EGFR transactivation.

Ligand-independent, intracellular mechanisms without ROS production have also been suggested. Src family kinases have been shown to be associated with GPCRs. This association may be through the direct interaction between the Src SH3 domain and the GPCR cytoplasmic domain that contains consensus prolinerich motifs in its C-terminal tail or third intracellular loop. This interaction could also be through the binding to GPCR-associated proteins, including the G-protein subunits and β-arrestins. This interaction activates Src family kinases, which phosphorylate the EGFR on its intracellular domain [191–193].

8 Implications in Human Cancer

ErbB2 and EGFR have been implicated in the development of many types of human cancer. Genetic changes that have been detected in human tumors include gene amplification leading to receptor overexpression, activating kinase domain mutations mainly in EGFR, but also in ErbB2, in-frame deletions in the extracellular domain of EGFR (EGFR vIII), and coexpression of ErbB ligands and receptors in tumors. Each alteration promotes constitutive receptor activation, a process that stimulates cancer development [10, 47, 48].

8.1 EGFR

Dysregulation of the EGFR signaling cascade due to overexpression or constitutively activating mutations is well established in many cancer types, including breast cancer, lung cancer, colorectal and esophageal cancers, head and neck cancers, glioma, anal cancers, and glioblastoma [10, 39, 48].

EGFR plays an important role in the pathogenesis of many lung cancers. EGFR mutations in the kinase domain occur in a range of 10–40% of lung cancer samples [48, 198]. The incidence of EGFR kinase-domain mutations is about 10% in Caucasian patients and 30–40% in Asian patients. EGFR gene amplification occurs in about 15% of adenocarcinomas and 30% of squamous cell carcinomas, while ERBB2 amplification occurs in 6% of adenocarcinomas and 2% of squamous cell carcinomas [48, 198]. These alterations are rarely, if ever, observed in small cell lung cancers. The overexpression of EGFR occurs in about 60% of NSCLCs, as measured by immunohistochemistry [48, 198]. An increase in EGFR levels as a result of gene amplification, increased biosynthesis, or decreased degradation in a large percentage of NSCLCs has prompted the development of therapies that inhibit EGFR activity [48, 199].

In contrast to lung cancer, mutation of the EGFR in breast cancer is rare [200, 201]. However, the overexpression of EGFR is observed in 15–30% of breast carcinomas and is associated with large tumor size and poor clinical outcomes [48, 201]. Most notably, EGFR is frequently overexpressed and associated with poor prognosis in TNBC [202, 203], a breast cancer subtype for which many therapeutics targeting EGFR have been in develompent. The overexpression of EGFR is partly attributable to EGFR gene amplification.

8.2 ErbB2

ErbB2 overexpression or ERBB2 gene amplification occurs in 20–30% of breast cancers and ovarian cancers, while 10–20% of breast cancers are triple negative, lack hormone receptors and do not overexpress ErbB2/HER2 [7, 25, 26, 28, 39, 48, 204]. Targeting HER2 has proven to be an effective therapeutic strategy for HER2-positive breast cancer [205, 206]. Since its approval by the FDA in 1998, trastuzumab has changed the paradigm for HER2-positive breast cancer treatment [206, 207]. Despite the improved outcomes achieved so far, acquired resistance to trastuzumab eventually develops in patients and represents a challenge that must be overcome [206, 208, 209]. Under these circumstances, the recent approval of pertuzumab by the FDA, to be used in combination with trastuzumab and docetaxel to treat HER2-positive breast cancer has raised new hope for better outcomes [204, 206, 210].

Although wild-type ErbB2 overexpression occurs in 20–30% of breast cancers, approximately 1.6% of breast cancer patients possess an ERBB2 mutation [211]. The incidence of new cases of breast cancer with mutated ERBB2 in the United States is about 4000 per year. The mutations occur in the extracellular domain, carboxy-terminal tail, and most frequently within the protein kinase domain [211]. The most common mutation, which was observed in six patients, was a Leu755Ser mutation. ERBB2 mutations in the protein kinase domain are also found in a subset of lung adenocarcinomas, and the incidence of these ERBB2 mutations in wild-type EGFR/KRAS and ALK-negative tumors is about 6% [212]. Moreover, the overexpression of ErbB2 has also been reported in gastric cancer patients, and the incidence of ErbB2 overexpression is higher in the intestinal subtype than in the diffuse cancer subtype [213].

8.3 ErbB3

Both ErbB3 mutation and overexpression have been reported in various cancers. Some early histological studies have shown that 83% of gastrointestinal tumours showed stronger staining for ErbB3 than the basal proliferating cells on normal mucosa [38]. A similarly high percentage of squamous cell carcinomas showed strong ErbB3 staining [38]. It has also been reported that the overexpression of ErbB3 occurs in about 20% of breast, ovarian, stomach and bladder tumors [39]. The overexpression of ErbB3 has also been reported in cervical, prostate, and head and neck cancers [39]. The mechanism underlying this overpression is likely due to increased transcription of the ERBB3 gene [39, 214].

It has been reported that ERBB3 mutations occur in 11% of colorectal tumors, 12% of gastric cancers, 1% of NSCLC adenocarcinomas and 1% of squamous cell NSCLCs [42, 48]. Although the mutations may occur in the kinase domain, the majority of the ERBB3 mutations occur in the extracellular domain [42, 48]. It has also been reported that when wild-type or mutant ErbB3 was expressed alone, it did not promote anchorage-independent growth. The majority of ErbB3 mutants promoted such growth when expressed with ErbB2. Mice receiving cells expressing wild-type ErbB3 and ErbB2 developed a leukemia-like disease with a longer median latency of 39 days. Mice implanted with wild type ERBB3, but not ERBB2, were alive at the end of the 60-day test period [42, 48].

8.4 ErbB4

In contrast to EGFR and ErbB2, which are well-established oncogenes and cancer drug targets, both oncogenic and tumor suppressor functions have been proposed for ErbB4 [43, 215–217]. Both the upregulation and downregulation of ErbB4 expression has been reported in several cancer types [43, 217]. Both significant overexpression and downregulation of ERBB4 mRNA has been observed in various cancer cell samples as compared with normal tissues. The median ERBB4 expression was upregulated in ovarian, uterine and breast tumors as compared with their respective normal tissues, while substantial variation was observed between individuals. Significant downregulation of ERBB4 in cancer was, however, observed more frequently than overexpression [43, 218, 219]. Importantly, some cancer types, such as lung cancer, generally demonstrated ERBB4 downregulation, but also included a small subgroup with abnormally high expression. In breast cancer, ErbB4 signaling is associated with differentiation. Attempts to downregulate or inactivate endogenous ErbB4 in different cancer cells have generally led to suppressed tumor growth [220]. Recently, work on the potential of ErbB4 as a cancer drug target has received new attention after the discovery of somatic ERBB4 mutations in approximately 5% of patients with non-small cell lung cancer and 19% of patients with metastatic melanoma [43, 220–222].

9 Conclusion

ErbB receptors are activated after homodimerization or heterodimerization. Like other RTKs, ErbB receptors control the cellular communication network and function as master switches. Their regulated signaling is pivotal to normal cell development and survival. ErbB receptors are overexpressed in many cancers, especially in breast cancer, ovarian cancer, and non-small cell lung cancer. The overexpression of ErbB receptors is correlated with poor prognosis, drug resistance, cancer metastasis, and lower survival rate. ErbB receptors, especially EGFR and ErbB2 have been the primary choices as targets for developing cancer therapies.

Acknowledgments

This work was supported in part by grants from the Canadian Institutes of Health Research (CIHR) and the Canadian Breast Cancer Foundation (CBCF).

References

1. Yarden Y, Sliwkowski MX (2001) Untangling the ErbB signalling network. Nat Rev Mol Cell Biol 2(2):127–137
2. Citri A, Yarden Y (2006) EGF-ERBB signalling: towards the systems level. Nat Rev Mol-Cell Biol 7(7):505–516
3. Heldin CH (1995) Dimerization of cell surface receptors in signal transduction. Cell 80 (2):213–223
4. Marmor MD, Skaria KB, Yarden Y (2004) Signal transduction and oncogenesis by ErbB/HER receptors. Int J Radiat Oncol Biol Phys 58(3):903–913
5. Carraway KL III, Cantley LC (1994) A neu acquaintance for erbB3 and erbB4: a role for receptor heterodimerization in growth signaling. Cell 78(1):5–8
6. Pinkas-Kramarski R, Soussan L, Waterman H, Levkowitz G, Alroy I, Klapper L, Lavi S, Seger R, Ratzkin BJ, Sela M et al (1996) Diversification of Neu differentiation factor and epidermal growth factor signaling by combinatorial receptor interactions. EMBO J 15 (10):2452–2467
7. Hynes NE, Lane HA (2005) ERBB receptors and cancer: the complexity of targeted inhibitors. Nat Rev Cancer 5(5):341–354
8. Yarden Y, Pines G (2012) The ERBB network: at last, cancer therapy meets systems biology. Nat Rev Cancer 12(8):553–563
9. Avraham R, Yarden Y (2011) Feedback regulation of EGFR signalling: decision making by early and delayed loops. Nat Rev Mol Cell Biol 12(2):104–117
10. Arteaga CL, Engelman JA (2014) ERBB receptors: from oncogene discovery to basic science to mechanism-based cancer therapeutics. Cancer Cell 25(3):282–303
11. Lemmon MA, Schlessinger J (2010) Cell signaling by receptor tyrosine kinases. Cell 141 (7):1117–1134
12. Kovacs E, Zorn JA, Huang Y, Barros T, Kuriyan J (2015) A structural perspective on the regulation of the epidermal growth factor receptor. Annu Rev Biochem 84:739–764
13. Ward CW, Hoyne PA, Flegg RH (1995) Insulin and epidermal growth factor receptors contain the cysteine repeat motif found in the tumor necrosis factor receptor. Proteins 22(2):141–153
14. Carpenter G, Lembach KJ, Morrison MM, Cohen S (1975) Characterization of the binding of 125-I-labeled epidermal growth factor to human fibroblasts. J Biol Chem 250 (11):4297–4304
15. Ullrich A, Coussens L, Hayflick JS, Dull TJ, Gray A, Tam AW, Lee J, Yarden Y, Libermann TA, Schlessinger J (1984) Human epidermal growth factor receptor cDNA sequence and aberrant expression of the amplified gene in A431 epidermoid carcinoma cells. Nature 309(5967):418–425
16. Velu TJ, Beguinot L, Vass WC, Willingham MC, Merlino GT, Pastan I, Lowy DR (1987) Epidermal-growth-factor-dependent transformation by a human EGF receptor proto-oncogene. Science (New York, NY) 238 (4832):1408–1410
17. Kim YT, Park SW, Kim JW (2002) Correlation between expression of EGFR and the prognosis of patients with cervical carcinoma. Gynecol Oncol 87(1):84–89
18. Kopp R, Rothbauer E, Ruge M, Arnholdt H, Spranger J, Muders M, Pfeiffer DG, Schildberg FW, Pfeiffer A (2003) Clinical implications of the EGF receptor/ligand system for tumor progression and survival in gastrointestinal carcinomas: evidence for new therapeutic options. Recent Results Cancer Res 162:115–132
19. Nicholson RI, Gee JM, Harper ME (2001) EGFR and cancer prognosis. Eur J Cancer (Oxford 1990) 37(Suppl 4):S9–15
20. Coussens L, Yang-Feng TL, Liao YC, Chen E, Gray A, McGrath J, Seeburg PH, Libermann TA, Schlessinger J, Francke U (1985) Tyrosine kinase receptor with extensive homology to EGF receptor shares chromosomal location with neu oncogene. Science (New York, NY) 230(4730):1132–1139
21. Schlessinger J, Ullrich A (1992) Growth factor signaling by receptor tyrosine kinases. Neuron 9(3):383–391
22. Pawson T (1995) Protein modules and signalling networks. Nature 373(6515):573–580

23. Shih C, Padhy LC, Murray M, Weinberg RA (1981) Transforming genes of carcinomas and neuroblastomas introduced into mouse fibroblasts. Nature 290(5803):261–264
24. Bargmann CI, Hung MC, Weinberg RA (1986) Multiple independent activations of the neu oncogene by a point mutation altering the transmembrane domain of p185. Cell 45(5):649–657
25. Slamon DJ, Clark GM, Wong SG, Levin WJ, Ullrich A, McGuire WL (1987) Human breast cancer: correlation of relapse and survival with amplification of the HER-2/neu oncogene. Science (New York, NY) 235 (4785):177–182
26. Slamon DJ, Godolphin W, Jones LA, Holt JA, Wong SG, Keith DE, Levin WJ, Stuart SG, Udove J, Ullrich A (1989) Studies of the HER-2/neu proto-oncogene in human breast and ovarian cancer. Science (New York, NY) 244(4905):707–712
27. Natali PG, Nicotra MR, Bigotti A, Venturo I, Slamon DJ, Fendly BM, Ullrich A (1990) Expression of the p185 encoded by HER2 oncogene in normal and transformed human tissues. Int J Cancer 45(3):457–461
28. Reese DM, Slamon DJ (1997) HER-2/neu signal transduction in human breast and ovarian cancer. Stem Cells 15(1):1–8
29. Yonemura Y, Ninomiya I, Ohoyama S, Kimura H, Yamaguchi A, Fushida S, Kosaka T, Miwa K, Miyazaki I, Endou Y et al (1991) Expression of c-erbB-2 oncoprotein in gastric carcinoma. Immunoreactivity for c-erbB-2 protein is an independent indicator of poor short-term prognosis in patients with gastric carcinoma. Cancer 67(11):2914–2918
30. Press MF, Pike MC, Hung G, Zhou JY, Ma Y, George J, Dietz-Band J, James W, Slamon DJ, Batsakis JG et al (1994) Amplification and overexpression of HER-2/neu in carcinomas of the salivary gland: correlation with poor prognosis. Cancer Res 54(21):5675–5682
31. Hudziak RM, Schlessinger J, Ullrich A (1987) Increased expression of the putative growth factor receptor p185HER2 causes transformation and tumorigenesis of NIH 3T3 cells. Proc Natl Acad Sci U S A 84 (20):7159–7163
32. Di Fiore PP, Pierce JH, Fleming TP, Hazan R, Ullrich A, King CR, Schlessinger J, Aaronson SA (1987) Overexpression of the human EGF receptor confers an EGF-dependent transformed phenotype to NIH 3T3 cells. Cell 51 (6):1063–1070
33. Pietras RJ, Arboleda J, Reese DM, Wongvipat N, Pegram MD, Ramos L, Gorman CM, Parker MG, Sliwkowski MX, Slamon DJ (1995) HER-2 tyrosine kinase pathway targets estrogen receptor and promotes hormone-independent growth in human breast cancer cells. Oncogene 10 (12):2435–2446
34. Kraus MH, Issing W, Miki T, Popescu NC, Aaronson SA (1989) Isolation and characterization of ERBB3, a third member of the ERBB/epidermal growth factor receptor family: evidence for overexpression in a subset of human mammary tumors. Proc Natl Acad Sci U S A 86(23):9193–9197
35. Plowman GD, Whitney GS, Neubauer MG, Green JM, McDonald VL, Todaro GJ, Shoyab M (1990) Molecular cloning and expression of an additional epidermal growth factor receptor-related gene. Proc Natl Acad Sci U S A 87(13):4905–4909
36. Cacace AM, Guadagno SN, Krauss RS, Fabbro D, Weinstein IB (1993) The epsilon isoform of protein kinase C is an oncogene when overexpressed in rat fibroblasts. Oncogene 8 (8):2095–2104
37. Rajkumar T, Majhi U, Malligarjuna V, Gullick W (1995) Prevalence of C-erbb3 expression in squamous-cell carcinomas of the cervix as determined by the monoclonal-antibody rtj2. Int J Oncol 6(1):105–109
38. Rajkumar T, Gullick WJ (1994) The type I growth factor receptors in human breast cancer. Breast Cancer Res Treat 29(1):3–9
39. Gullick WJ (1996) The c-erbB3/HER3 receptor in human cancer. *Cancer Surv* 27:339–349
40. Prigent SA, Gullick WJ (1994) Identification of c-erbB-3 binding sites for phophatidylinositol 3′-kinase and SHC using an EGF receptor/c-erbB-3 chimera. EMBO J 13:2831–2841
41. Shi F, Telesco SE, Liu Y, Radhakrishnan R, Lemmon MA (2010) ErbB3/HER3 intracellular domain is competent to bind ATP and catalyze autophosphorylation. Proc Natl Acad Sci U S A 107(17):7692–7697
42. Jaiswal BS, Kljavin NM, Stawiski EW, Chan E, Parikh C, Durinck S, Chaudhuri S, Pujara K, Guillory J, Edgar KA et al (2013) Oncogenic ERBB3 mutations in human cancers. Cancer Cell 23(5):603–617
43. Veikkolainen V, Vaparanta K, Halkilahti K, Iljin K, Sundvall M, Elenius K (2011) Function of ERBB4 is determined by alternative splicing. Cell Cycle (Georgetown TX) 10 (16):2647–2657
44. Kinugasa Y, Ishiguro H, Tokita Y, Oohira A, Ohmoto H, Higashiyama S (2004)

Neuroglycan C, a novel member of the neuregulin family. Biochem Biophys Res Commun 321(4):1045–1049
45. Kochupurakkal BS, Harari D, Di-Segni A, Maik-Rachline G, Lyass L, Gur G, Kerber G, Citri A, Lavi S, Eilam R et al (2005) Epigen, the last ligand of ErbB receptors, reveals intricate relationships between affinity and mitogenicity. J Biol Chem 280(9):8503–8512
46. Normanno N, Bianco C, Strizzi L, Mancino M, Maiello MR, De Luca A, Caponigro F, Salomon DS (2005) The ErbB receptors and their ligands in cancer: an overview. Curr Drug Targets 6(3):243–257
47. Hynes NE, MacDonald G (2009) ErbB receptors and signaling pathways in cancer. Curr Opin Cell Biol 21(2):177–184
48. Roskoski R Jr (2014) The ErbB/HER family of protein-tyrosine kinases and cancer. Pharmacol Res 79:34–74
49. Gullick WJ (2001) The type 1 growth factor receptors and their ligands considered as a complex system. Endocr Relat Cancer 8 (2):75–82
50. Blobel CP (2005) ADAMs: key components in EGFR signalling and development. Nat Rev Mol Cell Biol 6(1):32–43
51. Rocks N, Paulissen G, El Hour M, Quesada F, Crahay C, Gueders M, Foidart JM, Noel A, Cataldo D (2008) Emerging roles of ADAM and ADAMTS metalloproteinases in cancer. Biochimie 90(2):369–379
52. Wilson KJ, Gilmore JL, Foley J, Lemmon MA, Riese DJ II (2009) Functional selectivity of EGF family peptide growth factors: implications for cancer. Pharmacol Ther 122 (1):1–8
53. Celikel C, Eren F, Gulluoglu B, Bekiroglu N, Turhal S (2007) Relation of neuroendocrine cells to transforming growth factor-alpha and epidermal growth factor receptor expression in gastric adenocarcinomas: prognostic implications. Pathol Oncol Res 13(3):215–226
54. Eckstein N, Servan K, Girard L, Cai D, von Jonquieres G, Jaehde U, Kassack MU, Gazdar AF, Minna JD, Royer HD (2008) Epidermal growth factor receptor pathway analysis identifies amphiregulin as a key factor for cisplatin resistance of human breast cancer cells. J Biol Chem 283(2):739–750
55. Ishikawa N, Daigo Y, Takano A, Taniwaki M, Kato T, Hayama S, Murakami H, Takeshima Y, Inai K, Nishimura H et al (2005) Increases of amphiregulin and transforming growth factor-alpha in serum as predictors of poor response to gefitinib among patients with advanced non-small cell lung cancers. Cancer Res 65(20):9176–9184
56. Ritter CA, Perez-Torres M, Rinehart C, Guix M, Dugger T, Engelman JA, Arteaga CL (2007) Human breast cancer cells selected for resistance to trastuzumab in vivo overexpress epidermal growth factor receptor and ErbB ligands and remain dependent on the ErbB receptor network. Clin Cancer Res 13 (16):4909–4919
57. Wang H, Grzywacz B, Sukovich D, McCullar V, Cao Q, Lee AB, Blazar BR, Cornfield DN, Miller JS, Verneris MR (2007) The unexpected effect of cyclosporin A on CD56+CD16- and CD56+CD16+ natural killer cell subpopulations. Blood 110 (5):1530–1539
58. Franklin MC, Carey KD, Vajdos FF, Leahy DJ, de Vos AM, Sliwkowski MX (2004) Insights into ErbB signaling from the structure of the ErbB2-pertuzumab complex. Cancer Cell 5(4):317–328
59. Lemmon MA (2009) Ligand-induced ErbB receptor dimerization. Exp Cell Res 315 (4):638–648
60. Ferguson KM (2008) Structure-based view of epidermal growth factor receptor regulation. Annu Rev Biophys 37:353–373
61. Sato K (2013) Cellular functions regulated by phosphorylation of EGFR on Tyr845. Int J Mol Sci 14(6):10761–10790
62. Rotin D, Honegger AM, Margolis BL, Ullrich A, Schlessinger J (1992) Presence of SH2 domains of phospholipase C gamma 1 enhances substrate phosphorylation by increasing the affinity toward the epidermal growth factor receptor. J Biol Chem 267 (14):9678–9683
63. Cohen BD, Kiener PA, Green JM, Foy L, Fell HP, Zhang K (1996) The relationship between human epidermal growth-like factor receptor expression and cellular transformation in NIH3T3 cells. J Biol Chem 271 (48):30897–30903
64. Zrihan-Licht S, Deng B, Yarden Y, McShan G, Keydar I, Avraham H (1998) Csk homologous kinase, a novel signaling molecule, directly associates with the activated ErbB-2-receptor in breast cancer cells and inhibits their proliferation. J Biol Chem 273 (7):4065–4072
65. Keilhack H, Tenev T, Nyakatura E, Godovac-Zimmermann J, Nielsen L, Seedorf K, Bohmer FD (1998) Phosphotyrosine 1173 mediates binding of the protein-tyrosine phosphatase SHP-1 to the epidermal growth factor receptor and attenuation of receptor

signaling. J Biol Chem 273 (38):24839–24846

66. Hellyer NJ, Kim MS, Koland JG (2001) Heregulin-dependent activation of phosphoinositide 3-kinase and Akt via the ErbB2/ ErbB3 co-receptor. J Biol Chem 276 (45):42153–42161
67. Schulze WX, Deng L, Mann M (2005) Phosphotyrosine interactome of the ErbB-receptor kinase family. Mol Syst Biol 1:2005.0008
68. Kaushansky A, Gordus A, Budnik BA, Lane WS, Rush J, MacBeath G (2008) System-wide investigation of ErbB4 reveals 19 sites of Tyr phosphorylation that are unusually selective in their recruitment properties. Chem Biol 15 (8):808–817
69. Mitsudomi T, Yatabe Y (2010) Epidermal growth factor receptor in relation to tumor development: EGFR gene and cancer. FEBS J 277(2):301–308
70. Pawson T (1997) New impressions of Src and Hck [news; comment]. Nature 385 (6617):582–583. 585
71. Pearson G, Robinson F, Gibson TB, BE X, Karandikar M, Berman K, Cobb MH (2001) Mitogen-activated protein (MAP) kinase pathways: regulation and physiological functions. Endocr Rev 22(2):153–183
72. Avruch J (2007) MAP kinase pathways: the first twenty years. BBA-Mol Cell Res 1773 (8):1150–1160
73. Marshall CJ (1994) MAP kinase kinase kinase, MAP kinase kinase and MAP kinase. Curr Opin Genet Dev 4(1):82–89
74. Marshall CJ (1996) Cell signalling. Raf gets it together. Nature 383(6596):127–128
75. Wu P, Wee P, Jiang J, Chen X, Wang Z (2012) Differential regulation of transcription factors by location-specific EGF receptor signaling via a spatio-temporal interplay of ERK activation. PLoS One 7(9):e41354
76. Wahl M, Carpenter G (1991) Selective phospholipase C activation. BioEssays 13 (3):107–113
77. Kamat A, Carpenter G (1997) Phospholipase C-gamma1: regulation of enzyme function and role in growth factor-dependent signal transduction. Cytokine Growth Factor Rev 8 (2):109–117
78. Wells A, Gupta K, Chang P, Swindle S, Glading A, Shiraha H (1998) Epidermal growth factor receptor-mediated motility in fibroblasts. Microsc Res Tech 43(5):395–411
79. Gual P, Giordano S, Williams TA, Rocchi S, Van Obberghen E, Comoglio PM (2000) Sustained recruitment of phospholipase C-gamma to Gab1 is required for HGF-induced branching tubulogenesis. Oncogene 19 (12):1509–1518
80. Meyer RD, Latz C, Rahimi N (2003) Recruitment and activation of phospholipase Cgamma1 by vascular endothelial growth factor receptor-2 are required for tubulogenesis and differentiation of endothelial cells. J Biol Chem 278(18):16347–16355
81. Bivona TG, Perez DCI, Ahearn IM, Grana TM, Chiu VK, Lockyer PJ, Cullen PJ, Pellicer A, Cox AD, Philips MR (2003) Phospholipase Cgamma activates Ras on the Golgi apparatus by means of RasGRP1. Nature 424 (6949):694–698
82. Chou J, Burke NA, Iwabu A, Watkins SC, Wells A (2003) Directional motility induced by epidermal growth factor requires Cdc42. Exp Cell Res 287(1):47–56
83. Choi JH, Park JB, Bae SS, Yun S, Kim HS, Hong WP, Kim IS, Kim JH, Han MY, Ryu SH et al (2004) Phospholipase C-gamma1 is a guanine nucleotide exchange factor for dynamin-1 and enhances dynamin-1-dependent epidermal growth factor receptor endocytosis. J Cell Sci 117(Pt 17):3785–3795
84. Thomas SM, Coppelli FM, Wells A, Gooding WE, Song J, Kassis J, Drenning SD, Grandis JR (2003) Epidermal growth factor receptor-stimulated activation of phospholipase Cgamma-1 promotes invasion of head and neck squamous cell carcinoma. Cancer Res 63(17):5629–5635
85. Wells A, Grandis JR (2003) Phospholipase C-gamma1 in tumor progression. Clin Exp Metastasis 20(4):285–290
86. Wahl MI, Daniel TO, Carpenter G (1988) Antiphosphotyrosine recovery of phospholipase C activity after EGF treatment of A-431 cells. Science (New York, NY) 241 (4868):968–970
87. Margolis B, Rhee SG, Felder S, Mervic M, Lyall R, Levitzki A, Ullrich A, Zilberstein A, Schlessinger J (1989) EGF induces tyrosine phosphorylation of phospholipase C-II: a potential mechanism for EGF receptor signaling. Cell 57(7):1101–1107
88. Meisenhelder J, Suh PG, Rhee SG, Hunter T (1989) Phospholipase C-gamma is a substrate for the PDGF and EGF receptor protein-tyrosine kinases in vivo and in vitro. Cell 57 (7):1109–1122
89. Anderson D, Koch CA, Grey L, Ellis C, Moran MF, Pawson T (1990) Binding of SH2 domains of phospholipase C gamma 1,

GAP, and Src to activated growth factor receptors. Science (New York, NY) 250 (4983):979–982

90. Margolis B, Zilberstein A, Franks C, Felder S, Kremer S, Ullrich A, Rhee SG, Skorecki K, Schlessinger J (1990) Effect of phospholipase C-gamma overexpression on PDGF-induced second messengers and mitogenesis. Science (New York, NY) 248(4955):607–610
91. Kim HK, Kim JW, Zilberstein A, Margolis B, Kim JG, Schlessinger J, Rhee SG (1991) PDGF stimulation of inositol phospholipid hydrolysis requires PLC-gamma 1 phosphorylation on tyrosine residues 783 and 1254. Cell 65(3):435–441
92. Ronnstrand L, Mori S, Arridsson AK, Eriksson A, Wernstedt C, Hellman U, Claesson-Welsh L, Heldin CH (1992) Identification of two C-terminal autophosphorylation sites in the PDGF beta-receptor: involvement in the interaction with phospholipase C-gamma. EMBO J 11(11):3911–3919
93. Jang IH, Lee S, Park JB, Kim JH, Lee CS, Hur EM, Kim IS, Kim KT, Yagisawa H, Suh PG et al (2003) The direct interaction of phospholipase C-gamma 1 with phospholipase D2 is important for epidermal growth factor signaling. J Biol Chem 278 (20):18184–18190
94. Wang Y, Wu J, Wang Z (2006) Akt Binds to and Phosphorylates Phospholipase C-{gamma}1 in Response to Epidermal Growth Factor. Mol Biol Cell 17(5):2267–2277
95. Ye K, Aghdasi B, Luo HR, Moriarity JL, FY W, Hong JJ, Hurt KJ, Bae SS, Suh PG, Snyder SH (2002) Phospholipase C gamma 1 is a physiological guanine nucleotide exchange factor for the nuclear GTPase PIKE. Nature 415(6871):541–544
96. Li S, Wang Q, Wang Y, Chen X, Wang Z (2009) PLC-gamma1 and Rac1 coregulate EGF-induced cytoskeleton remodeling and cell migration. Mol Endocrinol 23 (6):901–913
97. Wang Z, Gluck S, Zhang L, Moran MF (1998) Requirement for phospholipase C-gamma1 enzymatic activity in growth factor-induced mitogenesis. Mol Cell Biol 18 (1):590–597
98. Wells A, Kassis J, Solava J, Turner T, Lauffenburger DA (2002) Growth factor-induced cell motility in tumor invasion. Acta Oncol 41 (2):124–130
99. Downward J (1998) Mechanisms and consequences of activation of protein kinase B/Akt. Curr Opin Cell Biol 10(2):262–267
100. Castellano E, Downward J (2011) RAS interaction with PI3K: more than just another effector pathway. Genes Cancer 2 (3):261–274
101. Burgering BM, Coffer PJ (1995) Protein kinase B (c-Akt) in phosphatidylinositol-3-OH kinase signal transduction [see comments]. Nature 376(6541):599–602
102. Okano J, Gaslightwala I, Birnbaum MJ, Rustgi AK, Nakagawa H (2000) Akt/protein kinase B isoforms are differentially regulated by epidermal growth factor stimulation. J Biol Chem 275(40):30934–30942
103. Hanada M, Feng J, Hemmings BA (2004) Structure, regulation and function of PKB/AKT–a major therapeutic target. Biochim Biophys Acta 1697(1-2):3–16
104. Quesnelle KM, Boehm AL, Grandis JR (2007) STAT-mediated EGFR signaling in cancer. J Cell Biochem 102(2):311–319
105. Yeatman TJ (2004) A renaissance for SRC. Nat Rev Cancer 4(6):470–480
106. Summy JM, Gallick GE (2006) Treatment for advanced tumors: SRC reclaims center stage. Clin Cancer Res 12(5):1398–1401
107. Amata I, Maffei M, Pons M (2014) Phosphorylation of unique domains of Src family kinases. Front Genet 5:181
108. Maa MC, Leu TH, McCarley DJ, Schatzman RC, Parsons SJ (1995) Potentiation of epidermal growth factor receptor-mediated oncogenesis by c-Src: implications for the etiology of multiple human cancers. Proc Natl Acad Sci U S A 92(15):6981–6985
109. Tice DA, Biscardi JS, Nickles AL, Parsons SJ (1999) Mechanism of biological synergy between cellular Src and epidermal growth factor receptor. Proc Natl Acad Sci U S A 96 (4):1415–1420
110. Egloff AM, Grandis JR (2008) Targeting epidermal growth factor receptor and SRC pathways in head and neck cancer. Semin Oncol 35(3):286–297
111. Scaltriti M, Baselga J (2006) The epidermal growth factor receptor pathway: a model for targeted therapy. Clin Cancer Res 12 (18):5268–5272
112. Lurje G, Lenz HJ (2009) EGFR signaling and drug discovery. Oncology 77(6):400–410
113. Leu TH, Maa MC (2003) Functional implication of the interaction between EGF receptor and c-Src. Front Biosci 8:s28–s38
114. Cohen S, Fava RA (1985) Internalization of functional epidermal growth factor: receptor/kinase complexes in A-431 cells. JBiolChem 260(22):12351–12358

115. Kay DG, Lai WH, Uchihashi M, Khan MN, Posner BI, Bergeron JJ (1986) Epidermal growth factor receptor kinase translocation and activation in vivo. J Biol Chem 261 (18):8473–8480

116. Lai WH, Cameron PH, Doherty JJ II, Posner BI, Bergeron JJ (1989) Ligand-mediated autophosphorylation activity of the epidermal growth factor receptor during internalization. J Cell Biol 109(6 Pt 1):2751–2760

117. Di Guglielmo GM, Baass PC, WJ O, Posner BI, Bergeron JJ (1994) Compartmentalization of SHC, GRB2 and mSOS, and hyperphosphorylation of Raf-1 by EGF but not insulin in liver parenchyma. EMBO J 13 (18):4269–4277

118. Wang Z, Tung PS, Moran MF (1996) Association of p120 ras GAP with endocytic components and colocalization with epidermal growth factor (EGF) receptor in response to EGF stimulation. Cell Growth Differ 7 (1):123–133

119. Fukazawa T, Miyake S, Band V, Band H (1996) Tyrosine phosphorylation of Cbl upon epidermal growth factor (EGF) stimulation and its association with EGF receptor and downstream signaling proteins. J Biol Chem 271(24):14554–14559

120. Levkowitz G, Waterman H, Zamir E, Kam Z, Oved S, Langdon WY, Beguinot L, Geiger B, Yarden Y (1998) c-Cbl/Sli-1 regulates endocytic sorting and ubiquitination of the epidermal growth factor receptor. Genes Dev 12 (23):3663–3674

121. Clague MJ, Urbe S (2001) The interface of receptor trafficking and signalling. J Cell Sci 114(Pt 17):3075–3081

122. Haugh JM, Huang AC, Wiley HS, Wells A, Lauffenburger DA (1999) Internalized epidermal growth factor receptors participate in the activation of p21(ras) in fibroblasts. J Biol Chem 274(48):34350–34360

123. Kuruvilla R, Ye H, Ginty DD (2000) Spatially and functionally distinct roles of the PI3-K effector pathway during NGF signaling in sympathetic neurons. Neuron 27 (3):499–512

124. Slepnev VI, De Camilli P (2000) Accessory factors in clathrin-dependent synaptic vesicle endocytosis. Nat Rev Neurosci 1(3):161–172

125. Burke P, Schooler K, Wiley HS (2001) Regulation of epidermal growth factor receptor signaling by endocytosis and intracellular trafficking. Mol Biol Cell 12(6):1897–1910

126. Miller WE, Lefkowitz RJ (2001) Expanding roles for beta-arrestins as scaffolds and adapters in GPCR signaling and trafficking. Curr Opin Cell Biol 13(2):139–145

127. Vieira AV, Lamaze C, Schmid SL (1996) Control of EGF receptor signaling by clathrin-mediated endocytosis. Science (New York, NY) 274(5295):2086–2089

128. Ahn S, Maudsley S, Luttrell LM, Lefkowitz RJ, Daaka Y (1999) Src-mediated tyrosine phosphorylation of dynamin is required for beta2-adrenergic receptor internalization and mitogen-activated protein kinase signaling. JBiolChem 274(3):1185–1188

129. Maudsley S, Pierce KL, Zamah AM, Miller WE, Ahn S, Daaka Y, Lefkowitz RJ, Luttrell LM (2000) The beta(2)-adrenergic receptor mediates extracellular signal-regulated kinase activation via assembly of a multi-receptor complex with the epidermal growth factor receptor. J Biol Chem 275(13):9572–9580

130. Shen Y, Xu L, Foster DA (2001) Role for phospholipase D in receptor-mediated endocytosis. Mol Cell Biol 21(2):595–602

131. Johannessen LE, Ringerike T, Molnes J, Madshus IH (2000) Epidermal growth factor receptor efficiently activates mitogen-activated protein kinase in HeLa cells and Hep2 cells conditionally defective in clathrin-dependent endocytosis. Exp Cell Res 260(1):136–145

132. Wong ES, Fong CW, Lim J, Yusoff P, Low BC, Langdon WY, Guy GR (2002) Sprouty2 attenuates epidermal growth factor receptor ubiquitylation and endocytosis, and consequently enhances Ras/ERK signalling. EMBO J 21(18):4796–4808

133. Ceresa BP, Kao AW, Santeler SR, Pessin JE (1998) Inhibition of clathrin-mediated endocytosis selectively attenuates specific insulin receptor signal transduction pathways. Mol Cell Biol 18(7):3862–3870

134. Di Fiore PP, De Camilli P (2001) Endocytosis and signaling. an inseparable partnership. Cell 106(1):1–4

135. Wiley HS, Burke PM (2001) Regulation of receptor tyrosine kinase signaling by endocytic trafficking. Traffic 2(1):12–18

136. Wang Y, Pennock S, Chen X, Wang Z (2002) Internalization of inactive EGF receptor into endosomes and the subsequent activation of endosome-associated EGF receptors. Epidermal growth factor. Sci STKE 2002(161):L17

137. Wang Y, Pennock S, Chen X, Wang Z (2002) Endosomal signaling of epidermal growth factor receptor stimulates signal transduction pathways leading to cell survival. Mol Cell Biol 22(20):7279–7290

138. Pennock S, Wang Z (2003) Stimulation of cell proliferation by endosomal epidermal growth factor receptor as revealed through two distinct phases of signaling. Mol Cell Biol 23(16):5803–5815
139. Di Guglielmo GM, Le RC, Goodfellow AF, Wrana JL (2003) Distinct endocytic pathways regulate TGF-beta receptor signalling and turnover. Nat Cell Biol 5(5):410–421
140. Dobrowolski R, De Robertis EM (2012) Endocytic control of growth factor signalling: multivesicular bodies as signalling organelles. Nat Rev Mol Cell Biol 13(1):53–60
141. Sorkin A, Waters CM (1993) Endocytosis of growth factor receptors. BioEssays 15 (6):375–382
142. Lee FS, Rajagopal R, Chao MV (2002) Distinctive features of Trk neurotrophin receptor transactivation by G protein-coupled receptors. Cytokine Growth Factor Rev 13 (1):11–17
143. Ni CY, Murphy MP, Golde TE, Carpenter G (2001) gamma -Secretase cleavage and nuclear localization of ErbB-4 receptor tyrosine kinase. Science (New York, NY) 294 (5549):2179–2181
144. Medina M, Dotti CG (2003) RIPped out by presenilin-dependent gamma-secretase. Cell Signal 15(9):829–841
145. Rio C, Buxbaum JD, Peschon JJ, Corfas G (2000) Tumor necrosis factor-alpha-converting enzyme is required for cleavage of erbB4/HER4. J Biol Chem 275(14):10379–10387
146. Omerovic J, Santangelo L, Puggioni EM, Marrocco J, Dall'Armi C, Palumbo C, Belleudi F, Di Marcotullio L, Frati L, Torrisi MR et al (2007) The E3 ligase Aip4/Itch ubiquitinates and targets ErbB-4 for degradation. FASEB J 21(11):2849–2862
147. Omerovic J, Puggioni EM, Napoletano S, Visco V, Fraioli R, Frati L, Gulino A, Alimandi M (2004) Ligand-regulated association of ErbB-4 to the transcriptional coactivator YAP65 controls transcription at the nuclear level. Exp Cell Res 294(2):469–479
148. Lanning CC, Daddona JL, Ruiz-Velasco R, Shafer SH, Williams CL (2004) The Rac1 C-terminal polybasic region regulates the nuclear localization and protein degradation of Rac1. J Biol Chem 279(42):44197–44210
149. Linggi B, Carpenter G (2006) ErbB-4 s80 intracellular domain abrogates ETO2-dependent transcriptional repression. J Biol Chem 281(35):25373–25380
150. Zhu Y, Sullivan LL, Nair SS, Williams CC, Pandey AK, Marrero L, Vadlamudi RK, Jones FE (2006) Coregulation of estrogen receptor by ERBB4/HER4 establishes a growth-promoting autocrine signal in breast tumor cells. Cancer Res 66(16):7991–7998
151. Sardi SP, Murtie J, Koirala S, Patten BA, Corfas G (2006) Presenilin-dependent ErbB4 nuclear signaling regulates the timing of astrogenesis in the developing brain. Cell 127(1):185–197
152. Gilmore-Hebert M, Ramabhadran R, Stern DF (2010) Interactions of ErbB4 and Kap1 connect the growth factor and DNA damage response pathways. Mol Cancer Res 8 (10):1388–1398
153. Sundvall M, Veikkolainen V, Kurppa K, Salah Z, Tvorogov D, van Zoelen EJ, Aqeilan R, Elenius K (2010) Cell death or survival promoted by alternative isoforms of ErbB4. Mol Biol Cell 21(23):4275–4286
154. Lopez-Bigas N, Audit B, Ouzounis C, Parra G, Guigo R (2005) Are splicing mutations the most frequent cause of hereditary disease? FEBS Lett 579(9):1900–1903
155. Cai Z, Zhang H, Liu J, Berezov A, Murali R, Wang Q, Greene MI (2010) Targeting erbB receptors. Semin Cell Dev Biol 21 (9):961–966
156. Riese DJ II, Stern DF (1998) Specificity within the EGF family/ErbB receptor family signaling network. BioEssays 20(1):41–48
157. Gusterson B, Cowley G, McIlhinney J, Ozanne B, Fisher C, Reeves B (1985) Evidence for increased epidermal growth factor receptors in human sarcomas. Int J Cancer 36 (6):689–693
158. Kamio T, Shigematsu K, Sou H, Kawai K, Tsuchiyama H (1990) Immunohistochemical expression of epidermal growth factor receptors in human adrenocortical carcinoma. Hum Pathol 21(3):277–282
159. Marti U, Burwen SJ, Wells A, Barker ME, Huling S, Feren AM, Jones AL (1991) Localization of epidermal growth factor receptor in hepatocyte nuclei. Hepatology (Baltimore, MD) 13(1):15–20
160. Tervahauta A, Syrjanen S, Syrjanen K (1994) Epidermal growth factor receptor, c-erbB-2 proto-oncogene and estrogen receptor expression in human papillomavirus lesions of the uterine cervix. Int J Gynecol Pathol 13(3):234–240
161. Lin SY, Makino K, Xia W, Matin A, Wen Y, Kwong KY, Bourguignon L, Hung MC (2001) Nuclear localization of EGF receptor and its potential new role as a transcription factor. Nat Cell Biol 3(9):802–808

162. Carpenter G, Liao HJ (2009) Trafficking of receptor tyrosine kinases to the nucleus. Exp Cell Res 315(9):1556–1566
163. Wang YN, Yamaguchi H, Hsu JM, Hung MC (2010) Nuclear trafficking of the epidermal growth factor receptor family membrane proteins. Oncogene 29(28):3997–4006
164. Jones SM, Kazlauskas A (2001) Growth factor-dependent signaling and cell cycle progression. FEBS Lett 490(3):110–116
165. Massague J (2004) G1 cell-cycle control and cancer. Nature 432(7015):298–306
166. Jones SM, Kazlauskas A (2001) Growth-factor-dependent mitogenesis requires two distinct phases of signalling. Nat Cell Biol 3 (2):165–172
167. Mardin BR, Isokane M, Cosenza MR, Kramer A, Ellenberg J, Fry AM, Schiebel E (2013) EGF-induced centrosome separation promotes mitotic progression and cell survival. Dev Cell 25(3):229–240
168. Astuti P, Pike T, Widberg C, Payne E, Harding A, Hancock J, Gabrielli B (2009) MAPK pathway activation delays G2/M progression by destabilizing Cdc25B. J Biol Chem 284 (49):33781–33788
169. Nam HJ, Kim S, Lee MW, Lee BS, Hara T, Saya H, Cho H, Lee JH (2008) The ERK-RSK1 activation by growth factors at G2 phase delays cell cycle progression and reduces mitotic aberrations. Cell Signal 20 (7):1349–1358
170. Park YY, Nam HJ, Lee JH (2007) Hepatocyte growth factor at S phase induces G2 delay through sustained ERK activation. Biochem Biophys Res Commun 356(1):300–305
171. Dangi S, Chen FM, Shapiro P (2006) Activation of extracellular signal-regulated kinase (ERK) in G(2) phase delays mitotic entry through p21(CIP1). Cell Prolif 39 (4):261–279
172. Liu X, Yan S, Zhou T, Terada Y, Erikson RL (2004) The MAP kinase pathway is required for entry into mitosis and cell survival. Oncogene 23(3):763–776
173. Roberts EC, Shapiro PS, Nahreini TS, Pages G, Pouyssegur J, Ahn NG (2002) Distinct cell cycle timing requirements for extracellular signal-regulated kinase and phosphoinositide 3-kinase signaling pathways in somatic cell mitosis. Mol Cell Biol 22(20):7226–7241
174. Hanahan D, Weinberg RA (2011) Hallmarks of cancer: the next generation. Cell 144 (5):646–674
175. Marzo I, Naval J (2013) Antimitotic drugs in cancer chemotherapy: promises and pitfalls. Biochem Pharmacol 86(6):703–710
176. Kiyokawa N, Lee EK, Karunagaran D, Lin SY, Hung MC (1997) Mitosis-specific negative regulation of epidermal growth factor receptor, triggered by a decrease in ligand binding and dimerization, can be overcome by overexpression of receptor. J Biol Chem 272 (30):18656–18665
177. Gomez-Cambronero J (1999) p42-MAP kinase is activated in EGF-stimulated interphase but not in metaphase-arrested HeLa cells. FEBS Lett 443(2):126–130
178. Dangi S, Shapiro P (2005) Cdc2-mediated inhibition of epidermal growth factor activation of the extracellular signal-regulated kinase pathway during mitosis. J Biol Chem 280(26):24524–24531
179. Liu L, Shi H, Chen X, Wang Z (2011) Regulation of EGF-stimulated EGF receptor endocytosis during M phase. Traffic (Copenhagen) 12(2):201–217
180. Wee P, Shi H, Jiang J, Wang Y, Wang Z (2015) EGF stimulates the activation of EGF receptors and the selective activation of major signaling pathways during mitosis. Cell Signal 27(3):638–651
181. Daub H, Weiss FU, Wallasch C, Ullrich A (1996) Role of transactivation of the EGF receptor in signalling by G-protein-coupled receptors. Nature 379(6565):557–560
182. Daub H, Wallasch C, Lankenau A, Herrlich A, Ullrich A (1997) Signal characteristics of G protein-transactivated EGF receptor. EMBO J 16(23):7032–7044
183. Reichardt HM, Kaestner KH, Tuckermann J, Kretz O, Wessely O, Bock R, Gass P, Schmid W, Herrlich P, Angel P et al (1998) DNA binding of the glucocorticoid receptor is not essential for survival. Cell 93(4):531–541
184. Heeneman S, Haendeler J, Saito Y, Ishida M, Berk BC (2000) Angiotensin II induces transactivation of two different populations of the platelet-derived growth factor beta receptor. Key role for the p66 adaptor protein Shc. J Biol Chem 275(21):15926–15932
185. Lee FS, Chao MV (2001) Activation of Trk neurotrophin receptors in the absence of neurotrophins. Proc Natl Acad Sci U S A 98 (6):3555–3560
186. Tu H, Xu C, Zhang W, Liu Q, Rondard P, Pin JP, Liu J (2010) GABAB receptor activation protects neurons from apoptosis via IGF-1 receptor transactivation. J Neurosci 30 (2):749–759
187. Oligny-Longpre G, Corbani M, Zhou J, Hogue M, Guillon G, Bouvier M (2012) Engagement of beta-arrestin by transactivated insulin-like growth factor receptor is needed

for V2 vasopressin receptor-stimulated ERK1/2 activation. Proc Natl Acad Sci U S A 109(17):E1028–E1037

188. Tanimoto T, Jin ZG, Berk BC (2002) Transactivation of vascular endothelial growth factor (VEGF) receptor Flk-1/KDR is involved in sphingosine 1-phosphate-stimulated phosphorylation of Akt and endothelial nitric-oxide synthase (eNOS). J Biol Chem 277 (45):42997–43001
189. Seye CI, Yu N, Gonzalez FA, Erb L, Weisman GA (2004) The P2Y2 nucleotide receptor mediates vascular cell adhesion molecule-1 expression through interaction with VEGF receptor-2 (KDR/Flk-1). J Biol Chem 279 (34):35679–35686
190. Rauch BH, Millette E, Kenagy RD, Daum G, Clowes AW (2004) Thrombin- and factor Xa-induced DNA synthesis is mediated by transactivation of fibroblast growth factor receptor-1 in human vascular smooth muscle cells. Circ Res 94(3):340–345
191. Liebmann C (2011) EGF receptor activation by GPCRs: an universal pathway reveals different versions. Mol Cell Endocrinol 331(2):222–231
192. George AJ, Hannan RD, Thomas WG (2013) Unravelling the molecular complexity of GPCR-mediated EGFR transactivation using functional genomics approaches. FEBS J 280 (21):5258–5268
193. Cattaneo F, Guerra G, Parisi M, De Marinis M, Tafuri D, Cinelli M, Ammendola R (2014) Cell-surface receptors transactivation mediated by g protein-coupled receptors. Int J Mol Sci 15(11):19700–19728
194. Wang Z (2016) Transactivation of epidermal growth factor receptor by G protein-coupled receptors: recent progress, challenges and future research. Int J Mol Sci 17(1):pii: E95
195. Prenzel N, Fischer OM, Streit S, Hart S, Ullrich A (2001) The epidermal growth factor receptor family as a central element for cellular signal transduction and diversification. Endocr Relat Cancer 8(1):11–31
196. Ohtsu H, Dempsey PJ, Eguchi S (2006) ADAMs as mediators of EGF receptor transactivation by G protein-coupled receptors. Am J Physiol Cell Physiol 291(1):C1–10
197. Etkovitz N, Tirosh Y, Chazan R, Jaldety Y, Daniel L, Rubinstein S, Breitbart H (2009) Bovine sperm acrosome reaction induced by G-protein-coupled receptor agonists is mediated by epidermal growth factor receptor transactivation. Dev Biol 334(2):447–457
198. Herbst RS, Sandler A (2008) Bevacizumab and erlotinib: a promising new approach to the treatment of advanced NSCLC. Oncologist 13(11):1166–1176
199. Cataldo VD, Gibbons DL, Perez-Soler R, Quintas-Cardama A (2011) Treatment of non-small-cell lung cancer with erlotinib or gefitinib. N Engl J Med 364(10):947–955
200. Bhargava R, Gerald WL, Li AR, Pan Q, Lal P, Ladanyi M, Chen B (2005) EGFR gene amplification in breast cancer: correlation with epidermal growth factor receptor mRNA and protein expression and HER-2-status and absence of EGFR-activating mutations. Mod Pathol 18(8):1027–1033
201. Hsu JL, Hung MC (2016) The role of HER2, EGFR, and other receptor tyrosine kinases in breast cancer. Cancer Metastasis Rev 35 (4):575–588
202. Lehmann BD, Bauer JA, Chen X, Sanders ME, Chakravarthy AB, Shyr Y, Pietenpol JA (2011) Identification of human triple-negative breast cancer subtypes and preclinical models for selection of targeted therapies. J Clin Invest 121(7):2750–2767
203. Masuda H, Zhang D, Bartholomeusz C, Doihara H, Hortobagyi GN, Ueno NT (2012) Role of epidermal growth factor receptor in breast cancer. Breast Cancer Res Treat 136 (2):331–345
204. Rimawi MF, Schiff R, Osborne CK (2015) Targeting HER2 for the treatment of breast cancer. Annu Rev Med 66:111–128
205. Ross JS, Slodkowska EA, Symmans WF, Pusztai L, Ravdin PM, Hortobagyi GN (2009) The HER-2 receptor and breast cancer: ten years of targeted anti-HER-2 therapy and personalized medicine. Oncologist 14(4):320–368
206. De Mattos-Arruda L, Cortes J (2013) Use of pertuzumab for the treatment of HER2-positive metastatic breast cancer. Adv Ther 30(7):645–658
207. Dawood S, Broglio K, Buzdar AU, Hortobagyi GN, Giordano SH (2010) Prognosis of women with metastatic breast cancer by HER2 status and trastuzumab treatment: an institutional-based review. J Clin Oncol 28 (1):92–98
208. Nahta R, Yu D, Hung MC, Hortobagyi GN, Esteva FJ (2006) Mechanisms of disease: understanding resistance to HER2-targeted therapy in human breast cancer. Nat Clin Pract Oncol 3(5):269–280
209. Arteaga CL, Sliwkowski MX, Osborne CK, Perez EA, Puglisi F, Gianni L (2012) Treatment of HER2-positive breast cancer: current status and future perspectives. Nat Rev Clin Oncol 9(1):16–32
210. Baselga J, Cortes J, Kim SB, Im SA, Hegg R, Im YH, Roman L, Pedrini JL, Pienkowski T, Knott A et al (2012) Pertuzumab plus

trastuzumab plus docetaxel for metastatic breast cancer. N Engl J Med 366(2):109–119

211. Bose R, Kavuri SM, Searleman AC, Shen W, Shen D, Koboldt DC, Monsey J, Goel N, Aronson AB, Li S et al (2013) Activating HER2 mutations in HER2 gene amplification negative breast cancer. Cancer Discov 3(2):224–237
212. Arcila ME, Chaft JE, Nafa K, Roy-Chowdhuri S, Lau C, Zaidinski M, Paik PK, Zakowski MF, Kris MG, Ladanyi M (2012) Prevalence, clinicopathologic associations, and molecular spectrum of ERBB2 (HER2) tyrosine kinase mutations in lung adenocarcinomas. Clin Cancer Res 18(18):4910–4918
213. Bang YJ, Van Cutsem E, Feyereislova A, Chung HC, Shen L, Sawaki A, Lordick F, Ohtsu A, Omuro Y, Satoh T et al (2010) Trastuzumab in combination with chemotherapy versus chemotherapy alone for treatment of HER2-positive advanced gastric or gastro-oesophageal junction cancer (ToGA): a phase 3, open-label, randomised controlled trial. Lancet (London) 376(9742):687–697
214. Skinner A, Hurst HC (1993) Transcriptional regulation of the c-erbB-3 gene in human breast carcinoma cell lines. Oncogene 8 (12):3393–3401
215. Carpenter G (2003) ErbB-4: mechanism of action and biology. Exp Cell Res 284 (1):66–77
216. Jones FE (2008) HER4 intracellular domain (4ICD) activity in the developing mammary gland and breast cancer. J Mammary Gland Biol Neoplasia 13(2):247–258
217. Hollmen M, Elenius K (2010) Potential of ErbB4 antibodies for cancer therapy. Fut Oncol (London) 6(1):37–53
218. Junttila TT, Sundvall M, Lundin M, Lundin J, Tanner M, Harkonen P, Joensuu H, Isola J, Elenius K (2005) Cleavable ErbB4 isoform in estrogen receptor-regulated growth of breast cancer cells. Cancer Res 65 (4):1384–1393
219. Steffensen KD, Waldstrom M, Andersen RF, Olsen DA, Jeppesen U, Knudsen HJ, Brandslund I, Jakobsen A (2008) Protein levels and gene expressions of the epidermal growth factor receptors, HER1, HER2, HER3 and HER4 in benign and malignant ovarian tumors. Int J Oncol 33(1):195–204
220. Prickett TD, Agrawal NS, Wei X, Yates KE, Lin JC, Wunderlich JR, Cronin JC, Cruz P, Rosenberg SA, Samuels Y (2009) Analysis of the tyrosine kinome in melanoma reveals recurrent mutations in ERBB4. Nat Genet 41(10):1127–1132
221. Ding L, Getz G, Wheeler DA, Mardis ER, McLellan MD, Cibulskis K, Sougnez C, Greulich H, Muzny DM, Morgan MB et al (2008) Somatic mutations affect key pathways in lung adenocarcinoma. Nature 455 (7216):1069–1075
222. Kurppa KJ, Denessiouk K, Johnson MS, Elenius K (2016) Activating ERBB4 mutations in non-small cell lung cancer. Oncogene 35 (10):1283–1291

Chapter 2

New Insights from Drosophila into the Regulation of EGFR Signaling

Nicholas Harden

Abstract

Genetic analysis of Egfr signaling in Drosophila has a long-standing track record of making important contributions to our understanding of the Egfr pathway. While the central Ras/MAPK pathway has long been well defined, there is much to learn with regard to its cross talk with other pathways and how it is regulated. A better understanding of the regulation of Egfr signaling is of particular interest with regard to the participation of misregulated Egfr signaling in tumorigenesis. Recent studies in the fly have led to some surprising results, identifying regulators of Egfr acting in unexpected ways.

Key words EGFR, *Drosophila*, Signaling, Ras/MAPK, Tumorigenesis

1 Introduction

Genetic analysis in Drosophila has made tremendous contributions to our understanding of Egfr signaling and its regulation. In particular, the fly has been very informative in revealing details of the temporal and spatial regulation of Egfr signaling, including providing important insights into processing of the Egfr ligand Spitz and transcriptional induction of positive- and negative-feedback on the pathways [1, 2]. Positive regulators turned on by the pathway include the ligand Vein and Rhomboid, a protease that cleaves Spitz, whereas negative regulators include Argos, which forms nonfunctional heterodimers with Spitz, Kekkon, which forms heterodimers with Egfr, and Sprouty, an inhibitor of the Ras/MAPK pathway. Egfr signaling regulates the development of multiple tissues throughout the life cycle of the fly, including the eye and wing. Recent work in the latter two tissues has identified four new regulators of Egfr signaling, including the surprising findings that two well-known oncogenes antagonize the pathway, and thus Egfr signaling may have a tumor suppressor function (Fig. 1) [3–6].

Zhixiang Wang (ed.), *ErbB Receptor Signaling: Methods and Protocols*, Methods in Molecular Biology, vol. 1652, DOI 10.1007/978-1-4939-7219-7_2, © Springer Science+Business Media LLC 2017

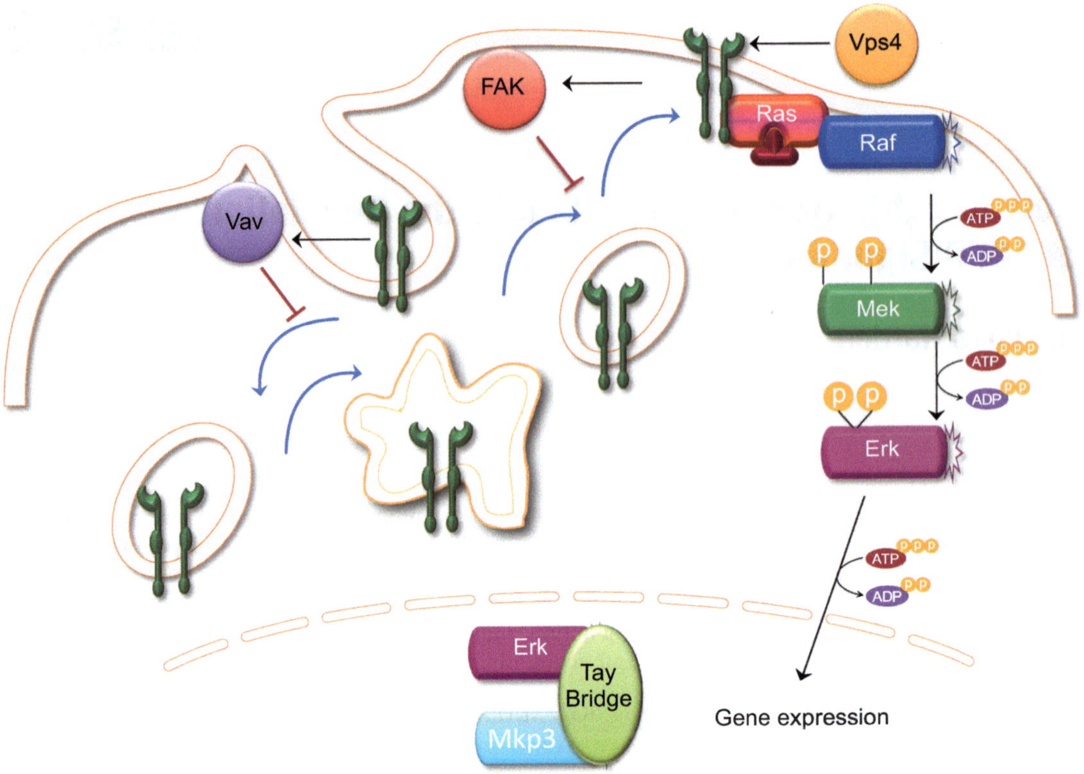

Fig. 1 Recent studies in Drosophila have identified four new regulators of Egfr signaling, with three of these, FAK, Vav, and Vps, likely acting upstream of pathway activation. FAK might inhibit Egfr signaling by repressing recycling of the receptor, whereas Vav might block endocytosis required for maximal Egfr signaling. Vps4 might be acting independently of protein trafficking at the receptor level, whereas Tay Bridge appears to negatively regulate Egfr signaling at the other end of the pathway in the nucleus

2 FAK

Focal Adhesion Kinase (FAK) is a non-receptor tyrosine kinase that interacts with integrins in focal adhesions and has been implicated in tumor cell survival and invasion [7]. FAK interacts with and is phosphorylated by several receptor tyrosine kinases including RET. This interaction was further investigated in Drosophila by Macagno and colleagues, who made the unexpected finding that FAK operates in a negative feedback loop, being activated by excessive RET activity but then inhibiting RET-driven phenotypes [4]. The authors went on to extend their study to FAK regulation of Egfr signaling, and the findings were similar: Egfr induced FAK activation and co-overexpression of FAK with Egfr in the eye suppressed the effects of excessive Egfr signaling. FAK's ability to inhibit signaling by two RTKs appears to be through impairment of the Ras/MAPK pathway, and this FAK function is conserved in a breast cancer cell line MDA-MB-231 [4]. Furthermore, in these cells FAK inhibits Egfr recycling to the membrane, which could be a mechanism for impairment of Egfr signaling. These various results indicate that FAK might act as a tumor suppressor in cells with elevated RTK activity, and suggest that caution should be observed in the use of FAK inhibitors in cancer therapies.

3 Vav

Vav is a guanine nucleotide exchange factor for the Rho subfamily small GTPases expressed in many cancers, which has been shown to be activated by Egfr [8]. Martin-Bermudo et al. recently studied the interaction of Vav with Egfr signaling using the Drosophila eye, and made the surprising discovery that rather than functioning as a oncogene and positive effector downstream of Egfr, Vav inhibits Egfr signaling [3]. Thus, as with FAK, Drosophila genetics has revealed potential tumor suppressor function for a well-known oncogene. *vav* mutant pupal retinas had increased cell numbers including extra photoreceptors. Several readouts for the Egfr pathway showed that consistent with the increased cell proliferation, Egfr signaling was elevated in *vav* mutant ommatidia. Strikingly, halving the dose of the Egfr ligand Spitz in *vav* mutants substantially rescued the eye defect, confirming that Vav antagonizes Egfr signaling in the eye. By creating new *vav* alleles using CRISPR/Cas9, the authors were able to show that the inhibitory function of Vav required both the GEF activity and binding to Egfr.

As with FAK, a possible area for Egfr regulation by Vav could be in protein trafficking, which has context-specific effects on Egfr signaling and might explain the unexpected finding that two well-established oncogenes could act as tumor suppressors with regard to Egfr signaling in a particular tissue. The regulation of Egfr signaling by endocytosis is complex and Egfr internalization can make both positive and negative contributions to Egfr signaling [9]. In Hela cells, Vav2 expression delays Egfr endocytosis, whereas knockdown of Vav2 enhances Egfr degradation and inhibits cell proliferation. In these cells, Vav has an oncogenic role, but in a tissue where endocytosis is required for maximal Egfr signaling Vav could function as a tumor suppressor. Consistent with this, work on the vesicle protein Myopic indicates that endocytosis is required for efficient Egfr signaling in the Drosophila eye [10].

4 Vsp4

Thus far, we have seen two oncogenes acting more like tumor suppressors with regard to Egfr signaling in Drosophila, possibly through effects on Egfr trafficking. The surprises continue with the recent finding that a known participant in endocytosis as part of the

ESCRT-III complex, Vps4, positively regulates Egfr signaling in an endocytosis-independent manner [6]. In studies performed in the Drosophila eye and wind discs, Legent showed that clones of *Vps4* mutant cells showed reduced expression of Egfr target genes despite having increased levels of Egfr protein. Epistasis analysis positioned Vps4 at the level of Egfr activation, with expression of the processed version of the Spitz ligand unable to rescue Egfr-dependent photoreceptor differentiation in *Vps4* mutant cells whereas activated Ras could. The effect of loss of Vps4 on Egfr signaling was opposite to that of mutations in other ESCRT-III complex genes, which prompted Legent et al. to determine if Vps4 was functioning in endocytosis in its regulation of Egfr signaling [6, 11]. The authors addressed this by performing epistasis analysis in the eye using clones of cells double mutant for *Vps4* and *shibire* (*shi*), which encodes dynamin, a GTPase that internalizes Egfr, and is a negative regulator of Egfr signaling [12, 13]. The increased Egfr signaling in *shi* mutant eye tissue causes excessive photoreceptor differentiation, but this is blocked when cells are additionally mutant for Vps4, suggesting that Vps4 is required for Egfr activation upstream of endocytosis. Further studies on the effects of Vps4 on Egfr signaling in the ovary were consistent with a model in which Vps4 favors Egfr signaling in an endocytosis-independent manner. It remains to be determined how Vps4 contributes to Egfr signaling; it does not appear to be at the level of transcription or translation of the receptor; one possibility proposed by Legent et al. is that it could be acting indirectly on Egfr at the cell surface through effects on lipid raft composition.

5 Tay Bridge

For our last new regulator of Egfr signaling, we move from the cell surface to the other end of the pathway in the nucleus. Tay Bridge (Tay) is a nuclear protein that was picked up in a gain-of-function screen for genes that when overexpressed cause effects on wing vein differentiation [5]. Overexpression of Tay caused the elimination of longitudinal veins, a phenotype similar to loss of Egfr signaling. Conversely, loss of Tay resulted in excess wing vein differentiation similar to excessive Egfr signaling. Genetic interaction studies were consistent with Tay functioning as a negative regulator of Egfr signaling, for example loss of Tay enhanced the extra vein formation caused by expression of RasV12 or ectopic Rhomboid. Furthermore, Tay overexpression enhanced the loss-of –vein phenotype caused by reduction of Egfr signaling. To determine if Tay was directly regulating the Egfr pathway, Molnar and de Celis looked at several readouts for the pathway: the accumulation of di-Phosphorylated Erk (dP-Erk) and expression of Egfr target genes. Consistent with Tay functioning as a negative regulator of the pathway, these readouts were reduced with Tay overexpression and increased with loss of Tay.

At what level is Tay regulating the Egfr pathway? Tay is a nuclear protein and the authors focused on nuclear events in Egfr signaling. Following its production by Egfr activation, dP-Erk moves into the nucleus where it regulates gene expression by phosphorylating transcription factors. The phosphatase Mkp3 dephosphorylates and inactivates dP-Erk in the nucleus, and the authors found that Tay binds both Erk and Mkp3, suggesting that it might negatively regulate Egfr signaling by sequestering Erk in the nucleus.

These four recent papers on Egfr signaling in the fly have revealed some unexpected findings about the regulation of this pathway that will be of interest to address in vertebrates. Of particular note are the findings that two well-known oncogenes, *fak* and *vav* are behaving as tumor suppressors with regard to Drosophila Egfr, suggesting that their functions with regard to oncogenic pathways may be context-dependent. Furthermore, proteins similar to Tay may not always negatively regulate Egfr signaling. Expression in the Drosophila wing disc of human AUTS2, which has a 250-amino acid stretch of homology to Tay, causes activation of Egfr signaling [5]. These various results add to a growing list of findings that many genes implicated in cancer exhibit an "antagonistic duality" with regard to the regulation of tumorigenesis, and indicate that they may not be good therapeutic targets [14].

References

1. Shilo BZ (2003) Signaling by the drosophila epidermal growth factor receptor pathway during development. Exp Cell Res 284 (1):140–149. doi:S0014482702000940 [pii]
2. Shilo BZ (2005) Regulating the dynamics of EGF receptor signaling in space and time. Development 132(18):4017–4027. doi:10.1242/dev.02006
3. Martin-Bermudo MD, Bardet PL, Bellaiche Y, Malartre M (2015) The vav oncogene antagonises EGFR signalling and regulates adherens junction dynamics during drosophila eye development. Development 142(8):1492–1501. doi:10.1242/dev.110585
4. Macagno JP, Diaz Vera J, Yu Y, MacPherson I, Sandilands E, Palmer R, Norman JC, Frame M, Vidal M (2014) FAK acts as a suppressor of RTK-MAP kinase signalling in *Drosophila melanogaster* epithelia and human cancer cells. PLoS Genet 10(3):e1004262. doi:10.1371/journal.pgen.1004262
5. Molnar C, de Celis JF (2013) Tay bridge is a negative regulator of EGFR signalling and interacts with Erk and Mkp3 in the *Drosophila melanogaster* wing. PLoS Genet 9(12):e1003982. doi:10.1371/journal.pgen.1003982
6. Legent K, Liu HH, Treisman JE (2015) Drosophila Vps4 promotes epidermal growth factor receptor signaling independently of its role in receptor degradation. Development 142 (8):1480–1491. doi:10.1242/dev.117960
7. Sulzmaier FJ, Jean C, Schlaepfer DD (2014) FAK in cancer: mechanistic findings and clinical applications. Nat Rev Cancer 14 (9):598–610. doi:10.1038/nrc3792
8. Lazer G, Katzav S (2011) Guanine nucleotide exchange factors for RhoGTPases: good therapeutic targets for cancer therapy? Cell Signal 23 (6):969–979. doi:10.1016/j.cellsig.2010.10.022
9. Tomas A, Futter CE, Eden ER (2014) EGF receptor trafficking: consequences for signaling and cancer. Trends Cell Biol 24(1):26–34. doi:10.1016/j.tcb.2013.11.002
10. Miura GI, Roignant JY, Wassef M, Treisman JE (2008) Myopic acts in the endocytic pathway to enhance signaling by the drosophila EGF receptor. Development 135(11):1913–1922. doi:10.1242/dev.017202

11. Vaccari T, Rusten TE, Menut L, Nezis IP, Brech A, Stenmark H, Bilder D (2009) Comparative analysis of ESCRT-I, ESCRT-II and ESCRT-III function in drosophila by efficient isolation of ESCRT mutants. J Cell Sci 122(Pt 14):2413–2423. doi:10.1242/jcs.046391
12. Henriksen L, Grandal MV, Knudsen SL, van Deurs B, Grovdal LM (2013) Internalization mechanisms of the epidermal growth factor receptor after activation with different ligands. PLoS One 8(3):e58148. doi:10.1371/journal.pone.0058148
13. Sousa LP, Lax I, Shen H, Ferguson SM, De Camilli P, Schlessinger J (2012) Suppression of EGFR endocytosis by dynamin depletion reveals that EGFR signaling occurs primarily at the plasma membrane. Proc Natl Acad Sci U S A 109(12):4419–4424. doi:10.1073/pnas.1200164109
14. Stepanenko AA, Vassetzky YS, Kavsan VM (2013) Antagonistic functional duality of cancer genes. Gene 529(2):199–207. doi:10.1016/j.gene.2013.07.047

Chapter 3

C. elegans Vulva Induction: An In Vivo Model to Study Epidermal Growth Factor Receptor Signaling and Trafficking

Kimberley Gauthier and Christian E. Rocheleau

Abstract

Epidermal growth factor receptor (EGFR)-mediated activation of the canonical Ras/MAPK signaling cascade is responsible for cell proliferation and cell growth. This signaling pathway is frequently overactivated in epithelial cancers; therefore, studying regulation of this pathway is crucial not only for our fundamental understanding of cell biology but also for our ability to treat EGFR-related disease. Genetic model organisms such as *Caenorhabditis elegans*, a hermaphroditic nematode, played a vital role in identifying components of the EGFR/Ras/MAPK pathway and delineating their order of function, and continues to play a role in identifying novel regulators of the pathway. Polarized activation of LET-23, the *C. elegans* homolog of EGFR, is responsible for induction of the vulval cell fate; perturbations in this signaling pathway produce either a vulvaless or multivulva phenotype. The translucent cuticle of the nematode facilitates in vivo visualization of the receptor, revealing that localization of LET-23 EGFR is tightly regulated and linked to its function. In this chapter, we review the methods used to harness vulva development as a tool for studying EGFR signaling and trafficking in *C. elegans*.

Key words EGFR, Ras GTPase, LET-23, Vulva, *C. elegans*

1 Introduction

Epidermal growth factor receptor (EGFR) is a receptor tyrosine kinase that stimulates signaling cascades responsible for numerous aspects of cell development and cell cycle regulation, including cell proliferation and growth. There are four homologous EGFR genes in humans: ErbB1 (HER1, EGFR), ErbB2 (HER2, Neu), ErbB3 (HER3), and ErbB4 (HER4). Upon binding to an extracellular growth factor ligand, the receptor dimerizes and activates four major intracellular signaling pathways mediated by Ras GTPase/ mitogen-activated protein kinase (MAPK), PI3K/Akt, JAK/STAT, and PLCγ/PKC [1].

The EGFR family and the downstream targets Ras GTPase and Raf kinase are frequently mutated or amplified in epithelial cancers, such as non-small cell lung cancer, breast cancer, and squamous cell

Zhixiang Wang (ed.), *ErbB Receptor Signaling: Methods and Protocols*, Methods in Molecular Biology, vol. 1652, DOI 10.1007/978-1-4939-7219-7_3, © Springer Science+Business Media LLC 2017

carcinoma [2–5]. EGFR expression in breast cancer is associated with higher proliferation, genomic instability, and risk of relapse [2]. Resistance to tyrosine kinase (TK) inhibitors is a growing challenge to treating EGFR-positive cancers [4, 6, 7]. Studying the numerous ways in which this pathway is positively and negatively regulated is necessary for understanding and treating EGFR-related diseases.

1.1 *Caenorhabditis elegans* Vulva Development as a Model for EGFR Signaling

The hermaphroditic nematode *Caenorhabditis elegans* serves as a powerful genetic model to identify and study regulators of signaling pathways. Its three-day life-cycle begins as a newly hatched L1 larva, progresses through three additional larval stages, then completes as an egg-laying adult. A nondisjunction event resulting in a loss of the X-chromosome in developing gametes can spontaneously produce XO males that can mate with their hermaphroditic counterparts. This quick life-cycle and ability to both self-fertilize and outcross facilitates genetic studies in this model organism. The translucent cuticle allows for direct, visual inspection of the worms' cells and organs through a Nomarski differential interference contrast (DIC) filter in a compound microscope [8–10]. There is also a considerable amount of conservation of genes and signaling pathways between *C. elegans* and humans [11–13], indicating that this species can serve as a useful model organism to increase our fundamental understanding of human cell function, development, and disease. Guidelines for worm rearing have been well described in text and online [14, 15].

C. elegans has a single EGFR homolog named LET-23 that shares homology with all four ErbB genes in humans, but is most closely related to EGFR/ErbB1 [16], and shares a C-terminal PDZ-binding motif also seen in HER4/ErbB4 [17]. LET-23 EGFR can stimulate the LET-60 Ras/LIN-45 Raf/MEK-2 MEK/MPK-1 MAPK signaling pathway to promote vulva development and duct cell fate specification [18], and can also activate developmental pathways mediated by PLCγ/PKC [19, 20].

Activation of the canonical Ras/MAPK pathway in *C. elegans* by LET-23 EGFR has been studied extensively for its role in vulva development. Basolateral localization of LET-23 EGFR in the vulva precursor cells (VPCs), a set of six polarized epithelial cells (named P3.p–P8.p) on the ventral side of the L3 larval worm, is required for vulva induction by the LIN-3 EGF-like ligand, released from the anchor cell in the overlying gonad. A recent study suggests that the extracellular region of LET-23 EGFR is constitutively dimerized, and the LIN-3 ligand induces allosteric changes that allow for activation of the intracellular tail of the receptor [21], ultimately leading to stimulation of the downstream LET-60 Ras/MPK-1 MAPK signaling pathway (Fig. 1).

The VPC in closest proximity to the anchor cell, P6.p, receives the greatest amount of LET-23 EGFR-mediated stimulation, and thereby assumes the primary vulval cell fate. The neighboring P5.

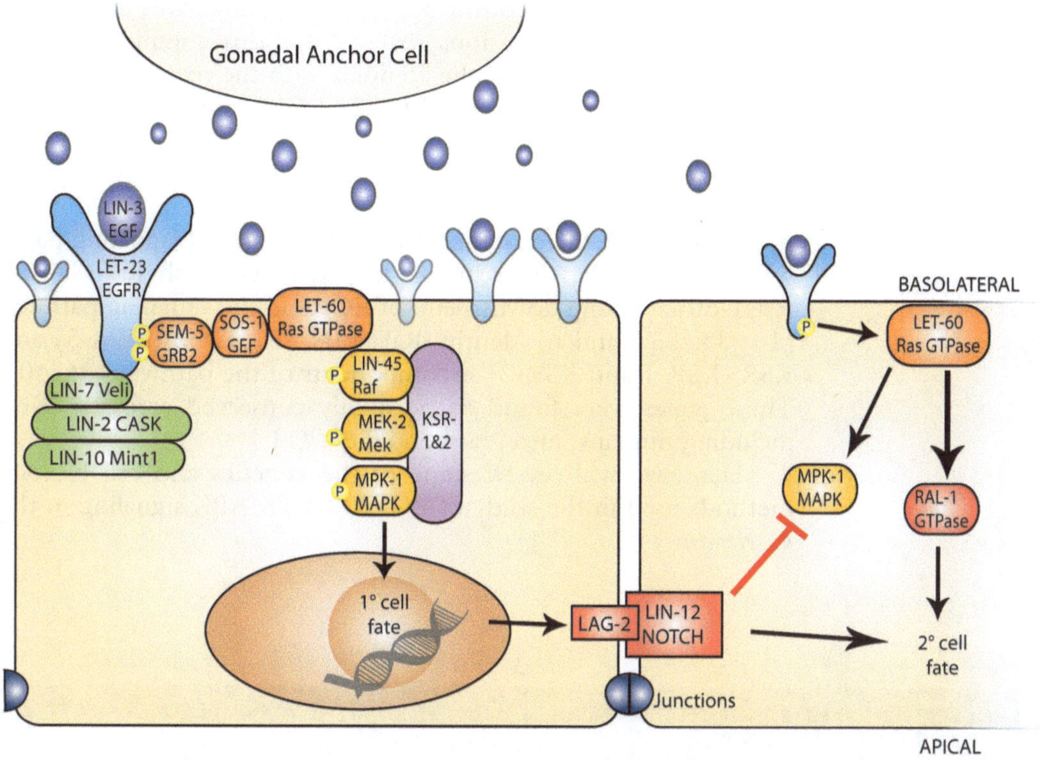

Fig. 1 LET-23 (epidermal growth factor receptor, EGFR) activates the canonical Ras/MAPK signaling pathway to specify the vulval cell fate. The LIN-3 EGF-like ligand, released from the gonadal anchor cell, is necessary to stimulate the canonical LET-23 EGFR/ LET-60 Ras/ LIN-45 Raf/MPK-1 MAPK signaling pathway in the vulva precursor cells (VPCs) of L3 *C. elegans* larvae to specify the vulval cell fate. Basolateral localization of LET-23 EGFR, mediated by an interaction with the LIN-2 CASK/ LIN-7 Veli/LIN-10 Mint1 complex, is necessary for activation of the pathway. The VPC closest to the anchor cell, P6.p, receives the most ligand and differentiates into the primary vulval cell fate. The two adjacent cells (P5.p and P7.p) receive inhibitory Notch signaling from the primary cell that downregulates MPK-1 MAPK. This, in addition to reduced EGF stimulation, promotes LET-60 Ras to signal through an alternate Ral GTPase-mediated pathway and ultimately causes these cells to differentiate into the secondary cell fate as referenced in the text

p and P7.p cells adopt a secondary vulval cell fate via a combination of graded LET-23 EGFR signaling through the Ras/Ral GTPase pathway and LIN-12 Notch signaling induced by the LAG-2 Notch ligand expressed on the surface of the P6.p cell [22–26] (Fig. 1). These three cells divide to produce the 22 cells of the vulva: eight cells from P6.p and seven cells each from P5.p and P7.p. The remaining VPCs adopt a tertiary nonvulval cell fate. P4.p and P8.p divide once and then fuse with the hypodermis, while P3.p has an equal chance of dividing or not, prior to fusing with the hypodermis [27, 28] (Fig. 2a).

A conserved tripartite complex consisting of the scaffold proteins LIN-2 CASK, LIN-7 Veli, and LIN-10 Mint-1 interact with LET-23 EGFR via its C-terminal PDZ binding motif and maintain basolateral localization of the receptor by an unknown mechanism (Figs. 1 and 2b). Mutations that disrupt this complex, or that prevent the interaction of the complex with the receptor, result in exclusive apical localization of LET-23 EGFR and a vulvaless (Vul) phenotype [29–33] (Fig. 2c).

Genetics studies on vulva induction in *C. elegans* played a central role in identifying the components and mapping out the specific order of the canonical EGFR/Ras/MAPK signaling pathway. For example, these studies helped to establish Ras GTPase (LET-60) as acting downstream of EGFR in the signaling pathway [16, 34, 35], and to identify Raf (LIN-45), Grb2 (SEM-5) and KSR (KSR-1 and KSR-2) as components of the pathway [36–40]. These genes were found to be highly conserved across species, including humans, mice, rats, and flies [41].

Here, we will review some of the genetics and cell biology methods used in the study of EGFR/Ras/MAPK signaling in the *C. elegans* VPCs.

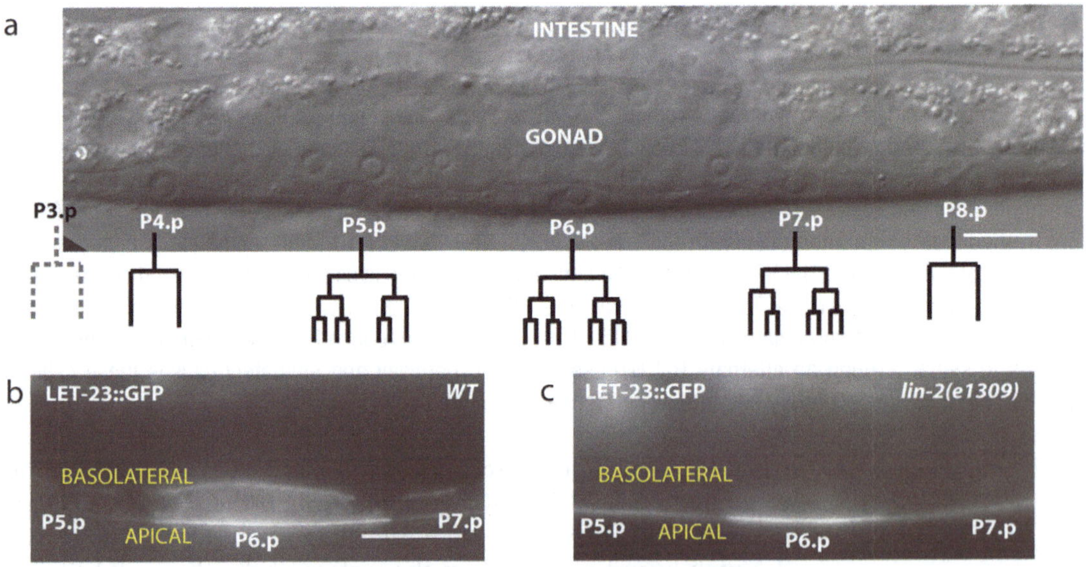

Fig. 2 Basolateral localization of LET-23 EGFR in the VPCs, mediated by the LIN-2/7/10 complex, is necessary for wild-type vulval cell fate induction. (**a**) DIC image of five of the six VPCs, with vulval lineages shown below. The three VPCs closest to the anchor cell (P5.p–P7.p) are induced to undergo three rounds of division to produce the 22 cells of the vulva. The remaining uninduced cells divide once and fuse with the hypodermis, with the exception of P3.p—whose nucleus is not visible in this image—which has an equal chance of fusing with the hypodermis with or without dividing. The VPCs can be identified by their location ventral to the gonad and their large oblong nucleus and round nucleolus. (**b**) In the L3 larva of wild-type worms, LET-23::GFP (EGFR) is most strongly expressed in P6.p. LET-23::GFP localizes to both the apical and basolateral membranes. (**c**) Disruption of the LIN-2/7/10 complex in a *lin-2(e1309)* mutant results in exclusive apical localization of LET-23::GFP and a vulvaless (Vul) phenotype. Scale bars 10 μm

2 Genetics Techniques for Studying EGFR Signaling and Vulva Induction

Development of the worm vulva can serve as a functional read-out of LET-23 EGFR signaling. Activation of the canonical Ras/MAPK signaling pathway by LET-23 EGFR is the first step towards vulval development, and perturbations in this pathway can lead to a vulvaless (Vul) or multivulva (Muv) phenotype. The signaling events that lead to vulva development have been well characterized and can serve as a template for genetic epistasis tests to determine where a new gene functions within the pathway.

2.1 Egg-Laying and Morphological Phenotypes

The vulva is needed for egg-laying; therefore, a simple way to identify potential perturbations in LET-23 EGFR signaling is to measure the frequency of egg-laying defective (Egl) worms. After reaching reproductive maturity, a wild-type adult worm (Fig. 3a) will lay several hundred eggs after 1–3 days, whereas mutant worms with a Vul phenotype will lay few or no eggs, causing the eggs to fill up inside the worm (Fig. 3b). The eggs then hatch inside the mother, producing a distinct "bag of worms" phenotype [33] (Fig. 3c).

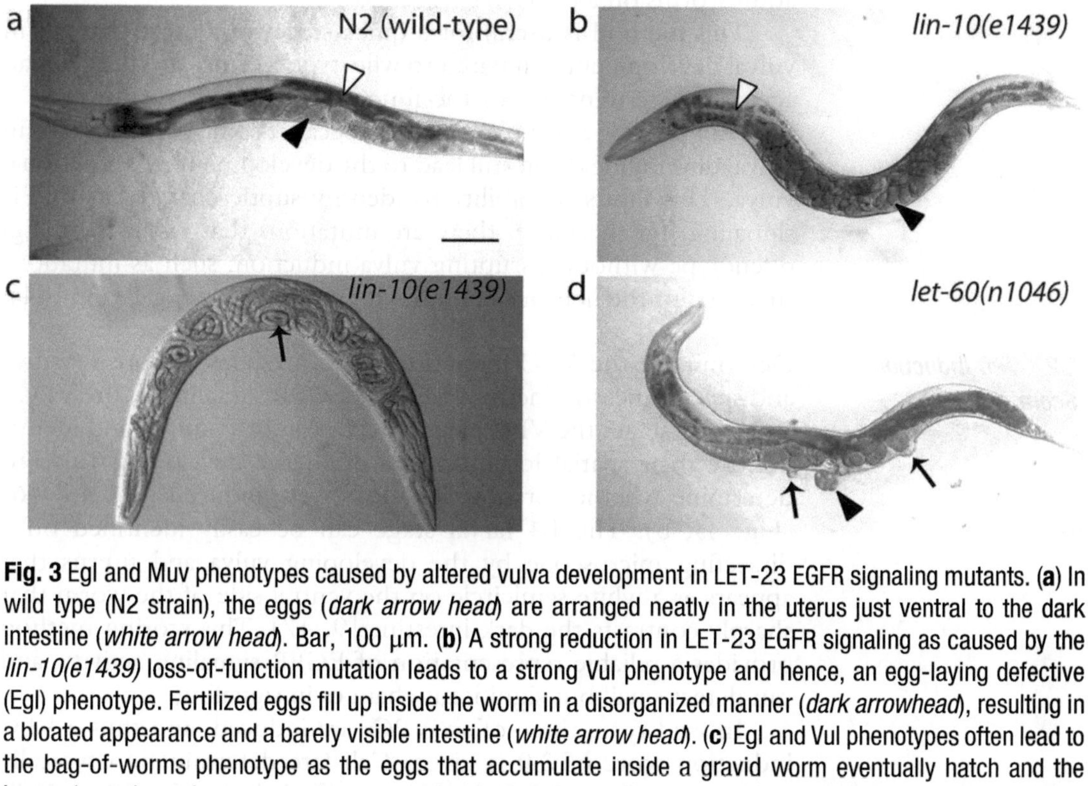

Fig. 3 Egl and Muv phenotypes caused by altered vulva development in LET-23 EGFR signaling mutants. (**a**) In wild type (N2 strain), the eggs (*dark arrow head*) are arranged neatly in the uterus just ventral to the dark intestine (*white arrow head*). Bar, 100 μm. (**b**) A strong reduction in LET-23 EGFR signaling as caused by the *lin-10(e1439)* loss-of-function mutation leads to a strong Vul phenotype and hence, an egg-laying defective (Egl) phenotype. Fertilized eggs fill up inside the worm in a disorganized manner (*dark arrowhead*), resulting in a bloated appearance and a barely visible intestine (*white arrow head*). (**c**) Egl and Vul phenotypes often lead to the bag-of-worms phenotype as the eggs that accumulate inside a gravid worm eventually hatch and the larvae (*arrow*) can be seen moving around inside their mother's carcass. (**d**) Overactivation of LET-23 EGFR/LET-60 Ras signaling pathway as caused by the *let-60(n1046)* gain-of-function mutation, leads to a multivulva (Muv) phenotype and the development of pseudovulvae (*arrows*) protruding from ventral side. The arrowhead points to a fertilized egg that was laid by the functioning vulva

This phenotype can be observed by quickly screening plates on a dissecting microscope for bloated worms or bags of worms. To get a more quantitative measurement, the development of a population of worms can be synchronized by soaking gravid mothers in a bleach and sodium hydroxide solution. This melts away the cuticle of the mothers and leaves the eggs intact and viable because they are more resistant to the effects of bleach. After washing the eggs to remove bleach, the eggs can be left to hatch in M9 buffer overnight. The absence of food (*E. coli*) will cause development of newly hatched L1 worms to arrest. Once the L1s are plated on bacteria-seeded agar in a petri dish, development will resume and the worms will grow at a similar pace [15, 42]. After growing on the plates for 4 days, the frequency of Egl worms can be scored. Glodowski et al. followed a similar method by measuring the number of animals that failed to lay eggs 48 h into adulthood [43].

Alternatively, strong overactivation of LET-23 EGFR signaling can be spotted by the occurrence of Muv worms. Mutant animals with this phenotype (e.g., gain-of-function mutations in *let-60 ras*) will develop ectopic vulvae (pseudovulvae) that protrude from the ventral epidermis. These protrusions can be spotted by observing adult worms on a plate [34] (Fig. 3d).

This method is useful for a quick screen to detect changes in vulval development compared to wild-type worms, and provides an accurate determination of the function of the vulva. However, it does not provide a complete picture, because some perturbations in vulval development can still lead to the development of a functional vulva. This limits the ability to identify subtle changes in EGFR signaling. Furthermore, there are mutations that result in an Egl phenotype without disrupting vulva induction, such as mutations that disrupt the neurons and muscles required for egg-laying [44].

2.2 VPC Induction Score

Determining the VPC induction score provides a more sensitive and specific measurement of LET-23 EGFR signaling in the VPCs. At the L4 stage, the VPC progeny cells can be counted and identified by their spatial localization around the developing vulva to determine whether or not the VPCs were induced [34, 45, 46] (Fig. 4a, b). The L4 larval stage can be easily identified on a dissecting microscope by the developing vulva and uterus that appears as a white semicircle on the ventral side of the worm that sharply contrasts the dark intestine [9, 47]. This scoring method provides a reliable representation of EGFR signaling activity, and can detect small but significant changes in signaling.

In wild type, three of six VPCs are induced, and thus a VPC induction score of 3.0 is given. Cell fate determination typically occurs at the Pn.p (VPC) stage; however, their daughter Pn.px cells are still competent for induction, and in mutants compromised for the inductive signal it is common to see only one of two Pn.p daughters being induced to produce three or four vulval cells.

a

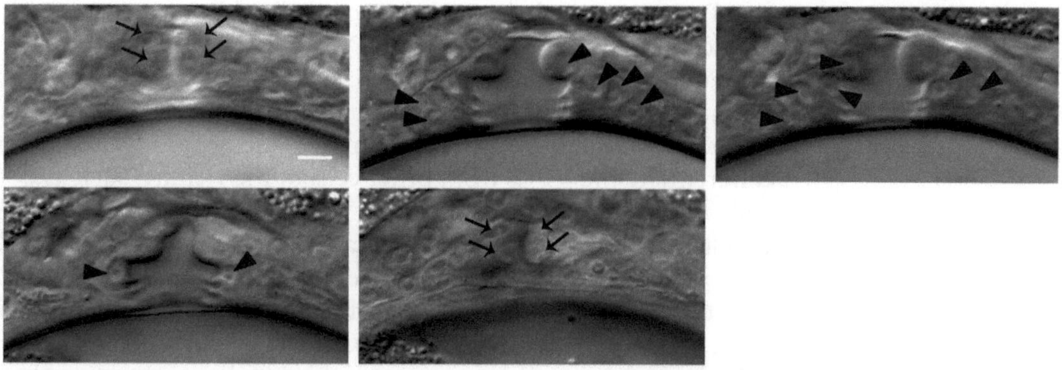

b

Phenotype	P3.p	P4.p	P5.p	P6.p	P7.p	P8.p	VPC Induction Score
Wild-type	✗✗	✗✗	✓✓	✓✓	✓✓	✗✗	3.0
Vul	✗✗	✗✗	✗✗	✓✓	✓✗	✗✗	1.5
Muv	✗	✓✓	✓✓	✓✓	✓✓	✓✗	4.5

c

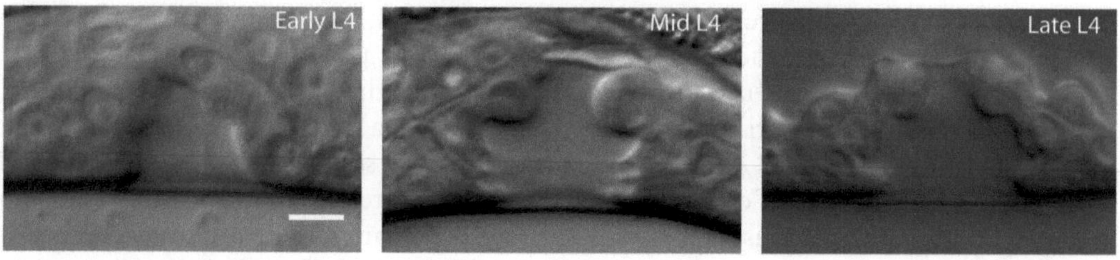

Fig. 4 Vulva induction serves as a readout of LET-23 EGFR signaling. (**a**) DIC images of the vulval cells at the mid-L4 larval stage. These five focal planes show the eight descendants of P6.p (*arrows*) and the seven descendants each from P5.p and P7.p (*arrow heads*) that make up the 22 cells of the wild-type vulva. See also Inoue et al. for a representation of the three-dimensional orientation of the vulval cells [45]. (**b**) A sample scoring chart of a wild-type, Vul, and Muv worm. Each checkmark (√) denotes an induced half-lineage and is given a score of 0.5. Uninduced cells (X) receive a score of 0. Note that in the sample Muv worm, P3.p did not divide prior to fusing with the hypodermis. The VPC induction score is calculated as the sum of all induced half-lineages, and the average score is used to measure LET-23 EGFR signaling for a given mutant. (**c**) DIC images showing the morphological changes of vulval cells that occur after tubulogenesis at later stages of the L4 larva (*right*) make determining the VPC induction score difficult. We recommend scoring the vulval cells at early (*left*) or mid (*center*) stages of vulval development in L4 larvae. Scale bars 5 μm

Therefore, we count half inductions as 0.5 cells induced [34, 46]. Animals with 0–2.5 cells induced (<3) are Vul, and animals with 3.5–6 cells induced (>3) are Muv, as demonstrated in Fig. 4b. The average number of VPCs induced is an indicator of the severity of the Muv or Vul phenotypes.

Identifying differentiated or undifferentiated vulval cells can be challenging, but can be learned with a little practice. VPC membranes and boundaries are not visible by DIC optics and thus cells are identified by the location, size, and morphology of the nucleus and nucleoli. Generally, the cell nuclei appear smaller and closer together with each subsequent division. Undifferentiated VPCs at the Pn.p stage each have a characteristic oblong nucleus with a round nucleolus (Fig. 2a). The VPCs can typically be found all in the same central focal plane on the ventral side of the animal along with the nerve cells in the ventral nerve cord that can be identified by their small circular nuclei and stippled nucleoli. This also helps to differentiate the VPCs from other cells in the vicinity, such as body wall muscle cells that flank the VPCs to the left and right. An excellent set of Nomarski DIC images identifying key cells of the *C. elegans* anatomy, including those described here, are freely available on Wormbook.org [47] and therefore will not be reproduced here. WormAtlas.org is another excellent resource of images and diagrams of the *C. elegans* anatomy that can be used as a guide.

When scoring the induction of the VPCs, we typically count all the Pn.p cells (P1.p–P12.p) to ensure that any phenotypes we see are not due to defects in the generation or location of the VPCs. Starting at the posterior of the animal, there are three cells just anterior to the rectum: two of these nuclei are parallel to the rectum and are derived from the U or Y cell, and the more ventral/posteriorly located nuclei belongs to P12.pa. Anterior to this cell is the larger nucleus of P11.p. Together they provide an easily recognizable triangle of cell nuclei to use as a landmark [47]. As a side note, EGFR/Ras/MAPK signaling also specifies the P12.p cell fate as being different from the P11.p fate [48], so often mutants that decrease signaling will have two larger P11.p-like nuclei and no P12.pa nucleus. U/Y should not be affected. In contrast, mutations that increase signaling can cause P11.p to adopt a P12.p fate, resulting in two smaller P12.pa nuclei and no P11.p nucleus. This phenotype can easily be scored as a second readout of LET-23 EGFR signaling.

Moving anteriorly, the P10.p nuclei can be found roughly half way between the P11.p cell and the bend of the posterior gonad arm and looks similar to P11.p. Closer to the bend of the posterior gonad is P9.p. Next, P8.pp—which can be challenging to get in focus due to nearby muscle cells—and P8.pa nuclei can be found.

Even with overactivation or underactivation of the EGFR/Ras/MAPK pathway, the vulva is generally easy to spot and is located just ventral to the developing uterus. The primary and secondary fates are easily identified by the location and plane of division within the wild-type vulva. The final division of the outer secondary cells P5.pa and P7.pp are lateral and result in four cells clustered at the outer sides of the vulva in the central focal plane of the vulva. The inner secondary cells, P5.pp and P7.pa, give rise to three cells each. The innermost cells P5.ppp and P7.paa do not divide and almost look like a pair of eyes in the central focal plane (Fig. 4a). Their siblings, P5.ppa and P7.pap, divide transversely such that they can be seen when focusing up and down as nuclei flanking sides of the vulva lumen. P6.p gives rise to eight cells that also results from a transverse final division plane such that four nuclei on each side of the vulva can be seen at a time by focusing up and down (Fig. 4a) [45].

The P4.pp cell nuclei is just anterior to the vulva, and P4.pa is sometimes a little obscure, hiding anterior and a little dorsal. The last VPC, P3.p, normally divides once about half the time, and the size does not differ much between P3.p or P3.px. If only three more anterior Pn.p cells can be found after P4.p, then it is inferred that P3.p did not divide. If four anterior Pn.p cells are identified, then P3.p must have divided. The round shape of P3.p progeny cells can be differentiated from P2.p, which often appears smaller or oblong and is squished between the cuticle and the rest of the animal. Finally, P1.p can be found near the base of the pharynx. Identifying the non-VPC Pn.p cells can be helpful in knowing which VPCs were induced or not induced, especially in mutants with strong Vul or Muv phenotypes, or in rare scenarios in which the vulva has shifted due to P5.p or P7.p adopting the primary cell fate.

When scoring, we mark which VPCs are induced or not (induced meaning that a half lineage divided to produce three or four vulval cells), as demonstrated in Fig. 4b. Each induced half-lineage will receive a score of 0.5. Mutants in which fates are lost or gained are harder to decipher in older L4 animals, which therefore may be deemed unscorable. To avoid any bias in your scoring, we recommend using younger L4 animals (Fig. 4c). Scoring between 20 and 30 animals should be sufficient to know whether a particular mutation or RNAi treatment is significantly altering VPC induction and LET-23 EGFR signaling. Combining the percentage of Vul and Muv animals with the average VPC induction score provides a quantitative readout of LET-23 EGFR signaling. We use Fisher's exact test for statistical analysis of categorical data to compare the number of VPCs induced vs. non-induced, Vul vs. non-Vul, or Muv vs. non-Muv between two genotypes.

2.3 Transcriptional Reporters

Several transcriptional targets of LET-23 EGFR signaling in P6.p have been identified and can be used as transcriptional reporters to identify vulval cell fates. Two in particular, *egl-17* and *lag-2,* encode ligands for fibroblast growth factor receptor and Notch, respectively [24, 45, 49]. They are specifically expressed in P6.p in a LET-23 EGFR signaling-dependent manner and can be used to differentiate primary vs. secondary vulval cell fates. Expression of *egl-17* provides a more sensitive assay for signaling as the amount of signaling needed to activate reporter expression is less than what is needed for primary cell fate induction [50]. This transcriptional reporter was vital in demonstrating that the secondary cell fate was dependent on both Notch signaling and a graded inductive signal acting through a Ras/Ral GTPase-mediated signaling pathway [22, 50]. Transcriptional reporters facilitate the identification of changes in cell fate, such as if P5.p or P7.p assumes a primary cell fate as a result of an upregulation of the EGFR/Ras/MAPK signaling pathway or loss of Notch signaling from the adjacent primary cell [22, 50, 51]. Transcriptional reporters can complement vulva induction scoring by monitoring their expression in induced cells.

Single molecule fluorescence in situ hybridization (smFISH) has also been used as a way of monitoring expression of cell fate markers [52]. This method allows for labeling and imaging of individual mRNA molecules using highly specific oligonucleotide fluorescent probes [53]. smFISH can reveal subtle changes in gene expression and has been used to visualize the dynamics of *lag-2* expression in the VPCs of wild-type and mutant animals [54].

2.4 Sensitized Genetic Backgrounds and Epistasis

Genetic screens for Muv and Vul phenotypes have identified numerous mutations in core components of the canonical EGFR/Ras/MAPK signaling pathway. These gain-of-function and loss-of-function mutations, both weak and strong, can be used as tools for genetic analysis of the pathway. Using these mutations for suppressor or enhancer screens can be useful in identifying new mutants in genes that are essential for fertility/viability and thus may have been previously overlooked, or identifying genes that are otherwise wild type for vulva development. For example, loss of negative regulators such as SLI-1 and GAP-1 alone do not cause vulval phenotypes; however, mutations in these genes in a sensitized background cause a strong Muv phenotype [55, 56]. Furthermore, enhancer/suppressor screens provide specificity in identifying mutations that affect EGFR/Ras/MAPK signaling rather than other aspects of vulva development. Genetic epistasis with various Muv and Vul mutants can be used to determine where a newly identified gene functions within the canonical EGFR/Ras/MAPK pathway.

Suppressors of Muv screens have been effective at identifying positive regulators of EGFR/Ras/MAPK. Several mutations result in penetrant Muv phenotypes at distinct steps of the pathway. Screens for suppressors of the constitutively active *let-60 ras* mutant allele *n1046gf* have identified many downstream components of the pathway that did not come out in the initial screens for Vul and Muv mutants. For example, MPK-1 was initially identified as a **su**ppressor of **R**as and was originally named SUR-1 before being identified as an Erk1/2 homolog [57, 58]. The MAPK kinase MEK-2 followed a similar history [39, 59]. Finally, the **k**inase **s**uppressor of **R**as gene *ksr-1* (formerly SUR-3), which codes for a scaffold protein that functions redundantly with KSR-2 during vulva induction, was also identified for its ability to suppress the *let-60 ras* Muv phenotype. A similar suppressor screen in *Drosophila* eye development also identified KSR as a suppressor of Ras [38, 39, 60].

Synthetic Muv (SynMuv) mutants are organized into two groups (SynMuv A and B) that are redundantly required to limit LIN-3 ligand expression to the anchor cell in the gonad [61, 62].

Loss of both SynMuv A and B genes results in overactivation of the pathway by inducing ectopic LIN-3 EGF expression and a potent Muv phenotype. A notable example is the *lin-15* locus that was found to affect the two distinct adjacent genes *lin-15a* and *lin-15b* that are SynMuv A and B genes, respectively [63, 64]. Screens for suppressors of the *lin-15* Muv identified LIN-45 Raf and SEM-5 Grb2 as components of the VPC induction pathway [34, 36, 37].

Screens for **su**ppressors of the **V**ul phenotypes of mutations in *let-23 egfr* and mutations in the *lin-2/7/10* complex have been used for the identification of negative regulators of EGFR/Ras/MAPK signaling. Novel negative regulators of the EGFR/Ras/MAPK pathway identified for their ability to suppress the Vul phenotypes of the *sy1* and *sy97* mutants of *let-23* include SLI-1, homologous to the Cbl family of ubiquitin ligases; and UNC-101, a subunit of the AP-1 clathrin adaptor complex that directs trafficking from the Golgi complex [55, 65, 66]. The *lin-2* gene locus is particularly useful in these genetic screens: its position on the X-chromosome and its lack of physiological perturbations when mutated other than the Vul phenotype facilitate genetic crosses. The Ras GTPase activating protein GAP-1 was initially identified as a suppressor of *lin-2* and *lin-10* Vul phenotypes in *C. elegans* [56]. Furthermore, using screens for suppressors of *lin-2*, our lab has been able to identify negative regulators of LET-23 EGFR signaling that regulate LET-23 EGFR trafficking, including the late endosome-associated RAB-7 GTPase, AGEF-1 (a putative Arf guanine exchange factor), the Golgi-associated ARF-1.2 and 3 GTPases, and the dynein heavy chain DHC-1 [67–69].

Enhancer screens are an effective method for the identification of new positive and negative regulators of EGFR/Ras/MAPK signaling. For example, ARK-1 ACK kinase or the CDT-2 ubiquitin ligase were identified as novel negative regulators for their ability to induce a Muv phenotype when mutated in a *sli-1 Cbl* genetic background [70, 71]. **E**nhancers **o**f **R**af (the EOR genes) were identified for their ability to enhance the mild developmental defects in the vulva and excretory system of hypomorphic (weak) alleles of *lin-45 raf* [38, 72]. Screens for enhancers of *lin-45 raf* phenotypes identified the novel proteins EOR-1 (a BTB-zinc finger transcription factor targeted by ERK) and EOR-2 (an EOR-1 binding partner) as positive regulators of EGFR/Ras/MAPK signaling [72, 73] as well as novel mutations in core components of the pathway. Hypomorphic *lin-45 raf* alleles provide a sensitized background to test candidate positive regulators [74].

Chemical mutagenesis allows for the generation of novel mutants that can be screened for a desired phenotype. Alternatively, RNAi provides a quick way to knock down genes of interest to screen for enhancement or suppression of vulval phenotypes. To test for cell autonomous effects of mutants, VPC-specific RNAi can be achieved by expressing a wild-type argonaute *rde-1* transgene under a *lin-31* promoter in an *rde-1*-deficient mutant background [69, 75, 76]. However, RNAi causes much weaker phenotypes than genetic mutations in the VPCs, suggesting that VPCs may be partly refractory to RNAi [24, 76]. Still, RNAi has been a useful tool in identifying novel regulators of EGFR/Ras/MAPK in sensitized backgrounds [77, 78].

Screens and the identified genes described in this section are not exhaustive and are only meant to exemplify the types of screens that can be designed and carried out. Many of these screens are not yet saturated and future screening will continue to uncover new and interesting regulators of LET-23 EGFR signaling. The variety of Muv and Vul mutants identified through these screens provide a rich set of tools for dissecting where new regulators function through genetic epistasis experiments. With the advent of CRISPR/Cas9 genome editing, there are new opportunities to design specific mutants for functional analyses.

3 Cell Biology Techniques for Studying EGFR Localization and Trafficking

The localization of LET-23 EGFR in the VPCs is polarized: although it is present on both the apical and basolateral membranes, it is expressed more strongly on the apical membrane for reasons unknown (Fig. 2b) [79]. The basolateral localization of LET-23 EGFR is necessary for its ability to be activated and stimulate the Ras/MAPK pathway. The localization of the receptor can be monitored in vivo as a means to identify and test regulators of LET-23 EGFR trafficking.

3.1 Identifying Worms at the Right Stage

Signaling begins in L3 larvae; therefore, this is the optimal developmental stage to visualize LET-23 EGFR localization. We recommend letting synchronized L1 worms grow for 30–35 h at 20 °C, or longer if using a mutant or RNAi clone with developmental delays, to catch worms ranging between early Pn.p and Pn.pxx stages. The VPCs are organized neatly in a row, which facilitates visual monitoring of the cells. The VPCs can be readily observed with a compound light microscope using a DIC filter. They can be identified by their spatial positioning ventral to the developing gonad, and their oblong nuclei and round nucleoli (Fig. 2a). Using a LET-23::GFP reporter, the VPCs stands out with bright GFP expression. In early L3 larvae, LET-23 EGFR is expressed in similar levels throughout the VPCs; however, shortly before the first cell division, expression of the receptor diminishes rapidly in the secondary cells and more slowly in the tertiary cells (Fig. 2b) [32, 79, 80].

3.2 Antibody Staining

Immunostaining can be used to find the subcellular localization of LET-23 EGFR and of other components and regulators of the signaling pathway. A rabbit polyclonal antibody (not commercially available) specific for LET-23 EGFR was used to determine the wild-type localization of LET-23 EGFR, and to establish the requirement for the C-terminal tail of LET-23 as well as the LIN-2/7/10 complex in regulating LET-23 EGFR basolateral localization in the VPCs [31, 32, 79]. There are several detailed protocols available for whole mount immunostaining of *C. elegans* [42, 81–84]. To catch worms at the same stage, they should first be synchronized. Worms can be examined on a microscope at 63× or 100× in order to verify that they have reached the developmental stage of choice (e.g., L3 larvae at the Pn.p stage) before fixing a batch of larvae. Once the worms have reached the desired stage, worms need to be rinsed off of plates, fixed right away, permeabilized and then treated overnight with the primary antibody. After treating with the secondary antibody, the worms can be mounted on a glass slide. Since fixation can make identifying the VPCs and their stage of development difficult, co-staining with the MH27 antibody for AJM-1, the apical cell junction protein that stains the VPCs and their descendants can make VPC identification easier [32, 85, 86].

3.3 Live Imaging

With the use of modern imaging and cloning techniques, the subcellular localization and trafficking of LET-23 EGFR can be viewed in vivo in the VPCs on a confocal or epifluorescence microscope. Observations of LET-23::GFP in live worms provide a clearer image of the receptor's localization pattern at all stages of vulva development [68, 80].

Live-imaging also helps to further characterize the method by which regulators influence receptor signaling. Two fluorescently tagged LET-23 transgenic lines *zhIs035* and *zhIs038* were generated by the Hajnal lab in Zurich, Switzerland using Mos1 transposon-mediated single copy insertion (MosSCI) cloning [80, 87, 88]. The GFP tag in these transgenes is upstream of the intracellular C-terminal tail so as not to disrupt the receptor's interaction with LIN-7 [80]. These strains differ in the intensity of GFP fluorescence; however, both strains express LET-23::GFP at comparable levels to endogenous LET-23 EGFR and they recapitulate the receptor localization found using immunostaining techniques. We have found that these strains are able to rescue the *lin-2* Vul phenotype, despite a lack of visible basolateral receptor localization, indicating that the amount of receptor on the basolateral membrane required for signaling may be below detection levels [68].

A major advantage of imaging live worms over immunostaining is the ability to observe LET-23 EGFR localization and trafficking in real time. For example, using time-lapse imaging and fluorescence recovery after photobleaching, Haag et al. found that ERM-1, the Ezrin/Radixin/Moelin homolog, restricts the mobility of LET-23 EGFR on the basolateral membrane [80].

These LET-23::GFP lines were pivotal in our analysis of *lin-2* suppressor mutants. We found that an AGEF-1/Arf GTPase/AP-1 clathrin adaptor complex either promotes apical localization of LET-23 EGFR or antagonizes basolateral localization to regulate LET-23 EGFR signaling [68]. Additionally we found that LET-23::GFP accumulates in intracellular foci in *rab-7* and *dhc-1* mutants consistent with defects in endosome trafficking and degradation of LET-23 EGFR. We found that the localization phenotypes of these mutants were more pronounced at the Pn.px stage compared to Pn.p [68, 69], suggesting that localization defects in the Pn.px stage may alter signaling.

There are numerous well-described cloning techniques to achieve fluorescently tagged transgenic lines. An expression vector injected into an adult worm's gonad can be inherited in a mosaic fashion to progeny worms in order to establish an extrachromosomal array strain [89]. Extrachromosomal arrays can be inserted into the genome by UV or gamma-ray irradiation [90, 91], or gold-particle bombardment [92]. For single-copy insertion of a transgene, MosSCI transposable elements are an ideal tool [87, 88]. VPC-specific expression of fluorescently tagged proteins can be achieved by expressing these transgenes under the *lin-31* promoter [93]. Finally, it is also possible to insert a fluorescent tag at endogenous gene loci using CRISPR/Cas9 genome engineering [94–97] for endogenous expression of transgenic proteins. This would be an ideal method to visualize and modify LET-23 EGFR moving forward.

4 Discussion

Perturbations in the EGFR/Ras/MAPK pathway have a tremendous impact on human health. Therefore, understanding the diverse mechanisms by which this pathway is regulated is crucial for the development of novel, effective, and safe treatments and therapies of EGFR-related diseases.

C. elegans vulva development has been a vital in vivo resource for elucidating the components and order of the EGFR/Ras/MAPK signaling pathway, and continues to play a role in identifying new positive and negative regulators of the pathway. Studies on the subcellular localization of LET-23 EGFR have provided key insight into the role of intracellular trafficking in the propagation of signaling pathways. However, our model remains incomplete. Although the receptor appears to be more concentrated on the apical membrane (Fig. 2b), its function in this domain is unknown. Furthermore, the method by which the LIN-2/7/10 complex promotes basolateral localization of the receptor is unclear.

In addition to identifying new regulators, this model system may have further applications for in vivo anticancer drug screening.

A LET-23 chimeric receptor expressing wild-type human EGFR tyrosine kinase (TK) domain in place of its own TK domain can rescue *let-23* mutant phenotypes [98]. Activating mutations in the TK domain seen in human cancers cause a Muv phenotype in *C. elegans* that can be suppressed by known TK inhibitors gefinitib and erlotinib. Bae et al. developed this humanized *C. elegans* model to screen for compounds that suppress the Muv phenotype associated with more aggressive mutations in the TK domain that are resistant to TK inhibitors.

The identification of missense mutations can also contribute to our understanding of how the many components of EGFR/Ras/MAPK pathway function and how they can be targeted. Ras-suppressing missense mutations in *ksr* genes identified in *C. elegans* and *Drosophila* were used as a guide in designing drugs to target human KSR as an alternative method to treating Ras-overexpressing cancers [99].

Vulva induction provides a simple readout for the activation of EGFR/Ras/MAPK, and endogenous expression of fluorescently labeled LET-23 EGFR coupled with high resolution imaging techniques afford in vivo real-time visualization of receptor trafficking. The simple genetics and low cost of maintaining worm stocks make *C. elegans* an ideal system in which to screen for EGFR genetic and therapeutic regulators. This will in turn provide us with novel therapeutic targets and treatments to improve human health and outcomes of disease.

Acknowledgments

We thank Meera Sundaram and Simon Rousseau for providing comments and suggestions on the manuscript. Research by the authors is supported by a Natural Sciences and Engineering Research Council of Canada Doctoral award to K.G., a Canadian Institutes of Health Research operating grant (MOP-114935), and a Natural Sciences and Engineering Research Council of Canada discovery grant (RGPIN-341579-13) to C.E.R. The Research Institute of the McGill University Health Centre is funded in part by the Fonds de la recherche en santé du Québec.

References

1. Roskoski R Jr (2014) The ErbB/HER family of protein-tyrosine kinases and cancer. Pharmacol Res 79:34–74. doi:10.1016/j.phrs.2013.11.002
2. Rimawi MF, Shetty PB, Weiss HL, Schiff R, Osborne CK, Chamness GC, Elledge RM (2010) Epidermal growth factor receptor expression in breast cancer association with biologic phenotype and clinical outcomes. Cancer 116(5):1234–1242. doi:10.1002/cncr.24816
3. Herbst RS, Heymach JV, Lippman SM (2008) Lung cancer. N Engl J Med 359 (13):1367–1380. doi:10.1056/NEJMra0802714

4. Shigematsu H, Gazdar AF (2006) Somatic mutations of epidermal growth factor receptor signaling pathway in lung cancers. Int J Cancer 118(2):257–262. doi:10.1002/ijc.21496
5. Yarden Y, Sliwkowski MX (2001) Untangling the ErbB signalling network. Nat Rev Mol Cell Biol 2(2):127–137. doi:10.1038/35052073
6. Ahsan A (2016) Mechanisms of resistance to EGFR tyrosine kinase inhibitors and therapeutic approaches: an update. Adv Exp Med Biol 893:137–153. doi:10.1007/978-3-319-24223-1_7
7. Kuwano M, Sonoda K, Murakami Y, Watari K, Ono M (2016) Overcoming drug resistance to receptor tyrosine kinase inhibitors: learning from lung cancer. Pharmacol Ther 161:97–110. doi:10.1016/j.pharmthera.2016.03.002
8. Sulston JE (1976) Post-embryonic development in the ventral cord of *Caenorhabditis elegans.* Philos Trans R Soc Lond Ser B Biol Sci 275(938):287–297
9. Corsi AK, Wightman B, Chalfie M A transparent window into biology: a primer on *Caenorhabditis elegans.* WormBook. doi:10.1895/wormbook.1.177.1. WormBook
10. Sulston JE, Horvitz HR (1977) Post-embryonic cell lineages of the nematode, *Caenorhabditis elegans.* Dev Biol 56(1):110–156
11. Shaye DD, Greenwald I (2011) OrthoList: a compendium of *C. elegans* genes with human orthologs. PLoS One 6(5):e20085. doi:10.1371/journal.pone.0020085
12. Kaletta T, Hengartner MO (2006) Finding function in novel targets: *C. elegans* as a model organism. Nat Rev Drug Discov 5 (5):387–398. doi:10.1038/nrd2031
13. Culetto E, Sattelle DB (2000) A role for *Caenorhabditis elegans* in understanding the function and interactions of human disease genes. Hum Mol Genet 9(6):869–877
14. Brenner S (1974) The genetics of *Caenorhabditis elegans.* Genetics 77(1):71–94
15. Stiernagle T Maintenance of *C. elegans.* WormBook. doi:10.1895/wormbook.1.101.1. WormBook
16. Aroian RV, Koga M, Mendel JE, Ohshima Y, Sternberg PW (1990) The let-23 gene necessary for *Caenorhabditis elegans* vulval induction encodes a tyrosine kinase of the EGF receptor subfamily. Nature 348(6303):693–699. doi:10.1038/348693a0
17. Shelly M, Mosesson Y, Citri A, Lavi S, Zwang Y, Melamed-Book N, Aroeti B, Yarden Y (2003) Polar expression of ErbB-2/HER2 in epithelia. Bimodal regulation by Lin-7. Dev Cell 5(3):475–486
18. Sundaram MV (2013) Canonical RTK-Ras-ERK signaling and related alternative pathways. WormBook:1–38. doi:10.1895/wormbook.1.80.2
19. Yin X, Gower NJ, Baylis HA, Strange K (2004) Inositol 1,4,5-trisphosphate signaling regulates rhythmic contractile activity of myoepithelial sheath cells in *Caenorhabditis elegans.* Mol Biol Cell 15(8):3938–3949. doi:10.1091/mbc.E04-03-0198
20. Van Buskirk C, Sternberg PW (2007) Epidermal growth factor signaling induces behavioral quiescence in *Caenorhabditis elegans.* Nat Neurosci 10(10):1300–1307. doi:10.1038/nn1981
21. Freed DM, Alvarado D, Lemmon MA (2015) Ligand regulation of a constitutively dimeric EGF receptor. Nat Commun 6:7380. doi:10.1038/ncomms8380
22. Zand TP, Reiner DJ, Der CJ (2011) Ras effector switching promotes divergent cell fates in *C. elegans* vulval patterning. Dev Cell 20 (1):84–96. doi:10.1016/j.devcel.2010.12.004
23. Sternberg PW, Horvitz HR (1989) The combined action of two intercellular signaling pathways specifies three cell fates during vulval induction in *C. elegans.* Cell 58(4):679–693
24. Chen N, Greenwald I (2004) The lateral signal for LIN-12/Notch in *C. elegans* vulval development comprises redundant secreted and transmembrane DSL proteins. Dev Cell 6 (2):183–192
25. Greenwald I (1985) lin-12, a nematode homeotic gene, is homologous to a set of mammalian proteins that includes epidermal growth factor. Cell 43(3 Pt 2):583–590
26. Greenwald IS, Sternberg PW, Horvitz HR (1983) The lin-12 locus specifies cell fates in *Caenorhabditis elegans.* Cell 34(2):435–444
27. Sternberg PW, Horvitz HR (1991) Signal transduction during *C. elegans* vulval induction. Trends Genet 7(11–12):366–371
28. Sternberg PW (2005) Vulval development. WormBook:1–28. doi:10.1895/wormbook.1.6.1
29. Kim SK (1997) Polarized signaling: basolateral receptor localization in epithelial cells by PDZ-containing proteins. Curr Opin Cell Biol 9 (6):853–859
30. Kornfeld K (1997) Vulval development in *Caenorhabditis elegans.* Trends Genet 13(2):55–61
31. Kaech SM, Whitfield CW, Kim SK (1998) The LIN-2/LIN-7/LIN-10 complex mediates basolateral membrane localization of the *C. elegans* EGF receptor LET-23 in vulval epithelial cells. Cell 94(6):761–771

32. Simske JS, Kaech SM, Harp SA, Kim SK (1996) LET-23 receptor localization by the cell junction protein LIN-7 during *C. elegans* vulval induction. Cell 85(2):195–204
33. Horvitz HR, Sulston JE (1980) Isolation and genetic characterization of cell-lineage mutants of the nematode *Caenorhabditis elegans*. Genetics 96(2):435–454
34. Han M, Aroian RV, Sternberg PW (1990) The let-60 locus controls the switch between vulval and nonvulval cell fates in *Caenorhabditis elegans*. Genetics 126(4):899–913
35. Han M, Sternberg PW (1990) let-60, a gene that specifies cell fates during *C. elegans* vulval induction, encodes a ras protein. Cell 63 (5):921–931
36. Clark SG, Stern MJ, Horvitz HR (1992) *C. elegans* cell-signalling gene sem-5 encodes a protein with SH2 and SH3 domains. Nature 356(6367):340–344. doi:10.1038/356340a0
37. Han M, Golden A, Han Y, Sternberg PW (1993) *C. elegans* lin-45 raf gene participates in let-60 ras-stimulated vulval differentiation. Nature 363(6425):133–140. doi:10.1038/363133a0
38. Sundaram M, Han M (1995) The *C. elegans* ksr-1 gene encodes a novel Raf-related kinase involved in Ras-mediated signal transduction. Cell 83(6):889–901
39. Kornfeld K, Guan KL, Horvitz HR (1995) The *Caenorhabditis elegans* gene mek-2 is required for vulval induction and encodes a protein similar to the protein kinase MEK. Genes Dev 9 (6):756–768
40. Ohmachi M, Rocheleau CE, Church D, Lambie E, Schedl T, Sundaram MV (2002) *C. elegans* ksr-1 and ksr-2 have both unique and redundant functions and are required for MPK-1 ERK phosphorylation. Curr Biol 12 (5):427–433
41. Chang C, Sternberg PW (1999) *C. elegans* vulval development as a model system to study the cancer biology of EGFR signaling. Cancer Metastasis Rev 18(2):203–213
42. Sulston J, Hodgkin J (1988) Methods. In: Wood WB (ed) The nematode *Caenorhabditis elegans*, Cold Spring Harbor monographs, 1st edn. Cold Spring Harbor, New York
43. Glodowski DR, Wright T, Martinowich K, Chang HC, Beach D, Rongo C (2005) Distinct LIN-10 domains are required for its neuronal function, its epithelial function, and its synaptic localization. Mol Biol Cell 16 (3):1417–1426. doi:10.1091/mbc.E04-10-0885
44. Trent C, Tsuing N, Horvitz HR (1983) Egg-laying defective mutants of the nematode *Caenorhabditis elegans*. Genetics 104(4):619–647
45. Inoue T, Sherwood DR, Aspock G, Butler JA, Gupta BP, Kirouac M, Wang M, Lee PY, Kramer JM, Hope I, Burglin TR, Sternberg PW (2002) Gene expression markers for *Caenorhabditis elegans* vulval cells. Mech Dev 119 (Suppl 1):S203–S209
46. Sternberg PW, Horvitz HR (1986) Pattern formation during vulval development in *C. elegans*. Cell 44(5):761–772
47. Yochem JK (2006) Nomarski images for learning the anatomy, with tips for mosaic analysis. WormBook:1–47. doi:10.1895/wormbook.1.100.1
48. Jiang LI, Sternberg PW (1998) Interactions of EGF, Wnt and HOM-C genes specify the P12 neuroectoblast fate in *C. elegans*. Development 125(12):2337–2347
49. Burdine RD, Branda CS, Stern MJ (1998) EGL-17(FGF) expression coordinates the attraction of the migrating sex myoblasts with vulval induction in *C. elegans*. Development 125(6):1083–1093
50. Yoo AS, Bais C, Greenwald I (2004) Crosstalk between the EGFR and LIN-12/Notch pathways in *C. elegans* vulval development. Science 303(5658):663–666. doi:10.1126/science.1091639
51. Berset TA, Hoier EF, Hajnal A (2005) The *C. elegans* homolog of the mammalian tumor suppressor Dep-1/Scc1 inhibits EGFR signaling to regulate binary cell fate decisions. Genes Dev 19(11):1328–1340. doi:10.1101/gad.333505
52. Ji N, van Oudenaarden A Single molecule fluorescent in situ hybridization (smFISH) of *C. elegans* worms and embryos. WormBook. doi:10.1895/wormbook.1.153.1. WormBook
53. Raj A, van den Bogaard P, Rifkin SA, van Oudenaarden A, Tyagi S (2008) Imaging individual mRNA molecules using multiple singly labeled probes. Nat Methods 5(10):877–879. doi:10.1038/nmeth.1253
54. van Zon JS, Kienle S, Huelsz-Prince G, Barkoulas M, van Oudenaarden A (2015) Cells change their sensitivity to an EGF morphogen gradient to control EGF-induced gene expression. Nat Commun 6:7053. doi:10.1038/ncomms8053
55. Jongeward GD, Clandinin TR, Sternberg PW (1995) sli-1, a negative regulator of let-23-mediated signaling in *C. elegans*. Genetics 139 (4):1553–1566
56. Hajnal A, Whitfield CW, Kim SK (1997) Inhibition of *Caenorhabditis elegans* vulval induction by gap-1 and by let-23 receptor tyrosine kinase. Genes Dev 11(20):2715–2728
57. Lackner MR, Kornfeld K, Miller LM, Horvitz HR, Kim SK (1994) A MAP kinase homolog,

mpk-1, is involved in ras-mediated induction of vulval cell fates in *Caenorhabditis elegans*. Genes Dev 8(2):160–173

58. Wu Y, Han M (1994) Suppression of activated Let-60 ras protein defines a role of *Caenorhabditis elegans* Sur-1 MAP kinase in vulval differentiation. Genes Dev 8(2):147–159
59. Wu Y, Han M, Guan KL (1995) MEK-2, a *Caenorhabditis elegans* MAP kinase kinase, functions in Ras-mediated vulval induction and other developmental events. Genes Dev 9 (6):742–755
60. Therrien M, Chang HC, Solomon NM, Karim FD, Wassarman DA, Rubin GM (1995) KSR, a novel protein kinase required for RAS signal transduction. Cell 83(6):879–888
61. Ferguson EL, Horvitz HR (1989) The multivulva phenotype of certain *Caenorhabditis elegans* mutants results from defects in two functionally redundant pathways. Genetics 123(1):109–121
62. Saffer AM, Kim DH, van Oudenaarden A, Horvitz HR (2011) The *Caenorhabditis elegans* synthetic multivulva genes prevent ras pathway activation by tightly repressing global ectopic expression of lin-3 EGF. PLoS Genet 7 (12):e1002418. doi:10.1371/journal.pgen.1002418
63. Clark SG, Lu X, Horvitz HR (1994) The *Caenorhabditis elegans* locus lin-15, a negative regulator of a tyrosine kinase signaling pathway, encodes two different proteins. Genetics 137 (4):987–997
64. Huang LS, Tzou P, Sternberg PW (1994) The lin-15 locus encodes two negative regulators of *Caenorhabditis elegans* vulval development. Mol Biol Cell 5(4):395–411
65. Lee J, Jongeward GD, Sternberg PW (1994) unc-101, a gene required for many aspects of *Caenorhabditis elegans* development and behavior, encodes a clathrin-associated protein. Genes Dev 8(1):60–73
66. Yoon CH, Lee J, Jongeward GD, Sternberg PW (1995) Similarity of sli-1, a regulator of vulval development in *C. elegans*, to the mammalian proto-oncogene c-cbl. Science 269 (5227):1102–1105
67. Skorobogata O, Rocheleau CE (2012) RAB-7 antagonizes LET-23 EGFR signaling during vulva development in *Caenorhabditis elegans*. PLoS One 7(4):e36489. doi:10.1371/journal.pone.0036489
68. Skorobogata O, Escobar-Restrepo JM, Rocheleau CE (2014) An AGEF-1/Arf GTPase/AP-1 ensemble antagonizes LET-23 EGFR basolateral localization and signaling during *C. elegans* vulva induction. PLoS Genet 10(10): e1004728. doi:10.1371/journal.pgen.1004728
69. Skorobogata O, Meng J, Gauthier K, Rocheleau CE (2016) Dynein-mediated trafficking negatively regulates LET-23 EGFR signaling. Mol Biol Cell 27(23):3771–3779. doi:10.1091/mbc.E15-11-0757
70. Hopper NA, Lee J, Sternberg PW (2000) ARK-1 inhibits EGFR signaling in *C. elegans*. Mol Cell 6(1):65–75
71. Poulin GB, Ahringer J (2010) The *Caenorhabditis elegans* CDT-2 ubiquitin ligase is required for attenuation of EGFR signalling in vulva precursor cells. BMC Dev Biol 10:109. doi:10.1186/1471-213X-10-109
72. Rocheleau CE, Howard RM, Goldman AP, Volk ML, Girard LJ, Sundaram MV (2002) A lin-45 raf enhancer screen identifies eor-1, eor-2 and unusual alleles of Ras pathway genes in *Caenorhabditis elegans*. Genetics 161 (1):121–131
73. Howell K, Arur S, Schedl T, Sundaram MV (2010) EOR-2 is an obligate binding partner of the BTB-zinc finger protein EOR-1 in *Caenorhabditis elegans*. Genetics 184(4):899–913. doi:10.1534/genetics.109.111591
74. Rocheleau CE, Ronnlund A, Tuck S, Sundaram MV (2005) *Caenorhabditis elegans* CNK-1 promotes Raf activation but is not essential for Ras/Raf signaling. Proc Natl Acad Sci U S A 102(33):11757–11762. doi:10.1073/pnas.0500937102
75. Qadota H, Inoue M, Hikita T, Koppen M, Hardin JD, Amano M, Moerman DG, Kaibuchi K (2007) Establishment of a tissue-specific RNAi system in *C. elegans*. Gene 400 (1–2):166–173. doi:10.1016/j.gene.2007.06.020
76. Barkoulas M, van Zon JS, Milloz J, van Oudenaarden A, Felix MA (2013) Robustness and epistasis in the *C. elegans* vulval signaling network revealed by pathway dosage modulation. Dev Cell 24(1):64–75. doi:10.1016/j.devcel.2012.12.001
77. Fisher K, Southall SM, Wilson JR, Poulin GB (2010) Methylation and demethylation activities of a *C. elegans* MLL-like complex attenuate RAS signalling. Dev Biol 341(1):142–153. doi:10.1016/j.ydbio.2010.02.023
78. Lehner B, Crombie C, Tischler J, Fortunato A, Fraser AG (2006) Systematic mapping of genetic interactions in *Caenorhabditis elegans* identifies common modifiers of diverse signaling pathways. Nat Genet 38(8):896–903. doi:10.1038/ng1844
79. Whitfield CW, Benard C, Barnes T, Hekimi S, Kim SK (1999) Basolateral localization of the *Caenorhabditis elegans* epidermal growth

factor receptor in epithelial cells by the PDZ protein LIN-10. Mol Biol Cell 10 (6):2087–2100

80. Haag A, Gutierrez P, Buhler A, Walser M, Yang Q, Langouet M, Kradolfer D, Frohli E, Herrmann CJ, Hajnal A, Escobar-Restrepo JM (2014) An in vivo EGF receptor localization screen in *C. elegans* identifies the Ezrin homolog ERM-1 as a temporal regulator of signaling. PLoS Genet 10(5):e1004341. doi:10.1371/journal.pgen.1004341
81. Duerr JS Immunohistochemistry. WormBook. doi:10.1895/wormbook.1.105.1. WormBook
82. Finney M, Ruvkun G (1990) The unc-86 gene product couples cell lineage and cell identity in *C. elegans*. Cell 63(5):895–905
83. Ruvkun G, Giusto J (1989) The *Caenorhabditis elegans* heterochronic gene lin-14 encodes a nuclear protein that forms a temporal developmental switch. Nature 338(6213):313–319. doi:10.1038/338313a0
84. Lant B, Derry WB (2014) Immunostaining for markers of apoptosis in the *Caenorhabditis elegans* germline. Cold Spring Harb Protoc 2014 (5). doi:10.1101/pdb.prot080242
85. Francis R, Waterston RH (1991) Muscle cell attachment in *Caenorhabditis elegans*. J Cell Biol 114(3):465–479
86. Koppen M, Simske JS, Sims PA, Firestein BL, Hall DH, Radice AD, Rongo C, Hardin JD (2001) Cooperative regulation of AJM-1 controls junctional integrity in *Caenorhabditis elegans* epithelia. Nat Cell Biol 3(11):983–991. doi:10.1038/ncb1101-983
87. Zeiser E, Frokjaer-Jensen C, Jorgensen E, Ahringer J (2011) MosSCI and gateway compatible plasmid toolkit for constitutive and inducible expression of transgenes in the *C. elegans* germline. PLoS One 6(5):e20082. doi:10.1371/journal.pone.0020082
88. Frokjaer-Jensen C (2015) Transposon-assisted genetic engineering with Mos1-mediated single-copy insertion (MosSCI). Methods Mol Biol 1327:49–58. doi:10.1007/978-1-4939-2842-2_5
89. Mello CC, Kramer JM, Stinchcomb D, Ambros V (1991) Efficient gene transfer in C. elegans: extrachromosomal maintenance and integration of transforming sequences. EMBO J 10(12):3959–3970
90. Mariol MC, Walter L, Bellemin S, Gieseler K (2013) A rapid protocol for integrating extrachromosomal arrays with high transmission rate into the *C. elegans* genome. J Vis Exp 82:e50773. doi:10.3791/50773
91. Evans TC (2006) Transformation and microinjection. In: WormBook (ed) The *C. elegans* Research Community, WormBook, doi: 10.1895/wormbook.1.108.1., http://www.wormbook.org
92. Praitis V, Casey E, Collar D, Austin J (2001) Creation of low-copy integrated transgenic lines in *Caenorhabditis elegans*. Genetics 157 (3):1217–1226
93. Tan PB, Lackner MR, Kim SK (1998) MAP kinase signaling specificity mediated by the LIN-1 Ets/LIN-31 WH transcription factor complex during *C. elegans* vulval induction. Cell 93(4):569–580
94. Dickinson DJ, Goldstein B (2016) CRISPR-based methods for *Caenorhabditis elegans* genome engineering. Genetics 202 (3):885–901. doi:10.1534/genetics.115.182162
95. Kim H, Ishidate T, Ghanta KS, Seth M, Conte D Jr, Shirayama M, Mello CC (2014) A co-CRISPR strategy for efficient genome editing in *Caenorhabditis elegans*. Genetics 197 (4):1069–1080. doi:10.1534/genetics.114.166389
96. Paix A, Folkmann A, Rasoloson D, Seydoux G (2015) High efficiency, homology-directed genome editing in *Caenorhabditis elegans* using CRISPR-Cas9 ribonucleoprotein complexes. Genetics 201(1):47–54. doi:10.1534/genetics.115.179382
97. Farboud B, Meyer BJ (2015) Dramatic enhancement of genome editing by CRISPR/Cas9 through improved guide RNA design. Genetics 199(4):959–971. doi:10.1534/genetics.115.175166
98. Bae YK, Sung JY, Kim YN, Kim S, Hong KM, Kim HT, Choi MS, Kwon JY, Shim J (2012) An in vivo *C. elegans* model system for screening EGFR-inhibiting anti-cancer drugs. PLoS One 7(9):e42441. doi:10.1371/journal.pone.0042441
99. Dhawan NS, Scopton AP, Dar AC (2016) Small molecule stabilization of the KSR inactive state antagonizes oncogenic Ras signalling. Nature 537(7618):112–116. doi:10.1038/nature19327

Chapter 4

Targeting HER2 in Advanced Breast Cancer

Xiaofu Zhu and Anil Abraham Joy

Abstract

Human epidermal growth factor receptor 2 (*HER2*) is a human oncogene that is amplified in approximately 20% of breast cancers, and portends a worse prognosis if not treated with anti-HER2 agents. The advent of targeted anti-HER2 therapies has dramatically improved disease control and survival in patients with metastatic HER2-positive breast cancer, and is now considered standard of care in the first-line setting and beyond. This review summarizes the currently available data on targeted anti-HER2 therapies from completed randomized phase III clinical trials, and briefly discusses emerging advances that will address unmet needs in metastatic HER2-positive breast cancer.

Key words HER2, Breast cancer, Therapies, Clinical trials

1 Biological Role of HER2 in Normal Cellular Function

Human epidermal growth factor receptor 2 (*HER2*), also known as *ERBB2*, is a human oncogene located on position 12 of the long arm of chromosome 17 (17q12), and belongs to the human epidermal growth factor receptor family of genes [1]. Like its sibling genes, which include *HER1* (also known as *EGFR*), *HER3*, and *HER4*, *HER2* encodes for a transmembrane tyrosine kinase receptor that is crucially involved in cell proliferation and survival [2].

Unlike other members of the HER family, the 1255-amino acid HER2 protein has no known ligands that bind to its extracellular domain to directly activate downstream signaling. Instead, induction of downstream signal activation is largely due to dimerization with its siblings and other known receptors. Ligands including epidermal growth factor (EGF), and other EGF-like molecules, can bind to HER1 and lead to subsequent HER1–HER2 dimerization [3]. Similarly, members of the neuregulin family of proteins can bind to and activate HER3 and HER4, leading to formation of HER2–HER3 and HER2–HER4 heterodimers [4]. In this way, HER2 functions as a co-receptor for many HER-family ligands. After dimerization, subsequent phosphorylation of tyrosine

Zhixiang Wang (ed.), *ErbB Receptor Signaling: Methods and Protocols*, Methods in Molecular Biology, vol. 1652, DOI 10.1007/978-1-4939-7219-7_4, © Springer Science+Business Media LLC 2017

residues located on the intracellular domain occurs, and allows for subsequent activation of other downstream signaling molecules via a signal transduction cascade [5]. Principal targets of HER2 signaling include the phosphatidylinositol-4,5-bisphosphate 3-kinase (PI3K) and mitogen-activated protein kinases (MAPK) pathways, which are regulators of important cell functions including proliferation, differentiation, growth, division, and cellular survival [6].

2 Biological Result of HER2 Overexpression in Breast Cancer

Approximately 20% of all breast cancers overexpress cell surface HER2 protein, primarily as a result of overamplification of the *HER2* gene, resulting in high levels of protein translation and cellular incorporation. Clinically, these HER2-positive (HER2+) breast cancers historically embodied a more aggressive phenotype, and patients experienced worse outcomes (recurrence and survival) compared to those with non-HER2-amplied breast cancers [7, 8]. The advent of trastuzumab (Herceptin®), the first targeted monoclonal antibody against the HER2 receptor in 1998, ushered in a new era of molecularly targeted therapies against HER2+ breast cancer, and dramatically improved the prognosis of this subset of patients with breast cancer in both the advanced and early stage settings [9].

In the nearly two decades since, a number of targeted agents against HER2, spanning several drug classes, have been developed. These classes include monoclonal antibodies [e.g., trastuzumab (Herceptin®) and pertuzumab (Perjeta®)] that bind to the extracellular portion of the HER2 oncoprotein; the small molecule intracellular tyrosine kinase inhibitors [e.g., lapatinib (Tykerb®) and neratinib] which block subsequent intracellular downstream receptor signaling; and the antibody-drug conjugate [e.g., TDM-1/ado-trastuzumab emtansine (Kadcyla®)] that combines both HER2 signaling disruption as well as targeted cytotoxicity.

3 Systemic Therapy for HER2+ Advanced Breast Cancer

A number of phase III trials in HER2+ advanced breast cancer (ABC) have demonstrated significant improvement in breast cancer outcomes with the addition of HER2-directed therapy in combination with chemotherapy (Table 1).

4 First-Line Treatment

4.1 Trastuzumab

Slamon et al. provided the first definitive evidence of efficacy for anti-HER2 agents in a pivotal clinical trial published in 2001 [10].

Table 1
First-line phase III trials incorporating anti-HER2 therapy and systemic therapy for HER2+ ABC

Anti-HER2 and chemotherapy					
Study	**Pts (*n*)**	**Treatment arms**	**Median TTP/ PFS (mos)**	**Median OS (mos)**	**Response rate (%)**
Slamon et al. [10]	469	Chemotherapy (various), Trastuzumab 4 mg/kg loading, then 2 mg/kg weekly	7.4 ($p < 0.001$)	25.1 ($p = 0.046$)	50 ($p < 0.001$)
		Chemotherapy (various)	4.6	20.3	32
Robert et al. [11]	196	Trastuzumab 4 mg/kg loading, then 2 mg/kg weeklyv, Paclitaxel 175 mg/m^2 q3w, Carboplatin AUC 6 q3w	10.7 ($p = 0.03$)	35.7 ($p = 0.76$)	52 ($p = 0.04$)
		Trastuzumab 4 mg/kg loading, then 2 mg/kg weekly, Paclitaxel 175 mg/m^2 q3w	7.1	32.2	36
BCIRG 007 [12]	263	Trastuzumab 4 mg/kg loading, then 2 mg/kg weekly, Docetaxel 75 mg/m^2 q3w, Carboplatin AUC 6 q3w	10.4 ($p = 0.57$)	37.4 ($p = 0.99$)	72 ($p = 0.97$)
		Trastuzumab 4 mg/kg loading, then 2 mg/kg weekly, Docetaxel 100 mg/m^2 q3w	10.4	37.1	72
HERNATA [13]	284	Trastuzumab 8 mg/kg loading, then 6 mg/kg q3w, Docetaxel 100 mg/m^2 q3w	12.4 ($p = 0.67$)	35.7 ($p = 0.98$)	59 ($p = 1.00$)
		Trastuzumab 8 mg/kg loading, then 6 mg/kg q3w, Vinorelbine 30–35 mg/m^2 day 1 and 8 q3w	15.3	38.8	59
Burstein et al. [14]	81	Trastuzumab 4 mg/kg loading, then 2 mg/kg weekly, Paclitaxel 80 mg/m^2 weekly *Or* Docetaxel 35 mg/m^2 weekly for 7 weeks q8w	6.0 ($p = 0.09$)	Not reported	58
		Trastuzumab 4 mg/kg loading then 2 mg/kg weekly, Vinorelbine 25 mg/m^2 weekly	8.5	Not reported	66
Guan et al. [15]	444	Paclitaxel 80 mg/m^2 weekly for 3 weeks q4w, Lapatinib 1500 mg daily	9.7 ($p < 0.001$)	27.8 ($p = 0.012$)	69 ($p < 0.001$)
		Paclitaxel 80 mg/m^2 weekly for 3 weeks q4w, Placebo daily	6.5	20.5	50

(continued)

Table 1 (continued)

Anti-HER2 and chemotherapy					
Study	**Pts (*n*)**	**Treatment arms**	**Median TTP/ PFS (mos)**	**Median OS (mos)**	**Response rate (%)**
NCIC MA.31 [16]	652	Paclitaxel 80 mg/m^2 weekly *Or* Docetaxel 75 mg/m^2 q3w, Lapatinib 1500 mg daily	8.8 ($p = 0.01$)	Not reported	Not reported
		Paclitaxel 80 mg/m^2 weekly *Or* Docetaxel 75 mg/m^2 q3w, Trastuzumab 2 mg/kg weekly or 6 mg/kg q3w with chemo, followed by 6 mg/kg q3w	11.4	Not reported	Not reported
CLEOPATRA [17, 18]	808	Docetaxel 75–100 mg/m^2 q3w, Trastuzumab 8 mg/kg loading, then 6 mg/kg q3w, Pertuzumab 840 mg loading, then 420 mg q3w	18.5 ($p < 0.001$)	56.5 ($p < 0.001$)	80 ($p = 0.001$)
		Docetaxel 75–100 mg/m^2 q3w, Trastuzumab 8 mg/kg loading, then 6 mg/kg q3w, Placebo q3w	12.4	40.8	69

Two hundred and thirty-four patients with HER2+ ABC who had not previously received systemic therapy for metastatic disease were randomized to chemotherapy with or without trastuzumab. Chemotherapy regimens allowed included doxorubicin and cyclophosphamide (AC) and paclitaxel. At a median follow up of 30 months, the addition of trastuzumab to chemotherapy led to an improvement in response rate (RR) (32% vs 50%, $p < 0.001$), time to progression (TTP) (4.6 vs 7.4 months, $p = 0.046$), as well as overall survival (OS) (median 20.3 vs 25.1 months, 1-year 22% vs 33%, $p < 0.001$).

In terms of toxicity, higher rates of cardiac dysfunction were seen in trastuzumab treated patients, and particularly so in patients receiving concurrent anthracyclines in the AC treatment arm (27% of patients experienced cardiac dysfunction in the combined treatment arm, compared with 8% of patients in the AC chemotherapy only arm). One patient in each of the anthracycline-containing arms died of cardiac dysfunction. Similarly, cardiac dysfunction rates were higher with paclitaxel plus trastuzumab compared to paclitaxel alone (8% vs 1%). Due to the more favorable cardiac safety profile, the combination of a taxane (such as paclitaxel or docetaxel) plus trastuzumab rapidly became the standard-of-care first-line combination therapy for patients with HER2+ ABC.

The HERNATA study compared taxane and non-taxane-based chemotherapy backbones in association with trastuzumab [13]. In this study, 284 patients were randomized to trastuzumab plus either docetaxel or vinorelbine. There was no significant difference in TTP, the primary endpoint (12.4 vs 15.3 months, $p = 0.67$). Overall survival was also similar with 35.7 months for docetaxel and 38.8 months for vinorelbine. Response rates were also identical in both arms (59%). Despite similar efficacy, however, vinorelbine was much better tolerated compared to docetaxel, and significantly more patients in the docetaxel arm experienced grade 3–4 toxicity and discontinued therapy. These findings were confirmed in a smaller study that combined trastuzumab with either vinorelbine or a weekly taxane [14]. The primary endpoint was response rate while TTP was a secondary endpoint. This trial was closed early due to poor accrual, and outcomes were not statistically different in either response rate (51% for vinorelbine and 40% for taxane) or TTP (8.5 months for vinorelbine and 6.0 months for taxane). Vinorelbine was associated with more hematologic toxicity while taxanes caused more skin toxicity, myalgias, and fluid retention.

Trastuzumab in conjunction with combination chemotherapy has also been investigated on multiple occasions [11, 12]. Disappointingly, despite increased tumor response rates, it is accompanied by significantly increased incidence of toxicity. Moreover, the incorporation of additional chemotherapy agents on top of the standard regimen of a taxane plus trastuzumab did not significantly improve overall survival. Thus, trastuzumab is rarely administered with more than one chemotherapy agent for ABC.

4.2 Lapatinib

Two phase III trials have explored the use of the small molecule, intracellular tyrosine kinase inhibitor, lapatinib in the first-line treatment setting [15, 19]. Guan et al. randomized 444 HER2 positive patients who had not been treated with chemotherapy for metastatic disease to weekly paclitaxel (80 mg/m^2 once a week for 3 weeks of every 4 weeks) plus either lapatinib (1500 mg per day) or placebo [15]. The primary endpoint, overall survival, was significantly improved with lapatinib (27.8 vs 20.5 months, $p = 0.012$). Secondary endpoints including both progression-free survival (PFS) (9.7 vs 6.5 months, $p < 0.001$) and response rate (69% vs 50%, $p < 0.001$) were improved as well. Lapatinib treatment was associated with high rates of diarrhea (grade 3–4: 20% vs <1%) as well as neutropenia (51% vs 20%).

A second trial, the Canadian-led NCIC CTG MA.31, used a taxane chemotherapy backbone in combination with HER2-directed therapy, either lapatinib or trastuzumab [19]. The study randomized 652 patients to receive a taxane (paclitaxel 80 mg/m^2 weekly or docetaxel 75 mg/m^2 every 3 weeks) plus either lapatinib or trastuzumab. PFS was found to be inferior in the lapatinib treatment arm compared to trastuzumab (9.0 vs 11.3 months, $p = 0.001$). Overall survival was comparable between the two groups in the entire study population, however, among the subset of patients with centrally confirmed HER2+ disease, lapatinib was associated with inferior OS. Similar to the first study, lapatinib was associated with significantly higher rates of diarrhea (grade 3–4: 19% vs 1%).

Taken together, these studies suggest that although lapatinib has activity against HER2 breast cancer and offers additional benefit over chemotherapy alone, trastuzumab-based regimens should still be preferred over lapatinib in the first-line treatment setting.

4.3 Trastuzumab and Pertuzumab

Dual HER2 blockade with multiple anti-HER2 antibodies has been shown to be more efficacious than chemotherapy plus trastuzumab alone. The CLEOPATRA trial investigated the addition of pertuzumab to the standard therapy of taxane plus trastuzumab [17]. In this study, 808 patients were randomized to docetaxel (75–100 mg/m^2) and trastuzumab (8 mg/kg loading dose followed by 6 mg/kg every 3 weeks) plus either pertuzumab (840 mg loading dose followed by 420 mg every 3 weeks) or placebo. At least six cycles of docetaxel were recommended. The primary endpoint was PFS by independent assessment. Key secondary endpoints included overall survival and response rate. One line of endocrine therapy for metastatic disease, as well as prior adjuvant or neoadjuvant chemotherapy, was allowed, as long as chemotherapy ended at least 12 months prior. The median PFS was 18.7 months with dual HER2 blockade compared with 12.4 months with trastuzumab alone. Higher response rates were also noted with dual HER2 blockade as well, with 80.2% of patients in the experimental group and 69.3% of patients in the control group achieving a response. Even more impressively, the final analysis with updated overall survival results demonstrated a statistically significant improvement in overall survival in favor of pertuzumab (56.5 vs 40.8 mos, $p < 0.001$) [20]. Dual HER2 inhibition did not lead to increased risk of cardiac toxicity, but rates of febrile neutropenia were higher (13.8% vs 7.6%), and there was also a small increase in the rate of grade 3–4 diarrhea (7.9% vs 5.0%) in the pertuzumab-containing arm. This regimen has now become the standard of care for patients with previously untreated HER2 + ABC.

4.4 Trastuzumab Emtansine (T-DM1)

The antibody drug conjugate trastuzumab emtansine (also known as ado-trastuzumab emtansine, T-DM1, Kadcyla™) consists of the cytotoxic tubulin inhibitor derivative emtansine (abbreviated DM1) linked to trastuzumab. Its use in the first-line setting was examined in the MARIANNE study, which compared T-DM1, as monotherapy or in combination with pertuzumab, against a taxane plus trastuzumab [21]. The primary endpoint was PFS. Reported at the 2015 ASCO Annual Meeting, PFS was nearly identical among the three arms (T-DM1 vs taxane + trastuzumab: 34.9 vs 34.8 mos, $p = 0.31$; T-DM1 + pertuzumab vs taxane + trastuzumab: 34.7 vs 34.8 mos, $p = 0.14$). T-DM1 was found to be noninferior to taxane plus trastuzumab, and was associated with less grade 3–5 toxicity, especially neutropenia and febrile neutropenia. So while the combination of a taxane, trastuzuamab, and pertuzumab remains the standard first-line treatment option for HER2+ ABC, T-DM1 alone may be considered an option for patients who are unable to tolerate cytotoxic chemotherapy.

5 Second-Line Therapy and Beyond

Multiple phase III clinical trials have shown that in patients whose disease has recurred or progressed on first-line trastuzumab-based therapy, continuation of anti-HER2 therapy in the second-line setting provides clinically meaningful improvements in outcomes including survival (Table 2). Evidenced based therapeutic options include continuation of trastuzumab with a different systemic therapy partner, switching to a regimen of capecitabine plus lapatinib, switching to the antibody drug conjugate T-DM1, or the addition of the mTOR pathway inhibitor everolimus to HER2-directed therapy.

5.1 Trastuzumab Treatment Beyond Progression

The strategy of continuing trastuzumab while switching its chemotherapy partner was evaluated in three phase III trials [22–24]. The GBG 26 study enrolled 156 previously treated patients, almost all of whom had received prior trastuzumab for metastatic disease [22]. Patients were then randomized to receive either capecitabine (1250 mg/m^2 bid on days 1–14 every 21 days) plus trastuzumab (6 mg/kg every 3 weeks) or capecitabine alone. With a median follow up of 15.6 months, the addition of trastuzumab to capecitabine was associated with improvement in TTP (8.2 vs 5.6 months, $p = 0.034$) and response rate (48% vs 27%, $p = 0.012$). Although overall survival was numerically superior with the combination, the effect was not statistically significant (25.5 vs 20.4 months, $p = 0.257$). With the exception of anemia (64% vs 44%), the combination arm was not associated with increased toxicity.

In an attempt to replicated the success seen with dual HER2-directed therapy in the first-line setting with CLEOPATRA, the PHEREXA trial randomized 452 patients to capecitabine (1250 mg/m^2 BID on days 1–14 every 21 days) and trastuzumab (8 mg/kg loading then 6 mg/kg every 3 weeks), with or without pertuzumab (840 mg loading then 420 mg every 3 weeks) [23]. Reported in abstract form in 2016, PFS was 11.1 months in the capecitabine, trastuzumab, and pertuzumab arm, and 9.0 months in the capecitabine and trastuzumab arm ($p = 0.743$). Although the

Table 2
Second-line and beyond phase III trials incorporating anti-HER2 therapy and chemotherapy for HER2+ ABC

Study	Pts (*n*)	Treatment arms	Median TTP/ PFS (mos)	Median OS (mos)	Response rate (%)
GBG 26 [22]	156	Capecitabine 1250 mg/m^2 bid on days 1–14 q3w, Trastuzumab 6 mg/kg q3w	8.2 ($p = 0.034$)	25.5 ($p = 0.257$)	48 ($p = 0.012$)
		Capecitabine 1250 mg/m^2 bid on days 1–14 q3w	5.6	20.4	27
PHAREXA [23]	452	Capecitabine 1250 mg/m^2 bid on days 1–14 q3w, Trastuzumab 8 mg/kg loading, then 6 mg/kg q3w	9.0 ($p = 0.743$)	28.1	Not reported
		Capecitabine 1250 mg/m^2 bid on days 1–14 q3w, Trastuzumab 8 mg/kg loading, then 6 mg/kg q3w, Pertuzumab 840 mg loading, then 420 mg q3w	11.1	36.1	Not reported
EGF104900 [24]	296	Lapatinib 1000 mg daily, Trastuzumab 4 mg/kg loading, then 2 mg/kg weekly,	2.8 ($p = 0.008$)	14.0 ($p = 0.026$)	10 ($p = 0.46$)
		Lapatinib 1500 mg daily	1.9	9.5	7
BOLERO-3 [25]	569	Vinorelbine 25 mg/m^2 weekly, Trastuzumab 2 mg/kg weekly, Everolimus 5 mg daily	7.0 ($p = 0.007$)	Not reported	41 ($p = 0.211$)
		Vinorelbine 25 mg/m^2 weekly, Trastuzumab 2 mg/kg weekly, Placebo daily	5.8	Not reported	37
Geyer et al. [26]	324	Capecitabine 1000 mg/m^2 bid on days 1–14 q3w, Lapatinib 1250 mg daily	8.4 ($p < 0.001$)	17.5 ($p = 0.206$)	22 ($p = 0.09$)
		Capecitabine 1000 mg/m^2 bid on days 1-14q3w	4.4	15.1	14
EMILIA [27]	991	T-DM1 3.6 mg/kg q3w	9.6 ($p < 0.001$)	30.9 ($p < 0.001$)	44 ($p < 0.001$)
		Capecitabine 1000 mg/m^2 bid on days 1–14 q3w, Lapatinib 1250 mg daily	6.4	25.1	31

PFS difference was not statistically significant, the pertuzumab-containing arm had numerically improved PFS as well as overall survival (36.1 vs 28.1 months, p = non significant). The adverse events profile and incidence were similar for both arms.

Continuation of trastuzumab in conjunction with lapatinib without cytotoxic chemotherapy was investigated in the EGF104900 study [24]. Blackwell et al. randomized 296 heavily pretreated patients (median 6 prior lines of systemic therapy, 3 lines of trastuzumab-containing regimens for ABC) to lapatinib plus trastuzumab or lapatinib alone. The combination of two HER2-directed agents led to an improvement of PFS (12.0 vs 8.1 weeks, $p = 0.008$). The improvement in response rate was not significant (10% vs 7%, $p = 0.46$) but there was a significant increase in clinical benefit rate (25% vs 12%). In the final updated analysis, the PFS benefit was maintained and improvement in survival was seen also seen (14.0 vs 9.5 months, $p = 0.026$) [28]. Combination therapy was associated with increased diarrhea, but rates of grade 3–4 diarrhea remained comparable between arms (7% in both).

5.2 Capecitabine and Lapatinib

Geyer et al. conducted a phase III study comparing capecitabine (1000 mg/m^2 bid on days 1–14 every 21 days) plus lapatinib (1250 mg daily) versus capecitabine (1250 mg/m^2 bid daily on days 1–14 every 21 days) alone in patients who have progressed on prior trastuzumab-based therapy [26]. The study enrolled a heavily pretreated population of 324 patients, with almost 100% of the patients having previously received anthracyclines and taxanes, and approximately 50% had also received fluorouracil and/or vinorelbine. At the interim analysis, there was a significant improvement in PFS associated with the combination therapy compared to capecitabine alone (8.4 vs 4.4 months, $p < 0.001$). The trial was stopped early which likely reduced statistical power and the ability to detect a significant overall survival benefit. Despite this, the final update showed a trend toward superior overall survival with combination therapy (17.5 vs 15.1 months, $p = 0.206$) [29]. The addition of lapatinib resulted in higher incidence of all-grade diarrhea (60% vs 39%), rash (27% vs 15%), and dyspepsia (11% vs 0%); however, the incidences of severe (grade 4 or higher) toxicity were comparable between the arms.

5.3 Trastuzumab Emtansine (T-DM1)

The superiority of the antibody-drug conjugate T-DM1 over capecitabine plus lapatinib in the second-line setting was established in the EMILIA trial [27]. Nine hundred and ninety-one patients with HER2+ ABC who were pretreated with trastuzumab and a taxane were randomized to T-DM1 (3.6 mg/kg every 3 weeks) or capecitabine (1000 mg/m^2 bid on days 1–14 every 21 days) plus lapatinib (1250 mg daily). Primary endpoints included independently assessed PFS, as well as overall survival and safety. At interim analysis, with a median follow up of 13 months, a significant improvement in PFS was found with T-DM1 compared to capecitabine plus lapatinib (9.6 vs 6.4 months, $p < 0.001$). An updated analysis, with a median follow up of 18 months, showed a significant improvement in overall survival with T-DM1 as well (30.9 vs

25.1 months, $p < 0.001$). Overall, T-DM1 was better tolerated and was associated with less grade 3–4 toxicities compared to capecitabine plus lapatinib (41% vs 57%). The most common grade 3–4 toxicities were thrombocytopenia (12.9%), elevated AST (4.3%), and elevated ALT (2.9%) for T-DM1, and diarrhea (79.7%), hand foot syndrome (58%), and nausea (44.7%) for capecitabine plus lapatinib.

The only phase III trial to specifically address the efficacy of T-DM1 HER2-directed therapy in the third-line setting is the TH3RESA study [30]. Krop et al. randomized patients pretreated with both trastuzumab and lapatinib for advanced disease in a 2:1 fashion to T-DM1 (3.6 mg/kg every 3 weeks) versus treatment of physician's choice. The latter was mostly composed of combination therapy including at least one HER2-directed agent (68% trastuzumab plus chemotherapy, 10% trastuzumab plus lapatinib). The T-DM1 treatment arm was associated with improvement in PFS (6.2 vs 3.3 months, $p < 0.001$). At the second interim analysis, presented at the San Antonio Breast Cancer Symposium in 2015, an improvement in overall survival was also seen (22.7 vs 15.8 months, $p < 0.001$). Lower grade 3–4 toxicity was reported with T-DM1 (32% vs 43%), with the most common toxicity being thrombocytopenia (4%).

Several previously mentioned trials conducted in the second-line setting enrolled patients with extensive pretreatment. In EGF104900, patients had a median of 3 prior trastuzumab-based therapies for metastatic disease. In EMILIA, 39% of patients in both arms received two or more prior lines of chemotherapy, and some received trastuzumab in both the adjuvant and metastatic setting.

Based on existing data, it is reasonable to continue to offer some form of anti-HER2 therapy to patients who are fit enough for additional systemic therapy. This is supported by the recent guideline from ASCO on HER2+ ABC, which recommended the use of anti-HER2 therapy in the third-line setting, including T-DM1, pertuzumab, and capecitabine plus lapatinib [31].

5.4 Everolimus

Resistance to HER2 inhibition is thought to be in part due to activation of the mTOR pathway, which induces cancer cell proliferation and survival [32]. Simultaneous blockade of both HER2 and the mTOR pathway was therefore thought to have the potential to reverse resistance to trastuzumab. This hypothesis was investigated formally in the BOLERO-3 trial, where 569 patients who previously received a taxane and trastuzumab were randomized to weekly vinorelbine (25 mg/m^2) and trastuzumab (2 mg/kg), plus either mTOR inhibitor everolimus (5 mg daily) or placebo [25]. Addition of everolimus modestly improved PFS (from 5.8 to 7.0 months, $p = 0.007$), with the greatest benefit seen in patients with hormone receptor-negative tumors. Survival data was immature at time of publication. Increased grade 3–4 neutropenia, mucositis, anemia, leucopenia, fatigue, and febrile neutropenia were all seen in the everolimus-containing arm. The modest improvement in PFS along with significant increase in toxicity limits the uptake of this therapeutic approach in the clinical setting.

6 Special Treatment Considerations: Brain Metastases

Metastasis to the brain represents a common and serious concern for patients with ABC, and imparts additional risk for significant morbidity and mortality. HER2 positivity has been found to be an independent risk factor for the occurrence of brain metastasis [33, 34]. Standard local therapies include surgery, stereotactic radiosurgery, and/or whole brain radiotherapy [35]. Treatment selection depends in part on the extent and location of the disease.

Historically, chemotherapy has had limited impact on brain metastasis, owing in part to the sanctuary provided by the blood brain barrier. In the HER2-directed therapy era, the combination of capecitabine plus lapatinib has emerged as having activity against brain metastases. In the Geyer trial, the addition of lapatinib to capecitabine reduced the risk of symptomatic CNS progression as first site of failure compared to capecitabine monotherapy (2% vs 6%) [36].

The CEREBEL trial attempted to address the question of whether there is a superior regimen in preventing brain metastases [37]. In this study, 540 patients with HER2+ ABC who did not have brain metastases were enrolled and randomized to receiving either capecitabine and lapatinib (capecitabine 1000 mg/m^2 bid on days 1–14 every 21 days, lapatinib 1250 mg daily) or capecitabine and trastuzumab (capecitabine 1000 mg/m^2 bid on days 1–14 every 21 days, trastuzumab loading dose of 8 mg/kg followed by an infusion of 6 mg/kg every 3 weeks). Primary endpoint was CNS metastases as first site of relapse, with PFS and OS being secondary endpoints. No significant difference was seen in the rate of CNS relapse between the two groups (3% vs 5%, $p = 0.360$). However, PFS was inferior in the group treated with capecitabine and lapatinib (6.6 vs 8.1 months, $p = 0.021$), especially among patients who did not previously receive trastuzumab (6.3 vs 10.9 months). Unfortunately, this study was not able to identify an optimal regimen against brain metastases.

7 Future Directions

Effective inhibition of the HER2 pathway has dramatically altered the landscape of systemic therapy for patients with HER2+ ABC. Median overall survival duration of nearly 5 years has now been documented with the addition of pertuzumab to docetaxel and trastuzumab. Despite these advances, the optimal selection of HER2-directed therapy and personalization of treatment for patients remain an active area of research. To date, no biomarkers that predict response to anti-HER2 therapy other than HER2 overexpression itself have been discovered. Even with dual HER2 inhibition, a proportion of patients do not respond to therapy, as was seen in the CLEOPATRA study. Recently, the companion biomarker analysis of prospectively banked tissue from CLEOPATRA has been published [38]. No new predictive factors were identified, but low HER2 protein and mRNA level, low HER3 mRNA level, and *PIK3CA* mutations were associated with worse prognosis. Interestingly, the latter finding was partly seen in the biomarker analysis from EMILIA, where mutated *PIK3CA* was associated with worse prognosis when patients were treated with lapatinib plus capecitabine but not T-DM1, suggesting that T-DM1 may overcome the negative implications of *PIK3CA* mutations [39]. Continued efforts in investigation of novel biomarkers may help refine and optimize therapy for specific subsets of patients in the future.

Antibody-drug conjugates (ADCs) represent an exciting frontier in cancer medicine. The marriage of exquisite target selectivity and potent cytotoxicity may lead to improved therapeutic index. In addition to T-DM1, several dozen ADCs have either been approved or are undergoing active development against a variety of hematologic and solid tumors [40, 41]. Second-generation anti-HER2 ADCs are currently being investigated, including SYD985, a trastuzumab-based ADC linked to the highly toxic alkylator antibiotic duocarmycin [42]. Preliminary evidence suggests that it may have superior efficacy compared to T-DM1, especially in FISH-negative, IHC 1–2+ tumors [43]. These agents may eventually add to the growing armamentarium of therapeutic agents for patients with metastatic HER2+ breast cancer.

8 Conclusions

The landscape of HER2+ ABC treatment continues to be refined, and evolves. The incorporation of HER2-directed agents as part of the standard of care for HER2+ ABC has led to dramatic improvements in outcome for patients with this aggressive disease. The introduction and availability of novel agents now offer clinicians the ability to provide deep and prolonged inhibition of the HER2 pathway across multiple lines of treatment. As even more potential therapies appear over the horizon, advancements in biomarker discovery will be critical in optimizing treatment selection and providing personalized therapy for patients.

Acknowledgments

The authors wish to thank Dr. Sunil Verma, who contributed to a previous version of a review on HER2-positive breast cancer. We also thank Larissa Vos for manuscript preparation and assistance.

References

1. Krishnamurti U, Silverman JF (2014) HER2 in breast cancer. Adv Anat Pathol 21:100–107
2. Yarden Y (2001) The EGFR family and its ligands in human cancer. Signalling mechanisms and therapeutic opportunities. Eur J Cancer 37(Suppl 4):S3–S8
3. Brennan PJ, Kumogai T, Berezov A, Murali R, Greene MI, Kumogai T (2000) HER2/Neu: mechanisms of dimerization/oligomerization. Oncogene 19:6093–6101
4. Rubin I, Yarden Y (2001) The basic biology of HER2. Ann Oncol Off J Eur Soc Med Oncol 12(Suppl 1):S3–S8
5. Moasser MM (2007) The oncogene HER2: its signaling and transforming functions and its role in human cancer pathogenesis. Oncogene 26:6469–6487
6. Kirouac DC, Du J, Lahdenranta J et al (2016) HER2+ cancer cell dependence on PI3K vs. MAPK signaling axes is determined by expression of EGFR, ERBB3 and CDKN1B. PLoS Comput Biol 12:e1004827
7. Winstanley J, Cooke T, Murray GD et al (1991) The long term prognostic significance of c-erbB-2 in primary breast cancer. Br J Cancer 63:447–450
8. Slamon DJ, Clark GM, Wong SG, Levin WJ, Ullrich A, McGuire WL (1987) Human breast cancer: correlation of relapse and survival with amplification of the HER-2/neu oncogene. Science 235:177–182
9. Hudis CA (2007) Trastuzumab—mechanism of action and use in clinical practice. N Engl J Med 357:39–51
10. Slamon DJ, Leyland-Jones B, Shak S et al (2001) Use of chemotherapy plus a monoclonal antibody against HER2 for metastatic breast cancer that overexpresses HER2. N Engl J Med 344:783–792
11. Robert N, Leyland-Jones B, Asmar L et al (2006) Randomized phase III study of trastuzumab, paclitaxel, and carboplatin compared with trastuzumab and paclitaxel in women with HER-2-overexpressing metastatic breast cancer. J Clin Oncol 24:2786–2792
12. Valero V, Forbes J, Pegram MD et al (2011) Multicenter phase III randomized trial comparing docetaxel and trastuzumab with docetaxel, carboplatin, and trastuzumab as first-line chemotherapy for patients with HER2-gene-amplified metastatic breast cancer (BCIRG 007 study): two highly active therapeutic regimens. J Clin Oncol 29:149–156
13. Andersson M, Lidbrink E, Bjerre K et al (2011) Phase III randomized study comparing docetaxel plus trastuzumab with vinorelbine plus trastuzumab as first-line therapy of metastatic or locally advanced human epidermal growth factor receptor 2-positive breast cancer: the HERNATA study. J Clin Oncol 29:264–271
14. Burstein HJ, Keshaviah A, Baron AD et al (2007) Trastuzumab plus vinorelbine or taxane chemotherapy for HER2-overexpressing metastatic breast cancer: the trastuzumab and vinorelbine or taxane study. Cancer 110:965–972
15. Guan Z, Xu B, DeSilvio ML et al (2013) Randomized trial of lapatinib versus placebo added to paclitaxel in the treatment of human epidermal growth factor receptor 2-overexpressing metastatic breast cancer. J Clin Oncol 31:1947–1953
16. Gelmon KA, Boyle F, Bella K, et al. (2012) Open-label phase III randomized controlled trial comparing taxane-based chemotherapy (Tax) with lapatinib (L) or trastuzumab (T) as first-line therapy for women with HER2+ metastatic breast cancer: Interim analysis (IA) of NCIC CTG MA.31/GSK EGF 108919. | [Internet]. [cited 2014 Oct 21]; Available from: http://meetinglibrary.asco.org/content/102986-114
17. Baselga J, Cortés J, Kim S-B et al (2012) Pertuzumab plus trastuzumab plus docetaxel for metastatic breast cancer. N Engl J Med 366:109–119
18. Swain SM, Kim S-B, Cortés J, et al. (2014) Final overall survival (OS) analysis from the CLEOPATRA study of first-line (1L) pertuzumab (Ptz), trastuzumab (T), and docetaxel (D) in patients (pts) with HER2-positive metastatic breast cancer (MBC) [Internet]. [cited 2014 Oct 12]; Available from: http://www.esmo.org/Conferences/ESMO-2014-Congress/News-Articles/Final-Overall-Survival-Analysis-from-the-CLEOPATRA-Study-in-Patients-with-HER2-Positive-Metastatic-Breast-Cancer

19. Gelmon KA, Boyle FM, Kaufman B et al (2015) Lapatinib or trastuzumab plus taxane therapy for human epidermal growth factor receptor 2-positive advanced breast cancer: final results of NCIC CTG MA.31. J Clin Oncol 33:1574–1583
20. Swain SM, Baselga J, Kim S-B et al (2015) Pertuzumab, trastuzumab, and docetaxel in HER2-positive metastatic breast cancer. N Engl J Med 372:724–734
21. Ellis P, Barrios C, Eiermann W, et al. (2015) Phase III, randomized study of trastuzumab emtansine (T-DM1) ± pertuzumab (P) vs trastuzumab + taxane (HT) for first-line treatment of HER2-positive MBC: Primary results from the MARIANNE study. | 2015 ASCO Annual Meeting | Abstracts | Meeting Library [Internet]. [cited 2017 Jan 23]; Available from: http://meetinglibrary.asco.org/content/147990-156
22. von Minckwitz G, du Bois A, Schmidt M et al (2009) Trastuzumab beyond progression in human epidermal growth factor receptor 2-positive advanced breast cancer: a german breast group 26/breast international group 03-05 study. J Clin Oncol 27:1999–2006
23. Urruticoechea A, Rizwanullah M, Im S-A, et al. PHEREXA: Benefit for Second-Line Pertuzumab Questionable - The ASCO Post [Internet]. [cited 2017 Jan 30]; Available from: http://www.ascopost.com/issues/july-25-2016/pherexa-benefit-for-second-line-per tuzumab-questionable/
24. Blackwell KL, Burstein HJ, Storniolo AM et al (2010) Randomized study of Lapatinib alone or in combination with trastuzumab in women with ErbB2-positive, trastuzumab-refractory metastatic breast cancer. J Clin Oncol 28:1124–1130
25. André F, O'Regan R, Ozguroglu M et al (2014) Everolimus for women with trastuzumab-resistant, HER2-positive, advanced breast cancer (BOLERO-3): a randomised, double-blind, placebo-controlled phase 3 trial. Lancet Oncol 15:580–591
26. Geyer CE, Forster J, Lindquist D et al (2006) Lapatinib plus capecitabine for HER2-positive advanced breast cancer. N Engl J Med 355:2733–2743
27. Verma S, Miles D, Gianni L et al (2012) Trastuzumab emtansine for HER2-positive advanced breast cancer. N Engl J Med 367:1783–1791
28. Blackwell KL, Burstein HJ, Storniolo AM et al (2012) Overall survival benefit with lapatinib in combination with trastuzumab for patients with human epidermal growth factor receptor 2-positive metastatic breast cancer: final results from the EGF104900 study. J Clin Oncol 30:2585–2592
29. Cameron D, Casey M, Oliva C, Newstat B, Imwalle B, Geyer CE (2010) Lapatinib plus capecitabine in women with HER-2-positive advanced breast cancer: final survival analysis of a phase III randomized trial. Oncologist 15:924–934
30. Krop IE, Kim S-B, González-Martín A et al (2014) Trastuzumab emtansine versus treatment of physician's choice for pretreated HER2-positive advanced breast cancer (TH3RESA): a randomised, open-label, phase 3 trial. Lancet Oncol 15:689–699
31. Giordano SH, Temin S, Kirshner JJ et al (2014) Systemic therapy for patients with advanced human epidermal growth factor receptor 2-positive breast cancer: american society of clinical oncology clinical practice guideline. J Clin Oncol 32:2078–2099
32. Wang L, Zhang Q, Zhang J et al (2011) PI3K pathway activation results in low efficacy of both trastuzumab and lapatinib. BMC Cancer 11:248
33. Gabos Z, Sinha R, Hanson J et al (2006) Prognostic significance of human epidermal growth factor receptor positivity for the development of brain metastasis after newly diagnosed breast cancer. J Clin Oncol 24:5658–5663
34. Pestalozzi BC, Zahrieh D, Price KN et al (2006) Identifying breast cancer patients at risk for central nervous system (CNS) metastases in trials of the international breast cancer study group (IBCSG). Ann Oncol 17:935–944
35. Leyland-Jones B (2009) Human epidermal growth factor receptor 2-positive breast cancer and central nervous system metastases. J Clin Oncol 27:5278–5286
36. Cameron D, Casey M, Press M et al (2008) A phase III randomized comparison of lapatinib plus capecitabine versus capecitabine alone in women with advanced breast cancer that has progressed on trastuzumab: updated efficacy and biomarker analyses. Breast Cancer Res Treat 112:533–543
37. Pivot X, Manikhas A, Zurawski B et al (2015) CEREBEL (EGF111438): a phase III, randomized, open-label study of lapatinib plus capecitabine versus trastuzumab plus capecitabine in patients with human epidermal growth factor receptor 2-positive metastatic breast cancer. J Clin Oncol 33:1564–1573
38. Baselga J, Cortes J, Im S-A et al (2014) Biomarker analyses in CLEOPATRA: a phase III,

placebo-controlled study of pertuzumab in human epidermal growth factor receptor 2-positive, first-line metastatic breast cancer. J Clin Oncol 32(33):3753–3761

39. Baselga J, Verma S, Ro J, et al. (2013) Relationship between tumor biomarkers (BM) and efficacy in EMILIA, a phase III study of trastuzumab emtansine (T-DM1) in HER2-positive metastatic breast cancer (MBC) [Internet]. [cited 2014 Nov 14]; Available from: http://www.abstractsonline.com/Plan/ViewAbstract.aspx?sKey=1e9ae007-6c50-46e2-951f-4b825865e743&cKey=15187a93-e3d3-443c-81b5-24d4f5e7bf0d

40. Lambert JM (2013) Drug-conjugated antibodies for the treatment of cancer. Br J Clin Pharmacol 76:248–262

41. Sievers EL, Senter PD (2013) Antibody-drug conjugates in cancer therapy. Annu Rev Med 64:15–29

42. Dokter W, Ubink R, van der Lee M et al (2014) Preclinical profile of the HER2-targeting ADC SYD983/SYD985: introduction of a new duocarmycin-based linker-drug platform. Mol Cancer Ther 13:2618–2629

43. Verheijden G, Beusker P, Ubink R, et al. (2014) Toward clinical development of SYD985, a novel HER2-targeting antibody-drug conjugate (ADC). 2014 ASCO Annual Meeting, Abstracts, Meeting Library [Internet]. [cited 2014 Nov 11]; Available from: http://meetinglibrary.asco.org/content/131605-144

Part II

Classical Methods in the Studies of ErbB Receptor Signaling

Chapter 5

Methods to Investigate EGFR Ubiquitination

Alexia Conte and Sara Sigismund

Abstract

Ubiquitination of the epidermal growth factor receptor (EGFR) is an important intracellular signal that occurs upon EGF stimulation and controls EGFR trafficking at multiple steps, finally destining the receptor to lysosomal degradation. In this chapter, we give an overview of the biochemical methods to investigate EGFR ubiquitination.

Firstly, we describe the in vitro ubiquitination assay, a method where, in the presence of the minimal ubiquitination machinery, the biological milieu for EGFR ubiquitination is reproduced in a test tube. In the second protocol, we explain how to immunoprecipitate the EGFR from total lysate and reveal its ubiquitinated form by western blot analysis. Then, with an ELISA-derived assay, we illustrate a robust and reliable method to assess EGFR ubiquitination from low amount of sample; lastly, we illustrate an immunofluorescence protocol to visualize ubiquitinated species (including the EGFR itself) within the EGFR-positive endocytic compartments upon EGF stimulation.

Key words Epidermal growth factor receptor, EGFR, Ubiquitination, Cbl, Phosphorylation, In vitro ubiquitination assay, Immunoprecipitation, ELISA, Immunofluorescence

1 Introduction

The epidermal growth factor receptor (EGFR) belongs to the receptor tyrosine kinase (RTK) family and its function is critical in several physiological processes in epithelial cells [1]. EGFR-dependent signaling deregulation leads to cellular dysfunctions and pathologies, including cancer [2]. Once the EGFR is activated by its cognate ligand EGF, it undergoes structural and biochemical modifications—including tyrosine phosphorylation (pY) [3] and ubiquitination of the intracytoplasmic tail [4]—leading to activation of the downstream signaling cascade. Importantly, active EGFRs are internalized in the cell through endocytosis. Depending on the EGF dose, the EGFR can be internalized through clathrin-mediated endocytosis (CME) or non-clathrin endocytosis (NCE), which are associated with distinct signaling outcomes. CME is active at low EGF concentrations ($\leq$3 ng/ml) and is crucial for sustaining EGFR signaling, while NCE is essential for EGFR

Zhixiang Wang (ed.), *ErbB Receptor Signaling: Methods and Protocols*, Methods in Molecular Biology, vol. 1652, DOI 10.1007/978-1-4939-7219-7_5,

degradation in the presence of high EGF concentrations (≥10 ng/ml), thereby safeguarding cells against excessive signaling.

Interestingly, the levels of EGFR ubiquitination correlate with NCE activation, being minimal at low EGF concentrations, but increasing sharply at higher EGF concentrations (between 3 and 10 ng/ml, [5]). Ubiquitination is also essential for EGFR entry into NCE and for its subsequent degradation in the lysosome [6]. Thus, ubiquitination acts as a molecular "switch" that commits receptors to degradation and consequently attenuates EGFR signaling. Recent evidence suggests that EGFR ubiquitination also plays a non-essential role in CME, redundantly with several other mechanisms [7].

EGFR ubiquitination is mediated by the E3 RING ligase Cbl [4, 8], which binds directly to pY1045 in the EGFR cytosplasmic domain [8]. In addition, the adaptor protein Grb2, which is recruited to the EGFR tail through pY1068/pY1086, was found to be an additional, indirect docking site for Cbl to the receptor [9, 10]. Cooperation between Cbl direct and indirect binding sites is responsible for the sharp increase in EGFR ubiquitination at higher EGF concentrations [11].

By quantitative mass spectrometry analysis, the EGFR was found to be modified mainly by K63-linked ubiquitin (Ub) chains and by mono-Ub [12]. These Ub modifications are known to be involved in endocytic trafficking and lysosomal sorting [13]. Notably, the major Ub acceptor lysine residues in EGFR reside within the kinase domain [14].

There are several well-established biochemical methods to investigate EGFR ubiquitination. The first approach that we describe is an EGFR in vitro ubiquitination assay (Subheading 3.1). In this assay, we recreate the biological environment for the EGFR to be ubiquitinated. All of the molecular players of the EGFR ubiquitination machinery are incubated with the EGFR tail as substrate: Ub, E1, and E2 enzymes, the E3 ligase Cbl, and the adaptor molecule Grb2 (required to optimize Cbl recruitment and EGFR ubiquitination). This assay relies on the availability of bacteria- or baculovirus-purified proteins (either commercial or in-house). Of note, the ubiquitination reaction is preceded by an EGFR autophosphorylation reaction (requiring the presence of ATP), to allow recruitment of Cbl and Grb2 to pY residues in EGFR.

In the second approach, immunoprecipitation (IP) of the EGFR protein from cell lysate (EGF stimulated) is described, followed by western blot (WB) analysis with antibodies against Ub (*see* Subheading 3.2). This approach requires a good antibody against the target (i.e., EGFR) for performing IP in denaturing conditions, a condition needed to avoid the presence of interacting proteins coprecipitating with the EGFR.

A more quantitative approach to detect EGFR-Ub in cells is an ELISA based-methodology: the DELFIA assay (PerkinElmer), which allows more samples and experimental conditions to be compared in parallel (*see* Subheading 3.3).

The last protocol we describe relies on immunofluorescence (IF) to visualize EGFR-Ub-containing complexes (*see* Subheading 3.4). This protocol requires specific Ub antibodies, which can recognize all Ub modifications or, more specifically, K63 Ub-linked chains (an antibody directed against K48 Ub-linked chains will be used for the negative control). The signal coming from anti-Ub is colocalized with active EGFR, by stimulating cells with fluorescently labelled EGF, so that it is possible to spatially localize the Ub signal within the EGF-positive endocytic compartments. The caveat of this approach is that not only EGFR-Ub is visualized but also ubiquitination of EGFR-associated proteins. Indeed, upon EGF stimulation, several endocytic accessory factors are ubiquitinated and cotrafficked with the EGFR along the endocytic pathway (e.g., eps15, epsin, CIN85, and Hrs; [15–17]).

Cross-validation with more than one approach is usually recommended to establish whether the protein of interest or a particular treatment is regulating/affecting EGFR ubiquitination.

2 Materials

2.1 *In Vitro* Ubiquitination Assay Reagents

1. GST lysis buffer: 50 mM Na-HEPES pH 7.5, 200 mM NaCl, 1 mM EDTA, 0.1% NP40, 5% glycerol, 1:500 Protease Inhibitor Cocktail (Calbiochem).
2. GST maintenance solution: 50 mM Tris–HCl pH 7.4, 100 mM NaCl, 1 mM EDTA, 10% glycerol, 1 mM DTT, 1:500 Protease Inhibitor Cocktail (Calbiochem).
3. 5× Kinase assay buffer: 40 mM MOPS/NaOH pH 7, 1 mM EDTA.
4. 5× Kinase reaction buffer: 10 mM ATP, 50 mM $MnCl_2$, 4 M $(NH_4)_2SO_4$.
5. 10× Ubiquitination assay buffer: 250 mM Tris–HCl pH 7.6, 50 mM $MgCl_2$, 1 M NaCl.
6. 1× YY buffer: 50 mM Hepes pH 7.5, 10% glycerol, 150 mM NaCl, 1% Triton X-100, 1 mM EDTA.

   Add fresh: 20 mM sodium pyrophosphate pH 7.5, 50 mM NaF, 2 mM PMSF (prepared in ethanol), 10 mM Na orthovanadate (0.5 M stock freshly made in 1 M Hepes pH 7.5), 1:500 Protease Inhibitor Cocktail (Calbiochem).
7. His-tagged E1 enzyme Uba1 (plasmid: Addgene clone #34965).

8. Baculovirus-produced GST-EGFR (*Homo sapiens*) cytoplasmic tail (aa 696-end) from Millipore.
9. Recombinant human ubiquitin from SIGMA.
10. ATP and ATP regeneration system (creatine phosphokinase and creatine phosphate) from SIGMA.
11. Isopropyl β-D-1-thiogalactopyranoside (IPTG) from SIGMA.

2.1.1 Production and Purification of Uba1 (as E1), UbcH5B (as E2) His-Fusion Proteins

His-tagged E1 enzyme Uba1 is produced in Rosetta cells according to the protocol described in [18]. His-tagged E2 enzyme Ube2D3 (UBCH5c) is produced in BL21 as described in [19].

Alternatively, E1 and E2 proteins are commercially available as recombinant active proteins from Biomol.

2.1.2 Production and Purification of Cbl (as E3) and Grb2 GST-Fusion Proteins

GST-Cbl and GST-Grb2 are produced in Rosetta cells cultured at 37 °C until an OD_{600} of 0.5 is reached and 1 mM IPTG is then added. IPTG induction is performed as follows:

- For GST-Grb2–37 °C for 3 h;
- For GST-Cbl–18 °C overnight (O/N).

Following IPTG induction:

1. Pellet cells and resuspend them in GST lysis buffer.
2. Sonicate samples and centrifuge to clear lysate.
3. Prewash glutathione-Sepharose beads in PBS.
4. Incubate supernatant from **step 2** with 600 μl–1 ml of glutathione-Sepharose beads per liter of bacterial culture at 4 °C for 4 h (for Grb2) or O/N (for Cbl) to allow GST proteins to bind to the beads.
5. Wash beads once in 1% Triton PBS and twice in PBS.
6. Resuspend beads in 1:1 volume of GST-maintenance solution.
7. To cleave off the GST tag, add 10 units of PreScission protease (GE Healthcare) per mg of GST protein in the presence of GST-maintenance solution and incubate O/N at 4 °C.

2.2 Immuno-precipitation Reagents

1. 1× RIPA buffer 1% SDS: 50 mM Tris–HCl pH 7.4, 150 mM NaCl, 1 mM EDTA, 1% Triton, 1% Na deoxycholate, 1% SDS.
2. 1× RIPA buffer w/o SDS: 50 mM Tris–HCl pH 7.4, 150 mM NaCl, 1 mM EDTA, 1% Triton, 1% Na deoxycholate.

   Add fresh to 1× RIPA buffers: 20 mM sodium pyrophosphate pH 7.5, 250 mM NaF, 2 mM PMSF (prepared in ethanol), 10 mM Na orthovanadate (0.5 M stock freshly made in 1 M Hepes pH 7.5), 1:500 Protease Inhibitor Cocktail (Calbiochem).

3. 4× Laemmli buffer: 200 mM Tris–HCl pH 6.8, 8% SDS, 40% glycerol, 0.004% saturated bromophenol blue, 2.4% β-mercaptoethanol (14 M).
4. Denaturing solution: 6 M guanidinium chloride, 20 mM Tris–HCl pH 7.5.

   Add fresh: 1 mM PMSF, 5 mM β-mercaptoethanol.
5. EGF from INALCO (cat. AF-100-15).

2.3 DELFIA Reagents

1. The commercial DELFIA kit (PerkinElmer) includes: microwell plates, wash buffer, assay buffer, Europium-labeled secondary antibodies (anti-mouse cat. AD0124; anti-rabbit cat. AD0105, anti-pY PT66 cat. AD0040), and enhancement solution.
2. Plate coating buffer: 3.03 g $NaCO_3$, 6.0 g $NaHCO_3$ pH 9.6 dissolved in 1 l H_2O.
3. EnVision Instrument (PerkinElmer).

2.4 Anti-K63 and K48 Staining Reagents

1. Binding Buffer: HeLa complete medium without serum [Minimum Essential Media (MEM), 1 mM Na pyruvate, 0.1 mM nonessential amino acids] containing 20 mM Hepes and 0.1% BSA.
2. Rhodamine-EGF (Thermo Fisher Scientific, cat. E3481).
3. Alexa 555-EGF (Thermo Fisher Scientific, cat. E35350).
4. Alexa 488 goat anti-human (Thermo Fisher Scientific, cat. A-11013).
5. Alexa 488 goat anti-mouse (Thermo Fisher Scientific, cat. A-11001).
6. Dako pen (Agilent Technologies).

2.5 Antibodies

Antibodies used for the protocols described in Subheadings 3.1–3.3 are listed in Table 1.

3 Methods

3.1 In Vitro Ubiquitination Assay

This method reproduces the biological reaction leading to ubiquitination of the EGFR cytoplasmic tail with the minimal purified components, as depicted in the schematic representation in Fig. 1a.

3.1.1 In Vitro Phosphorylation of the EGFR Tail

Since the EGFR must be phosphorylated to allow the binding of Cbl (and Grb2) to its docking sites in the receptor tail, prior to performing the in vitro ubiquitination assay the GST-EGFR tail is allowed to trans autophosphorylate:

1. Incubate the kinase reaction (250 ng GST-EGFR tail, kinase assay buffer 1×, kinase reaction buffer 1×, H_2O to a volume of 50 μl, *see* **Note 1**) for 1 h at 30 °C.

Table 1
List of antibodies used, and their applications within this chapter

	Company	Description	Application
Anti-EGFR antibodies			
Rabbit polyclonal anti-EGFR (intracellular)	In-house	Directed against aa 1172–1186 of human EGFR	WB, IP, ELISA
Commercial rabbit polyclonal anti-EGFR (intracellular)	Millipore, cat. 06-847	Directed against aa 1156–1186 of human EGFR	WB, IP
Commercial mouse monoclonal anti-EGFR (extracellular, clone 108)	m108 hybridoma, American Type Culture Collection (ATCC), cat. HB-9764	Directed against extracellular portion of human EGFR	IP, ELISA
Commercial mouse monoclonal anti-EGFR (extracellular)	clone Ab-1, Millipore, cat. GR01L	Directed against the extracellular portion of human EGFR	IP
Anti-Ubiquitin antibodies			
Anti-Ub P4D1 (mouse)	Santa Cruz Biotechnology, cat. sc-8017		WB
Anti-Ub FK2 (mouse)	Enzo Lifescience, cat. BML-PW8810–0100		ELISA
Anti-Ub K63 (human, Apu3 clone) Anti-Ub K48 (human, Apu2 clone)	Genentech; these antibodies can be obtained by completing a request form and MTA agreement on Genentech website (www.gene.com). These antibodies give the highest signal-to-noise ratio in immunofluorescence. Alternatively, the same clones are commercially available as monoclonal anti-Ub K63 and anti-Ub K48 rabbit antibodies Merck Millipore, cat. 05-1308 and 05-1307, respectively		IF
Other antibodies			
Anti-phosphotyrosine (pY) (mouse)	clone 4G10; Millipore, cat. 05-321		WB, ELISA
Anti-EGFR-pY1068 (mouse)	clone 1H12; Cell Signaling, cat. 2236		WB, ELISA
Anti-EGFR-pY1045 (rabbit)	clone 1H12; Cell Signaling, cat. 2237		WB

2. Preincubate glutathione beads in 0.02% BSA PBS for 1 h, wash beads in YY buffer and add them to the kinase reaction (10 μl 50% slurry/reaction) for the last 30 min to allow the GST-EGFR tail to be pulled down by the beads (*see* **Note 2**).
3. Upon completion of the kinase reaction, wash EGFR-bound beads 3 times with YY buffer and once with ubiquitination assay buffer 1× (*see* **Note 3**).

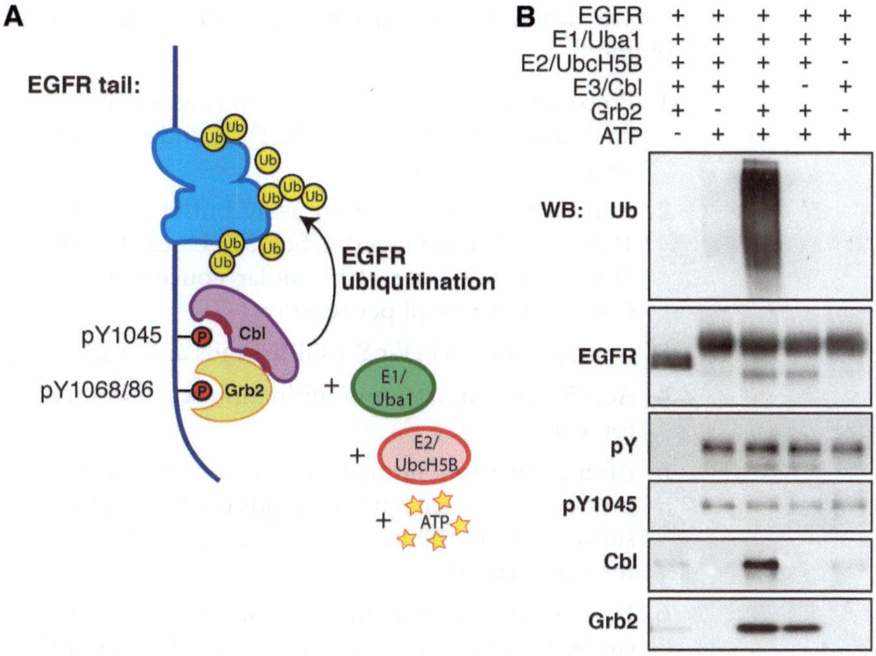

Fig. 1 In vitro ubiquitination assay. (**a**) Schematic representation of the in vitro ubiquitination reaction. Briefly, after in vitro phosphorylation of the intracytoplasmic tail of the EGFR, phosphorylation sites pY1045 and pY1068/1086 are competent for Cbl/Grb2 binding, and all the ubiquitination components are added to the reaction. Ubiquitination reaction requires ATP, as well as E1/Uba1, E2 UbcH5B, E3/Cbl, and Grb2 adaptor protein. (**b**) We report an example of an in vitro Ub assay on the EGFR intracytoplasmic tail. Only in the presence of all the components necessary for the ubiquitination reaction, the phosphorylated EGFR is ubiquitinated by Cbl. The presence of Grb2 in the reaction is required to achieve efficent EGFR ubiquitination. As control, reaction was run in absence of Cbl or E2, showing no ubiquitination

3.1.2 In Vitro Ubiquitination of the EGFR Tail

The in vitro ubiquitination assay requires the substrate and all the enzymes involved in the ubiquitination reaction. They are as follows:

1. Phosphorylated GST-EGFR tail.
2. El enzyme (bacterially purified Ubal, *see* Subheading 2.1.1).
3. E2 enzyme (bacterially purified UbcH5B, *see* Subheading 2.1.1).
4. Bacterially purified E3 enzyme Cbl (*see* Subheading 2.1.2).
5. Bacterially purified Grb2 (*see* Subheading 2.1.2).
6. Ubiquitin.
7. ATP (and ATP regeneration system).

The ubiquitination assay is performed in a 50 μl reaction volume.

1. Prepare the in vitro ubiquitination reaction mix 1×. Add to the final mix: 2 mM ATP, 0.2 μM DTT and the ATP regeneration system (*see* **Note 4**).
2. Add to the ubiquitination assay buffer 1×: ATP 2 mM, E1 100 ng, E2 His-tagged UbcH5c/Ube2D3 600 ng, E3 Cbl 500 ng, Grb2 1 μg (10× molar concentration compared to Cbl), H_2O to 50 μl per reaction.
3. Centrifuge the GST-pY-EGFR beads and discard supernatant.
4. Add 50 μl of in vitro ubiquitination reaction mix and incubate for 1 h at 30 °C.
5. After incubation, immediately transfer tubes on ice and centrifuge at 4 °C to separate the beads containing the ubiquitinated substrate from the free Ub/Ub chains and enzymes present in the supernatant.
6. Discard the supernatant (*see* **Note 5**), add 30 μl of Protein G (to better visualize the beads) and wash beads 4 times with YY buffer +0.1% SDS.
7. Add 2× Laemmli buffer (equal volume to beads), heat at 95 °C, and load on 8% SDS-PAGE gel.
8. Detect EGFR-Ub by WB analysis: half of the eluted proteins are analyzed by anti-Ub WB (*see* Subheading 3.2.5). The remaining half of the reactions are subsequently divided in two and analyzed by classical WB anti-EGFR or anti-pY, as described in the next section (Subheading 3.1.3).

3.1.3 Controls to Include in the In Vitro Ubiquitination Assay

Since this assay is composed of a series of catalytic and enzymatic reactions, to check step-by-step all the reactions, we suggest to include the following negative controls containing all the reaction components, except one:

1. No ATP.
2. No substrate.
3. No E2.
4. No E3.

Negative controls are treated exactly as described in Subheading 3.1.2.

To control for the presence of the EGFR system components in the reaction, half of the reaction is splitted in two and loaded on two 7% acrylamide gels, transferred onto a nitrocellulose membrane and blotted with:

- Gel 1: anti-EGFR to check for EGFR tail levels; a shift between nonphosphorylated and phosphorylated forms should be observed.
- Gel 2: anti-total pY 4G10 or anti-EGFR pY1045 to check the phosphorylation of the EGFR tail.

- Gel 1: (after anti-EGFR antibody): anti-Cbl to check for Cbl levels and anti-Grb2 to check for Grb2 levels.

In Fig. 1b, we show a representative in vitro ubiquitination assay experiment. In vitro ubiquitination reactions were analyzed by WB analysis, as indicated above and in Subheading 3.2.5.

3.2 Immunoprecipitation

This technique exploits the ability of an antibody to recognize and bind to its specific epitope, precipitating the protein of interest out of solution. For this experiment, HeLa cells are used as the model system since they have a relatively high number of surface EGFRs (~300,000 EGFRs/cell), which allows EGFR modifications to be easily detected. In addition, HeLa cells are well characterized for EGFR signaling, ubiquitination and endocytosis.

HeLa cells are cultured until they reach ~60–70% confluence on the day of the experiment. Prior to the experiment, HeLa cells are serum starved for 16 h, followed by stimulation with EGF (100 ng/ml, in order to have maximal EGFR activation and ubiquitination) for different time points. At 2 min post-stimulation, the EGFR is fully ubiquitinated, a status that persists at later times (~10–15 min) until the receptor is degraded in the lysosomal compartment (~30 min, [7]). Cellular lysates are then subjected to immunoprecipitation (IP) with an anti-EGFR antibody followed by SDS-PAGE and immunoblotting to reveal the ubiquitinated species with a specific anti-Ub antibody.

Following stimulation with high EGF concentrations (100 ng/ml), degradation of the EGFR is triggered. Therefore, a decrease of the ubiquitinated form of the receptor at late time points might reflect a decrease in the total level of the EGFR due to its degradation. To overcome this effect and to accumulate the ubiquitinated species, HeLa cells are treated with chloroquine, which prevents lysosomal acidification, and thus EGFR degradation. A pretreatment with 100 μM chloroquine for 1 h, followed by EGF stimulation in the continuous presence of chloroquine is sufficient to allow the accumulation of EGFR-Ub.

3.2.1 Experimental Controls to Include in Your Experiment

An important consideration when performing a biochemical experiment aimed at scoring a post-translational modification directed at a specific target (in this case the EGFR), is the fact that your result could be invalidated by the presence of other ubiquitinated substrates in the IP purification. This issue is partially overcome by performing all the experiments in stringent lysis conditions, achieved by increasing the SDS concentration in the standard RIPA buffer, used to lyse the cells, to 1% (*see* Subheading 2). SDS is subsequently diluted to 0.2% during the IP to avoid antibody denaturation.

An additional control for the specificity of the anti-EGFR IP/anti-Ub IB, is to perform the knockdown of the protein of interest, in our case the EGFR, and to perform a parallel IP under this condition. This control will, however, allow the exclusion of only those proteins unspecifically precipitating with the anti-EGFR antibodies, and not the proteins specifically coprecipitating with the EGFR.

3.2.2 Choice of Anti-EGFR Antibody

The choice of a good anti-EGFR antibody is essential for the success of this experiment, as well as for ELISA assay. A series of commercial anti-EGFR antibodies are available that work well both in IP and/or in ELISA (*see* Subheading 2.1, **item 5** and Table 2 below). Since the set of experiments herein described are performed in partial denaturing conditions (0.2% SDS in the IP), it is important to test in advance whether the antibody of your choice works under this setting.

Another characteristic to take into account when choosing the antibody is whether the epitope recognized by the antibody maps to a region close to the ubiquitinated kinase domain of the EGFR or to the phosphorylated residues in the tail. Ubiquitination might mask and interfere with epitope recognition of antibody, thus determining a bias or even the failure of the assay. The optimal choice would be an antibody directed against the extracellular domain of the EGFR (e.g., clone m108 or Ab-1). These antibodies are all mouse monoclonal and in some assays (e.g., the ELISA) they cannot be used in combination with the anti-Ub antibody, which is also a mouse monoclonal antibody.

In-house polyclonal antibodies can be a good choice, if available. In our experience, a good anti-EGFR in-house polyclonal antibody can be obtained by raising the host immune response

Table 2
Anti-EGFR antibodies used for IP and WB analysis

EGFR antibody	IP working conditions (set on HeLa cells)	Western blot conditions
Rabbit purified in-house polyclonal anti-EGFR directed against residues 1172–1186 of human EGFR (intracellular tail)	10 μg of antibody/mg of lysate	1:5000 in 5% milk/TBS-Tween
Rabbit polyclonal anti-EGFR directed against residues 1156–1186 of human EGFR (intracellular tail) Commercial: Millipore cat. 06–847	4 μg of antibody/mg of lysate	1:5000 in 5% milk/TBS-Tween
Mouse monoclonal anti-EGFR directed against the extracellular domain, clone Ab-1 Commercial: Millipore cat. GR01L	1 μg of antibody/mg of lysate	Not suitable
Mouse monoclonal anti-EGFR directed against the extracellular domain, m108 mouse hybridoma Commercial: ATCC, cat. HB-9764	15 μg of antibody/mg of lysate 2 cycles of IP	Not suitable

against the last 15 residues in the EGFR protein (residues 1172-1186). We have previously checked that the recognition of this antibody is not influenced by EGFR phosphorylation or ubiquitination (Sigismund 2013). A commercial antibody recognizing an overlapping epitope (residues 1156-1186) is also available (*see* Subheading 2.1, **item 5**).

Anti-Ub antibodies might also differ according to the question you want to address. P4D1 antibody (1:1000 in WB) and FK2 (1:1000 in WB) recognize both mono- and poly-Ub proteins, whereas FK1 (1:1000 in WB) recognizes only poly-Ub proteins. We routinely use P4D1 for WB analysis because it is the best available product to detect the ubiquitinated EGFR (that has both mono- and poly-Ub chains) and FK2 for ELISA assay (*see* Subheading 3.3) because it is the only one that works in a 0.2% SDS solution.

3.2.3 Cell Lysis

1. Immediately after EGF stimulation, place cells on ice.
2. Wash cells twice with cold PBS.
3. Accurately remove PBS by tilting plates for 2/3 min, keeping plates on ice.
4. Add 80–100 μl of RIPA 1% SDS plus inhibitors to each 10 cm plate (*see* **Note 6**) and scrape them. Collect the lysate (*see* **Note 7**) and let lysates rest on ice for 10 min.
5. Transfer lysate in a polycarbonate ultracentrifuge microtube and ultracentrifuge samples at 110,000 × *g* for 45 min at 4 °C.
6. Collect samples and dilute supernatants 1:5 in RIPA buffer w/o SDS to reach a final concentration of SDS equal to 0.2%.
7. Determine protein content using the BCA method (*see* **Note 8**).

3.2.4 IP Reaction

1. Incubate 500 μg of lysate in the presence of anti-EGFR antibody (*see* **Note 9**) on a wheel shaker at 4 °C for 2 h.
2. Add 30 μl of 50% slurry Protein G-conjugated Sepharose beads and incubate on a wheel shaker at 4 °C for an additional hour.
3. Centrifuge the beads for 1 min at 400 × *g*.
4. Wash the beads four times with 700 μl RIPA buffer 0.2% SDS for 1 min at 400 × *g*.
5. Dry bead pellets and add 30 μl of 2× Laemmli buffer.
6. Boil samples at 95 °C for 5 min and spin for 1 min at 400 × *g*.
7. Divide each sample as follows:
   - 1/10th of the IP to perform a WB anti-EGFR as normalizer of the IP reaction: load on a 7% acrylamide gel, transfer onto a nitrocellulose membrane (*see* **Note 10**) and blot with an anti-EGFR antibody following standard procedures (*see* Table 2 above for antibody dilution);

- 1/10th of the IP to perform a WB anti-total pY to assess the activation of the EGFR: load on a 7% acrylamide gel, transfer onto a nitrocellulose membrane and blot with total anti-pY 4G10 antibody (blocking and antibody incubation must be performed in 5% BSA/TBS-T, *see* **Note 11**);
- 8/10th of the IP to use in the anti-Ub WB protocol (next step).

Samples can now either be stored at −20 °C for later analysis or loaded onto a SDS-PAGE gel.

3.2.5 SDS-PAGE and Anti-Ub Western Blot

1. Load the IP on a 7% acrylamide gel, which will allow a good resolution of high molecular weight proteins (EGFR MW ~180 kDa).
2. Prior to blotting the gel onto a membrane, activate the polyvinylidene fluoride (PVDF) membrane by activating it through incubation in 100% methanol for 5 min at room temperature, followed by extensive washing in TBS-T (TBS with 0.1% Tween, *see* **Note 12**).
3. Transfer proteins to the PVDF membrane (*see* **Note 13**).
4. Extensively wash the membrane in TBS-T and incubate the filter in denaturing solution at 4 °C for 30 min (*see* **Note 14**).
5. Extensively wash the membrane in TBS-T and block filters in 5% BSA/TBS-T O/N.
6. Incubate the filter with primary antibody (anti-Ub P4D1, diluted 1:1000 in 5% BSA/TBS-T) for 1 h at room temperature.
7. Wash 3 times in TBS-T; change the buffer every 10 min.
8. Incubate the filter with the secondary antibody: anti-mouse horseradish peroxidase conjugated antibody in 5% BSA/TBS-T for 30 min at room temperature.
9. Wash 3 times in TBS-T; change the buffer every 10 min.
10. The EGFR-Ub signal can be revealed by a 1 min incubation with enhanced chemiluminescence (ECL) and by film exposures and development.

3.3 ELISA-Based Assay to Detect EGFR-Ub

For the protocol of this assay we refer to the DELFIA (Dissociation Enhanced Lanthanide Fluoroimmunoassay) technology from PerkinElmer. This technology is based on sandwich recognition of a target protein by a capture antibody and a detection antibody. After immobilization of the capture antibody (e.g., anti-Ub) on a solid surface (the bottom of a microwell), the analyte (appropriate cellular lysate) is incubated on the plate. After removal of the unbound analyte, incubation with the detection antibody against the substrate (e.g., the EGFR) will produce a specific signal in those samples where the substrate is ubiquitinated. The signal relies on a Europium (Eu)-conjugated antibody that is able to produce a fluorescent signal when incubated with acidic enhancement buffer. The lanthanide ions released into solution rapidly form a fluorescent species that is stable for up to 8 h.

To measure the ubiquitinated form of the EGFR a set of two antibodies is used, recognizing: (1) the EGFR and (2) the receptor-bound Ub. Two complementary approaches for this assay can be performed (Fig. 2a):

- The "forward" approach, where microwell plates are coated with anti-EGFR (intracellular/rabbit) antibody, which binds the receptor from the lysate. Detection is performed with the monoclonal anti-Ub (FK2) antibody followed by Eu-labeled secondary antibodies.
- The "reverse" approach, where microwell plates are coated with the monoclonal anti-Ub (FK2) that captures ubiquitinated proteins. Detection is performed with the anti-EGFR rabbit antibody followed by a Eu-labeled secondary antibody.

The best signal-to-noise ratio was achieved in our hands by applying the reverse approach, which we will illustrate in detail in the next section. Nonetheless, the forward approach is still a functional assay with a good signal-to-noise ratio that could be used to validate results obtained with the reverse approach (*see* Table 3 below for choosing the correct combinations of antibodies in the different approaches).

3.3.1 Lysate Preparation

For this experiment, prepare HeLa cells as previously described in Subheading 3.2 and proceed to cell lysis as in Subheading 3.2.3. Also in this case, lysates are prepared under denaturing conditions (1% SDS during lysis, 0.2% SDS during incubation) to minimize the presence of ubiquitinated proteins other than the EGFR in the final detection reaction.

3.3.2 Plate Coating and Analyte Incubation

1. Coat the flat bottomed DELFIA microwell plate with the capturing antibody: dilute the anti-Ub (FK2) antibody to 5 μg/ml in coating buffer (*see* **Note 15**); incubate plate O/N at 4 °C with gentle shaking. Consider working at least with duplicates.
2. Block the microwell plate by incubating with 2% BSA/PBS (*see* **Note 16**) for either 1–2 h at RT or O/N at 4 °C.
3. Wash plate once in wash buffer.
4. Prepare the cellular lysate as described in Subheading 3.2.3 and dilute samples in assay buffer (provided in the DELFIA kit) to reach a volume of 100 μl/well; incubate 25–50 μg of lysate/well (in triplicate) O/N at 4 °C (*see* **Note 17**).

A

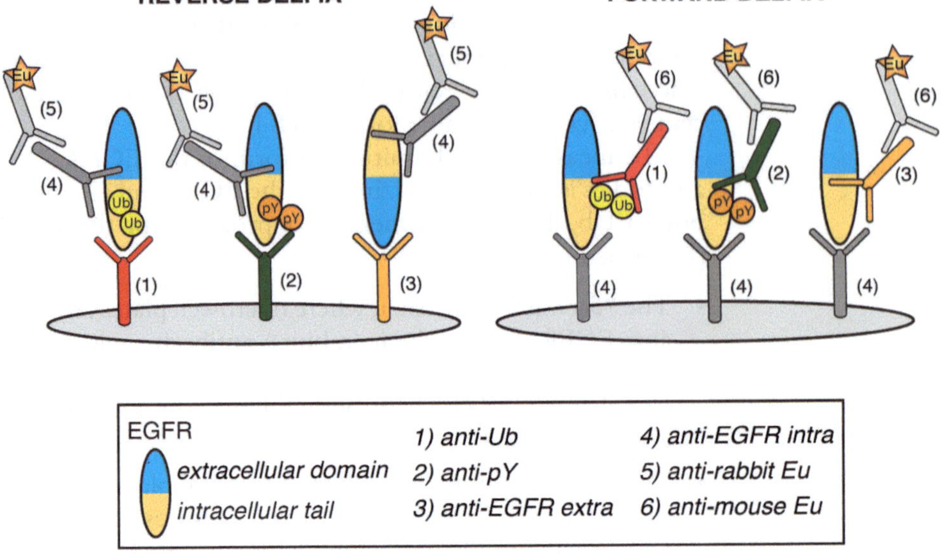

B

	no lysate		*SS*		*+EGF*	
Ub	5593	5206	31320	29850	262830	248863
pY	19128	21542	86927	84223	608279	612587
EGFR	2338	2184	5652	6120	5624	6135

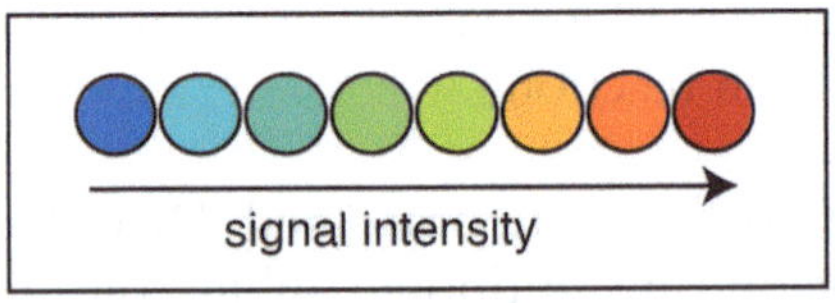

Fig. 2 DELFIA assay. (**a**) Schematic representation of the DELFIA assay described in this protocol, to measure EGFR phosphorylation and ubiquitination, and total level of EGFR as control. In the reverse approach, wells are coated in duplicate with the coating antibody: anti-Ub (1), anti-pY (2), or anti-EGFR extra (3). After incubation with lysate, detection antibody is added (anti-EGFR intra, 4) and revealed with a secondary anti-rabbit antibody conjugated with Europium, Eu (5). Addiction of enhancement solution will release the Eu^{3+} ions to generate the signal. In the forward approach, coating is performed with the anti-EGFR intra (4) to capture all the EGFR in the lysate, and the modification is revealed with the use of the specific antibodies [anti-Ub (1), anti-pY (2), and anti-EGFR extra (3)] followed by the Eu-labeled anti-mouse secondary antibody (6). (**b**) Representative raw data from a reverse DELFIA assay. Example of a DELFIA assay performed using cell lysate from HeLa cells that had been serum-starved (SS) or treated with 100 ng/ml EGF for 2 min. "No lysate" refers to the assay performed with all components except the lysate. Cell lysate was then added to 96-microwell plates coated as described in the main text (Subheading 3.3). Captured receptor was then detected using the detection antibody anti-EGFR intra (as in Fig.2a *upper panel*) coupled with the Eu-conjugated anti-rabbit antibody. Results are expressed as counts per second

Table 3
Antibodies used in ELISA assay. To perform the forward or reverse approach, use the indicated antibodies as capture or detection antibodies

	Capture antibodies	Detection antibodies
Forward ELISA	In-house polyclonal anti-EGFR directed against aa 1172–1186 of human EGFR (5 μg/ml)	• Anti-Ub (FK2) • EGFR (m108) • Anti-pY (clone 4G10) • Anti-EGFR-pY1068 All diluted at 1 μg/ml
Reverse ELISA	• Anti-Ub (FK2, 5 μg/ml) • EGFR (m108, 1 μg/ml) • Anti-pY (clone 4G10, 5 μg/ml) • Anti-EGFR-pY1068 (1 μg/ml)	Homemade polyclonal anti-EGFR directed against aa 1172–1186 of human EGFR (1 μg/ml)

3.3.3 Detection of Ubiquitinated EGFR

1. Wash plate 3 times in wash buffer.
2. Incubate the plate with detecting antibody for 1 h at room temperature: dilute the anti-EGFR (intracellular) antibody to 1 μg/ml in assay buffer.
3. Wash the plate 3 times in wash buffer.
4. Add the anti-rabbit Eu-labeled secondary antibody (1 μg/ml in assay buffer, 100 μl/well) and incubate for an additional 1 h at room temperature.
5. Wash the plate 3 times in wash buffer.
6. Add 100 μl/well of Enhancement Solution to release and activate EU3+ ions and incubate for 15 min without shaking or 5 min with shaking, at room temperature (*see* **Note 18**).
7. Measure signals with EnVision at $\lambda = 613$ nm emission (excitation $\lambda = 340$ nm).

3.3.4 Detection of Phosphorylated EGFR

With the DELFIA assay, it is also possible to measure EGFR phosphorylation. Since phosphorylation is a prerequisite of EGFR ubiquitination, you may add this as an additional control. Choose the appropriate combination of antibodies reported in Table 3, and follow protocols in Subheadings 3.3.2 and 3.3.3 to detect pY-EGFR.

3.3.5 How to Normalize the Signals

To normalize the signals of the Ub/pY-EGFR (e.g., in the case of a time course or an EGF dose titration experiment), we suggest to perform an additional DELFIA assay in order to obtain values for the total EGFR levels in the different samples. Total EGFR levels can then be used as internal normalizer. In this case, EGFR antibodies targeting different epitopes will be used as the capture and detection antibodies (Fig. 2).

According to the approach that had been used (forward vs. reverse), you can select the best combination of capture/detection EGFR antibodies as illustrated in Table 3.

3.4 Immunofluorescence

The last protocol we illustrate is the IF to visualize EGFR-Ub-containing complexes. This method allows to directly visualize ubiquitin chains in cells, through the use of specific Ub antibodies (*see* Subheading 2.1, **item 5**). Of note, the Ub signal is very plastic, rapidly rearranging within the cell depending on the stimuli. In the example reported in Fig. 3, EGF causes the rapid relocalization of K63 Ub from the cytosol to EGFR-positive foci (similarly to what happen in the nucleus following DNA damage inducing agent). By following the EGFR with fluorescent EGF in the same experiments, it is possible to co-localize the EGFR and K63 Ub chains, whereas an antibody directed against K48 Ub-linked chains will be used for the negative control (Fig. 3).

1. Plate HeLa cells on glass coverslips (*see* **Note 19**); cells should be at a ~60% confluence the day of the experiment.

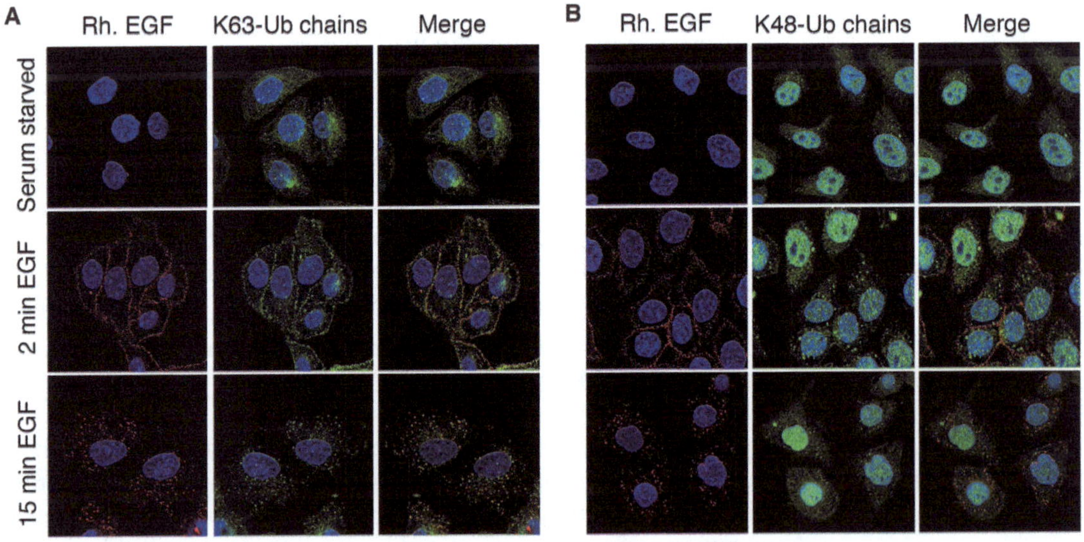

Fig. 3 Immunofluorescence experiment to detect Ub signal within the EGF-positive endocytic compartments. (**a**) Rhodamine-EGF (*red*) and K63 Ub chains (*green*) localization in unstimulated-serum starved HeLa cells (*upper panels*) or cells stimulated with Rodhamine-EGF (*red*) for 2 (*middle panels*) or 15 min (*lower panels*). In the serum starved cells, K63 staining is mainly diffused with no clear compartmentalized localization. After 2 min of EGF stimulation, K63 Ub signal relocalizes with EGF at the plasma membrane. At 15 min after stimulation, EGF and K63 colocalize in endocytic compartments (e.g., endosomes). (**b**) Rhodamine-EGF (*red*) and K48 Ub chains (*green*) localization in unstimulated-serum starved HeLa cells (*upper panels*) or cells stimulated with Rodhamine-EGF (*red*) for 2 (*middle panels*) or 15 min (*lower panels*). K48 Ub chain is the major signal to target proteins to proteasome degradation and it is not involved in endocytosis. Indeed, K48 staining is nuclear with diffused cytosolic staining both in unstimulated and stimulated cells, and no major relocalization is observed

2. Serum starved HeLa cells for 2 h in binding buffer, a starvation medium containing 0.1% BSA to saturate aspecific antibody binding sites.
3. Incubate cells in the presence of a fluorophore-conjugated EGF (e.g., Alexa 555-EGF 1:500 or Rhodamine-EGF 1:100, *see* **Note 20**) for different time points, between 2 and 15 min at 37 °C.
4. Immediately after EGF stimulation, place cells on ice.
5. Wash cells twice with cold PBS.
6. Fix cells for 8 min in 4% paraformaldehyde.
7. Permeabilize cells in 0.1% Triton for 10 min and wash 3 times with PBS.
8. Block in 10% goat serum diluted in 2% PBS/BSA for 30 min at RT.
9. Incubate coverslips with primary antibody in 1% BSA in PBS for 1 h at room temperature:
   - human K48 or K63 at 1 μg/ml (clone Apu2 and Apu3).
   - mouse FK2 at 1 μg/ml.
10. After washing twice with PBS, incubate coverslips with Alexa 488 goat anti-human or anti-mouse (1:400) in 1% BSA in PBS for an additional 45 min.
11. Wash twice in PBS and DAPI stain coverslips.
12. Mount coverslips in glycerol for confocal analysis:
    - Draw a circle on a microscope slide with a Dako pen to create a humidity chamber for the stained specimen so that is does not dry out;
    - Put 3 μl of glycerol in the center of the circle;
    - Dip the coverslip in distilled water;
    - Remove excess water with filter paper;
    - Carefully lower the coverslip onto the glycerol drop in such a way as to prevent the formation of bubbles;
    - Seal the edge of the coverslips with clear nail polish and let dry.

4 Notes

1. You can perform six reactions in the same volume of 50 μl. If you have to do more reactions, divide the reactions in two different tubes.
2. If you have less than three reactions, you may add more beads to reach a visible pellet (around 30 μl slurry), required for the subsequent washing step.

3. At this step, aliquot the kinase in the different tubes corresponding to the desired reactions.
4. ATP regeneration system is necessary to avoid ATP depletion by the active EGFR kinase.
5. You may want to keep an aliquot of supernatant for analysis of proteins present in the reaction, but not bound to the beads. To do so, run the supernatant (25 μl) with the samples as in Subheading 3.1.3.
6. Use the minimal volume of RIPA 1% SDS buffer to cover the plate, since in the next step the lysate will be diluted 1:5 to achieve the working concentration of 0.2% SDS to allow for antibody-epitope binding. For HeLa cells, we normally use 50 μl/10 cm plate.
7. After lysing the cells in RIPA 1% SDS buffer, samples will be viscous. We suggest to collect samples with a pipette tip that has been cut at the end.
8. Bradford protein quantification is not ideal in this case due to the high concentration of SDS that might interfere with the method.
9. Perform IP in a volume of ~200 μl to allow mixing of the tube when rotating.
10. For anti-Ub WB, PVDF membrane blotting is mandatory, since it allows the in situ denaturation step to be performed.
11. It is necessary to avoid blocking with milk since this may interfere with anti-pY recognition of phosphorylated proteins.
12. In this and in the following step, be aware not to allow the membranes to dry. If membranes dry, it is possible to hydrate them again by incubating in methanol for 2 min and washing in MilliQ water.
13. Ponceau staining is not recommended, because it could interfere with later antibody recognition. To check for an optimal transfer efficiency, prestained molecular weight markers can be used.
14. Denaturing step is necessary to completely unfold proteins blotted onto the membrane and prepare them for the incubation and recognition with the anti-Ub P4D1 antibody that we use in this assay. During this step, the nitrocellulose membrane will become completely transparent. If not, hydrate it again as in **Note 12** and denature again. Membranes will become white after the first wash in **step 19**.
15. Keep the working volume constant between the different steps; we suggest to use 100 μl/well.
16. Use 150–200 μl/well of wash buffer.

17. As a negative control, we suggest to include a "no-lysate" condition and a lysate of a cell line devoid of endogenous EGFR (e.g., NR6 or CHO cells).
18. Use Enhancement Solution at room temperature.
19. For HeLa cells, coverslip coating is dispensable; for other cell lines 0.5% porcine gelatin or fibronectin coating may be required.
20. EGF-Alexa 555 has a brighter signal since it has more fluorophores conjugated to the EGF molecules, while the rhodamine:EGF ratio is 1:1.

Acknowledgments

The authors gratefully acknowledge Rosalind Gunby for critically reviewing this manuscript and Roberta Pascolutti for the EGFR in vitro ubiquitination experiment. SS was supported by a grant from Worldwide Cancer Research (16-1245). AC was supported by FIRC Fellowship 14909.

References

1. Citri A, Yarden Y (2006) EGF-ERBB signalling: towards the systems level. Nat Rev Mol Cell Biol 7(7):505–516. doi:10.1038/nrm1962
2. Zwick E, Bange J, Ullrich A (2001) Receptor tyrosine kinase signalling as a target for cancer intervention strategies. Endocr Relat Cancer 8 (3):161–173
3. Kovacs E, Zorn JA, Huang Y, Barros T, Kuriyan J (2015) A structural perspective on the regulation of the epidermal growth factor receptor. Annu Rev Biochem 84:739–764. doi:10.1146/annurev-biochem-060614-034402
4. Levkowitz G, Waterman H, Zamir E, Kam Z, Oved S, Langdon WY, Beguinot L, Geiger B, Yarden Y (1998) C-Cbl/Sli-1 regulates endocytic sorting and ubiquitination of the epidermal growth factor receptor. Genes Dev 12 (23):3663–3674
5. Sigismund S, Algisi V, Nappo G, Conte A, Pascolutti R, Cuomo A, Bonaldi T, Argenzio E, Verhoef LG, Maspero E, Bianchi F, Capuani F, Ciliberto A, Polo S, Di Fiore PP (2013) Threshold-controlled ubiquitination of the EGFR directs receptor fate. EMBO J 32 (15):2140–2157. doi:10.1038/emboj.2013.149
6. Sigismund S, Argenzio E, Tosoni D, Cavallaro E, Polo S, Di Fiore PP (2008) Clathrin-mediated internalization is essential for sustained EGFR signaling but dispensable for degradation. Dev Cell 15(2):209–219. doi:10.1016/j.devcel.2008.06.012
7. Goh LK, Sorkin A (2013) Endocytosis of receptor tyrosine kinases. Cold Spring Harb Perspect Biol 5(5):a017459. doi:10.1101/cshperspect.a017459
8. Waterman H, Levkowitz G, Alroy I, Yarden Y (1999) The RING finger of c-Cbl mediates desensitization of the epidermal growth factor receptor. J Biol Chem 274(32):22151–22154
9. Waterman H, Katz M, Rubin C, Shtiegman K, Lavi S, Elson A, Jovin T, Yarden Y (2002) A mutant EGF-receptor defective in ubiquitylation and endocytosis unveils a role for Grb2 in negative signaling. EMBO J 21(3):303–313. doi:10.1093/emboj/21.3.303
10. Jiang X, Huang F, Marusyk A, Sorkin A (2003) Grb2 regulates internalization of EGF receptors through clathrin-coated pits. Mol Biol Cell 14(3):858–870. doi:10.1091/mbc.E02-08-0532
11. Capuani F, Conte A, Argenzio E, Marchetti L, Priami C, Polo S, Di Fiore PP, Sigismund S, Ciliberto A (2015) Quantitative analysis reveals how EGFR activation and downregulation are coupled in normal but not in cancer cells. Nat Commun 6:7999. doi:10.1038/ncomms8999

12. Huang F, Kirkpatrick D, Jiang X, Gygi S, Sorkin A (2006) Differential regulation of EGF receptor internalization and degradation by multiubiquitination within the kinase domain. Mol Cell 21(6):737–748. doi:10.1016/j.molcel.2006.02.018
13. Conte A, Sigismund S (2016) Chapter six – the ubiquitin network in the control of EGFR endocytosis and signaling. Prog Mol Biol Transl Sci 141:225–276. doi:10.1016/bs.pmbts.2016.03.002
14. Goh LK, Huang F, Kim W, Gygi S, Sorkin A (2010) Multiple mechanisms collectively regulate clathrin-mediated endocytosis of the epidermal growth factor receptor. J Cell Biol 189 (5):871–883. doi:10.1083/jcb.201001008
15. Sigismund S, Woelk T, Puri C, Maspero E, Tacchetti C, Transidico P, Di Fiore PP, Polo S (2005) Clathrin-independent endocytosis of ubiquitinated cargos. Proc Natl Acad Sci U S A 102(8):2760–2765. doi:10.1073/pnas.0409817102
16. Haglund K, Shimokawa N, Szymkiewicz I, Dikic I (2002) Cbl-directed monoubiquitination of CIN85 is involved in regulation of ligand-induced degradation of EGF receptors. Proc Natl Acad Sci U S A 99(19):12191–12196. doi:10.1073/pnas.192462299
17. Tomas A, Futter CE, Eden ER (2014) EGF receptor trafficking: consequences for signaling and cancer. Trends Cell Biol 24(1):26–34. doi:10.1016/j.tcb.2013.11.002
18. Berndsen CE, Wolberger C (2011) A spectrophotometric assay for conjugation of ubiquitin and ubiquitin-like proteins. Anal Biochem 418 (1):102–110. doi:10.1016/j.ab.2011.06.034
19. Maspero E, Polo S (2016) In vitro ubiquitination: self-ubiquitination, chain formation, and substrate ubiquitination assays. Methods Mol Biol 1449:153–160. doi:10.1007/978-1-4939-3756-1_7

Chapter 6

Dimerization Assessment of Epithelial Growth Factor Family of Receptor Tyrosine Kinases by Using Cross-Linking Reagent

Hamid Maadi, Babak Nami, and Zhixiang Wang

Abstract

Dimerization of the epithelial growth factor (EGF) family of receptor tyrosine kinases is a crucial step for activation of these receptors. Different chemicals such as BS^3 and DSS have been introduced to covalently bind the interacting receptors and fix the dimers. Unique properties of BS^3 including higher water solubility and membrane impermeability make it suitable for assessing receptor–receptor interactions in live cells. In this protocol, we aim to explain a method to evaluate the dimerization of EGF receptors family using BS^3 as a cross-linker reagent.

Key words EGF receptor family, Dimerization, Immunoblotting, BS^3

1 Introduction

The epithelial growth factor (EGF) family of receptor tyrosine kinases regulates different cellular processes such as survival, proliferation, and growth and consists of four members including epithelial growth factor receptor (EGFR), human epidermal growth factor receptor (HER2), HER3, and HER4. Except HER2, certain ligands can specifically bind to binding domain of EGF family receptors to form either homodimer or heterodimer. Dimerization of these receptors can induce certain tyrosine residues autophosphorylation or transphosphorylation which activates several intracellular signaling pathways including the phosphoinositide 3-kinase (PI3K)/protein kinase B (PKB or Akt), Ras/Raf mitogen-activated protein kinase (MAPK), and signal transduction and activator of transcription (STAT) pathways [1, 2]. Therefore, assessing the dimerization of EGF receptors family as a critical step for activation of this type of tyrosine kinase receptors is necessary in most studies.

Various methods have been introduced to evaluate the protein–protein interactions which can be applied to investigate

Zhixiang Wang (ed.), *ErbB Receptor Signaling: Methods and Protocols*, Methods in Molecular Biology, vol. 1652, DOI 10.1007/978-1-4939-7219-7_6,

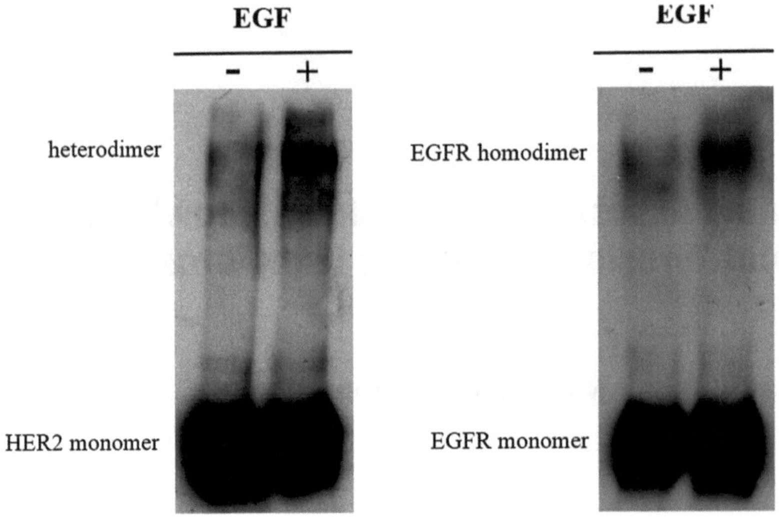

Fig. 1 Assessment of homodimerization and heterodimerization of EGFR receptors in BT474 breast cancer cells after stimulation with EGF ligands. EGF ligands can specifically bind to binding domain of EGFR receptors and induce the dimerization of these receptors with either EGFR receptors (*right*) or other members of EGF receptor family (*left*). The receptor dimers have twice the molecular weight of receptor monomers

receptors dimerization. Water-soluble and -insoluble chemicals such as bissulfosuccinimidyl suberate (BS^3) and disuccinimidyl suberate (DSS) are commonly used in several investigations to covalently bind interacting proteins [3]. Membrane impermeability and water-solubility of BS^3 are two important characteristics which makes it suitable for assessing membrane protein–protein interactions in live cells. In the buffers with the pH 7–9, two *N*-hydroxysulfosuccinimide (NHS) ester groups of BS^3 can form noncleavable amide bonds with primary amines on polypeptides' N-terminus region as well as the side chain of lysine residues [3]. In this chapter, we aim to explain a protocol to evaluate the homodimerization and heterodimerization of EGF receptors family by using BS^3 as a cross-linker reagent. As shown in Fig. 1, treatment of BT474 breast cancer cells with EGF ligand can induce the EGFR receptors homodimerization and heterodimerization which is easily detectable using this protocol.

2 Materials

2.1 Cell Culturing and EGF Treatment

1. Pipette and micropipette.
2. Micropipette tips.
3. Pasteur pipette.
4. Laboratory vacuum pump.
5. Optical microscope.

6. Neobar lam.
7. Mammalian cell culture incubator (providing 37 °C and 5% CO_2).
8. Sterile 6-cm plate.
9. Cell culture centrifuge.
10. Cell scraper.
11. Digital timer.
12. 1× Phosphate-Buffered Saline (PBS): dissolve the contents of one pouch of "Thermo Scientific BupH Phosphate Buffered Saline Packs" in 500 ml distilled water to obtain PBS solution with 0.1 M sodium phosphate and 0.15 M sodium chloride, pH 7.2.
13. Culturing medium: add 50 ml fetal bovine serum (FBS) and 5 ml penicillin–streptomycin mixtures (Pen-Strep) to 450 ml high glucose Dulbecco's Modified Eagle's Medium (DMEM) (from Sigma-Aldrich) containing L-glutamine and sodium bicarbonate to obtain 10% FBS, 1% Pen-Strep-containing DMEM.
14. Starvation medium: add 5 ml FBS to 495 ml DMEM containing L-glutamine and sodium bicarbonate to obtain 1% FBS-containing DMEM.
15. EGF working solution with a concentration of 50 ng/ml: add 1 μl from EGF stock solution with a concentration of 50 μg/ml to 1 ml DMEM starvation medium.

2.2 Cross-Linking Treatment

1. 1.5 ml microcentrifuge tube.
2. +4 °C Lab refrigerator.
3. BS^3 cross-linker.
4. BS^3 cross-linker reagent solution with a concentration of 0.75 mM: add 0.6 μl BS^3 cross-linker solution with a concentration of 125 mM to the 100 μl ice-cold 1× PBS.
5. Quench solution: Prepare 1 M Tris Base by dissolving 121.14 g Tris in 800 ml distilled water. Use concentrated HCl (12.1 M) to adjust the pH at 7.5. Add distilled water to the solution until it reaches to 1 l.

2.3 Protein Purification and Quantification

1. −80 °C Lab refrigerator.
2. Refrigerated ultracentrifuge.
3. Benchtop tube rotator/agitator.
4. Thermometer.
5. Spectrophotometer.
6. Protein extraction solution: Use the "Thermo Scientific M-PER Mammalian Protein Extraction Reagent" to purify the nuclear and cytoplasmic proteins from the cultured cells.

7. 1× bovine serum albumin (BSA) solution: prepare 10× solution of BSA by adding 5 mg BSA to 1 ml distilled water. This stock solution can be used to prepare 1× solution of BSA.
8. 1× Bradford solution: prepare 200 ml 5× solution of Bradford by adding 100 mg Coomassie Blue (G250) to 100 ml 85% phosphoric acid (H_3PO_4), 50 ml 95% ethanol, and 50 ml distilled water. The 1× Bradford solution can be prepared by diluting the stock solution.

2.4 Western Blotting

1. 90–100 °C water bath.
2. Essential accessories for casting gel, including combs, casting stand, glass plates, buffer tank, and so on.
3. Digital power supply.
4. Wash containers.
5. Filter paper.
6. Nitrocellulose paper.
7. Semidry transfer system.
8. Film cassette.
9. 4× Laemmli buffer (protein loading buffer): add 4.4 ml 0.5 M Tris base (pH 6.8) to 4.4 ml glycerol, 2.2 ml 20% sodium dodecyl sulfate (SDS) solution, 0.5 ml 1% Bromophenol Blue, and 0.5 ml 2-mercaptoethanol.
10. 5% separating gel: In 15 ml tube add 0.625 ml acrylamide–bis-acrylamide 40% solution to 3.125 ml double-distilled water (ddH_2O), 1.25 ml 1.5 M Tris–HCl buffer with pH 8.8 (lower gel buffer), 25 μl ammonium persulfate (APS), and 2.5 μl tetramethylethylenediamine (TEMED).
11. 4% stacking gel: In 15 ml tube add 250 μl acrylamide–bis-acrylamide 40% solution to 1.6 ml ddH_2O, 625 μl 0.5 M Tris–HCl buffer with pH 6.8 (higher gel buffer), 25 μl 10% SDS solution, 12.5 μl APS, and 2.5 μl TEMED.
12. 70% Ethanol.
13. Prestained high molecular weight protein ladder.
14. Running buffer: dilute 10× Tris–Glycine–SDS buffer solution from Bio Basic Canada Inc. to 1× buffer solution which contains 0.025 M Tris base, 0.192 M glycine, and 0.1% SDS.
15. Transfer buffer: to prepare 500 ml transfer buffer add 400 ml ddH_2O to 2.91 g Tris (48 mM), 1.47 g glycine (39 mM), 1.88 ml 10% SDS solution, and 100 ml methanol.
16. 1× Tris-buffered saline (TBS): add 475 ml ddH_2O to 25 ml 1 M Tris base solution (pH 7.5), 4.4 g NaCl.
17. 1× Tris-buffered saline–Tween 20 (TBST): add 250 μl Tween 20 to 1× TBS.

18. Blocking buffer: add 1 g skim milk powder to 30 ml TBS.
19. Diluted blocking buffer: dilute the blocking buffer in TBS at a 1:1 ratio.
20. Mouse monoclonal anti-EGFR antibody (A-10) (Santa Cruz Biotechnology, Dallas, TX, USA)
21. Mouse monoclonal anti-HER2 antibody (A-2) (Santa Cruz Biotechnology, Dallas, TX, USA)
22. Horseradish peroxidase (HRP)-conjugated mouse secondary antibody.
23. Thermo Scientific Pierce enhanced chemiluminescence (ECL) Western blotting substrate.
24. X-ray film.

3 Methods

3.1 Cell Culturing and EGF Treatment

1. Seed appropriate number of healthy cells expressing EGF receptors family in a 6-cm plate containing 3.5 ml culturing medium (*see* **Note 1**).
2. Incubate the cells in an incubator providing 37 °C and 5% CO_2 for 48 h.
3. After 48 h remove the culturing medium and wash the cells with 1× PBS and then pour 3.5 ml starvation medium to the 6-cm plate.
4. Incubate the cells at the mentioned condition for about 16 h (overnight).
5. Remove the starvation medium and wash the cells with 1× PBS and then add 1 ml EGF working solution to the 6-cm plate (*see* **Note 2**).
6. Incubate the cells at +4 °C for 1 h (*see* **Note 3**).

3.2 Cross-Linking Treatment

1. Put the 6-cm plate on the ice and wash the cells with ice-cold 1× PBS for three times (*see* **Note 4**).
2. Add appropriate amount of the ice-cold 1× PBS to the plate and scrape the cells and collect them into a 1.5 ml microcentrifuge tube (*see* **Note 5**).
3. Centrifuge the tube at 200 × *g* for 5 min.
4. Discard supernatant thoroughly and add fresh prepared 200 μl BS^3 cross-linker reagent solution to the tube and resuspend the cells gently by pipetting up and down (*see* **Notes 6–8**).
5. Incubate the tube at +4 °C for 2 h and resuspend the cells in the tube every half hour (*see* **Note 9**).
6. Add quench solution to a final concentration of 10–20 mM.

7. Incubate the tube on ice at +4 °C for 15 min.
8. Centrifuge the tube at 200 × *g* for 5 min.

3.3 Protein Purification

1. Discard the supernatant thoroughly and add the appropriate volume of protein extraction solution to the pellet of the cells.
2. Resuspend the cells by pipetting up and down and incubate the cells at +4 °C for 15 min with gentle agitation.
3. Centrifuge the tube with refrigerated ultracentrifuge (adjust it at +4 °C) at 8000 × *g* for 15 min.
4. Carefully transfer the supernatant containing extracted proteins into a precooled tube on ice.

3.4 Protein Quantification

1. Prepare protein standard solutions with different concentrations and protein samples in separate tubes according to Table 1.
2. Incubate the tubes at room temperature for 5 min.
3. Measure the absorbance of each protein standard solution and protein sample at 595 nm wavelength and write down the absorbance in the last row of Table 1.
4. Draw the standard curve by plotting the concentrations of each protein standard solutions against their absorbance.
5. Use the standard curve to calculate the concentrations of protein samples.

3.5 Western Blotting

1. Add 4× Laemmli buffer to each protein sample at a 1:3 ratio.
2. Incubate the protein samples in 90–100 °C water bath for 5 min (*see* **Note 10**).
3. Load 10–50 μg total protein sample to each well of 5% acrylamide gel.

Table 1
Preparation of protein standard solutions with different concentrations and protein samples for protein quantification

Groups	PSS1	PSS2	PSS3	PSS4	PSS5	PSS6	PSS7	PS
Distilled water	18 μl	15 μl	12 μl	9 μl	6 μl	3 μl	0 μl	18 μl
Protein extraction buffer	2 μl	2 μl	2 μl	2 μl	2 μl	2 μl	2 μl	–
BSA (1×)	0 μl	3 μl	6 μl	9 μl	12 μl	15 μl	18 μl	2 μl
Bradford (1×)	1000 μl	1000 μl	1000 μl	1000 μl	1000 μl	1000 μl	1000 μl	1000 μl
Concentration (mg/ml)	0	4.44	8.88	13.32	17.76	22.2	26.64	?
Absorbance	?	?	?	?	?	?	?	?

PSS: Protein standard solution
PS: Protein sample

4. Adjust the power supply at 100 V and run the electrophoresis for 185 min.
5. Put the gel in transfer buffer and incubate it at room temperature with gentle shaking for 15 min.
6. Prepare the protein transfer sandwich by putting the gel on the nitrocellulose paper, between two sets of filter papers.
7. Use the semidry transfer system at 15 V for 120 min to transfer proteins onto the nitrocellulose paper (*see* **Note 11**).
8. Incubate the membrane containing transferred proteins in blocking buffer at room temperature with gentle shaking for 1 h.
9. Dilute the primary antibodies (A-10 and A-2) in the blocking buffer at a 1:1000 ratio.
10. Incubate the membrane with the primary antibody solution at 4 °C for overnight.
11. Remove the primary antibody solution and wash the membrane three times, 10 min per wash, in TBST at room temperature with gentle shaking.
12. Dilute the HRP-conjugated mouse secondary antibody in the diluted blocking buffer at a 1:2000 ratio.
13. Incubate the membrane with the secondary antibody solution at room temperature for 1 h.
14. Remove the secondary antibody solution and wash the membrane three times, 10 min per wash, in TBST at room temperature with gentle shaking.
15. Incubate the membrane with ECL substrate solution for 5 min at room temperature with gentle shaking and then detect the protein bands on X-ray film.

4 Notes

1. The amount of seeded cells should be adjusted to have the 90–95% confluency on the day of treatment.
2. The concentration of the ligands may vary according to the cell types and experiments.
3. For assessment of receptor dimerization using BS^3 as a cross-linking reagent, it is mostly recommended to incubate the cells on ice to inhibit the receptor endocytosis.
4. Most of the cell culture media contain amine components which can interfere and quench the chemical reactions between BS^3 and polypeptides. Therefore, it is important to wash the cells carefully to be sure that all amine-containing components are removed from the 6-cm plate.

5. In the case of adding the BS^3 solution directly to the plate, it would be necessary to use considerable amount of BS^3; however, in this way much less amount of reagent will be used.
6. The reactivity of NHS ester groups of BS^3 can be easily lost after preparation; therefore, prepare the cross-linker solution just before the initiation of the experiment.
7. Adjust the BS^3 concentration according to the cell types and experiment. It is mostly recommended to use the BS^3 cross-linker solution with a final concentration of 1–5 mM.
8. The amount of cross-linker solution volume depends on pellet mass, and for higher pellet mass it would be necessary to add more fresh prepared BS^3 cross-linker reagent solution (e.g., to the cells collected from 10-cm plate add 500 μl cross-linker solution).
9. Some protocols suggest to incubate the cells with BS^3 cross-linker solution for 30 min at room temperature. However, incubate the cells at +4 °C for a longer time would decrease the BS^3 cell-mediated internalization and increase the efficiency of cross-linking assay.
10. Do not put the heated protein samples in too cold environment immediately. Leave them at room temperature for 2 min in order to cool down slowly.
11. Running the semidry transfer device for a long time increases the transfer buffer temperature and causes buffer evaporation. To avoid any possible damages to proteins, gel, and nitrocellulose paper, saturate the protein transfer sandwich with transfer buffer.

References

1. Ono M, Kuwano M (2006) Molecular mechanisms of epidermal growth factor receptor (EGFR) activation and response to gefitinib and other EGFR-targeting drugs. Clin Cancer Res 12(24):7242–7251
2. Hynes NE, MacDonald G (2009) ErbB receptors and signaling pathways in cancer. Curr Opin Cell Biol 21(2):177–184
3. Yakovlev AA (2009) Crosslinkers and their utilization for studies of intermolecular interactions. Neurochem J 3(2):139–144

Chapter 7

Application of Immunofluorescence Staining to Study ErbB Family of Receptor Tyrosine Kinases

Babak Nami and Zhixiang Wang

Abstract

Immunofluorescence staining is an effective method for visualizing ErbB family of receptor tyrosine kinases and to monitor their expression, phosphorylation, and localization in individual cells. However, immunofluorescence staining for membrane proteins is required preserving integrity of cell membrane intact by protecting the cells from organic solvents. Permeabilization of cells may increase nonspecific intracellular background signal as an inevitable result. This chapter describes procedures for the application of indirect immunofluorescence staining for ErbB receptors in adherent and suspended cells as well as in paraffin-embedded tissue sections.

Key words Immunofluorescence, ErbB, Receptor, Staining, EGFR, HER2

1 Introduction

The first use of immunofluorescence (IF) staining technique was reported in 1955 by scientists at Department of Bacteriology and Immunology of Harvard Medical School, who used it to study vaccine virus development in human cells [1]. After over 60 years of the first discovery of IF by the Harvard scientists [2], IF is still one of the most widely used methods in research and diagnostic laboratories to monitor eukaryotic cells, bacteria, organelles, viruses, and proteins. IF is desired so for its immense advantages. IF is a simple, rapid, sensitive, and safe technique. It allows us to investigate the presence of target antigens in specimen, utilizing the sensitivity and specificity of antibodies and conjugated fluorophores, without disruption of integrity of a cell or tissue. Another advantage of this technique becomes clearer when we intend to study expression and localization of target antigen in individual cells. Moreover, using double-IF, which refers to concurrent staining more than one target antigen with different fluorophores (*see* Fig. 1a), can reveal interaction and colocalization of the antigens in any site of a cell or tissue [3].

Zhixiang Wang (ed.), *ErbB Receptor Signaling: Methods and Protocols*, Methods in Molecular Biology, vol. 1652, DOI 10.1007/978-1-4939-7219-7_7, © Springer Science+Business Media LLC 2017

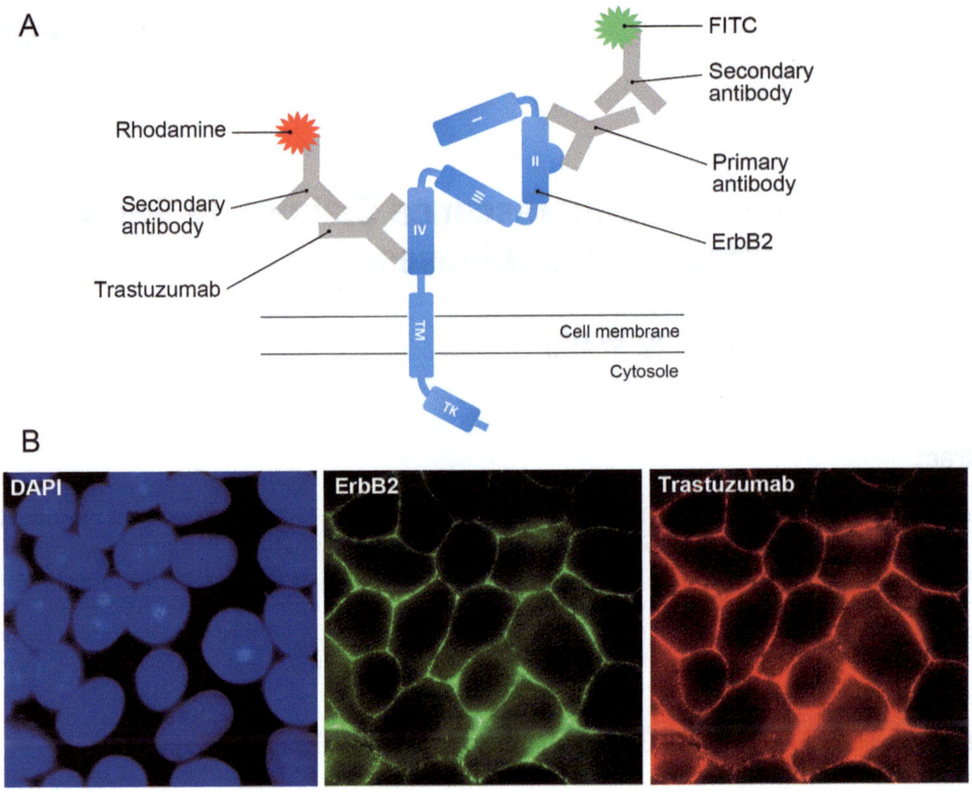

Fig. 1 Double indirect immunofluorescence staining of ErbB2 and trastuzumab. (**a**) Model illustrating the principle of double indirect immunofluorescence staining for ErbB2 (*Blue*) using trastuzumab (#DIN-02240692, Roche, Basel, Switzerland), a therapeutic monoclonal human IgG against domain IV of ErbB2 and a commercial rabbit anti-human ErbB2 primary antibody (9G6; #sc-08) against the extracellular domain of ErbB2. Trastuzumab was stained by rhodamine-conjugated rabbit anti-human IgG (#sc-3946) and 9G6 antibody was stained by FITC-conjugated mouse anti-rabbit IgG (#sc-2359) all from Santa Cruz Biotechnology Inc., Dallas, TX, USA. (**b**) Fluorescence microscope images of ErbB2 (FITC; *green*) and trastuzumab (rhodamine; *red*) as described in (**a**) in MCF7 cells. MCF7 cells were cultured on a coverslip and were treated with 10 μg/ml of trastuzumab for 30 min. Double indirect immunofluorescence staining of methanol-fixed cells was done as per the protocol described in this chapter. DAPI (*blue*) illustrates cell nucleus. The image reveals the specific binding of trastuzumab to ErbB2

ErbB family of receptor tyrosine kinases consists of four transmembrane proteins (ErErbB2/EGFR, ErbB2/HER2, ErbB3/HER3, and ErbB4/HER4) which play prominent role in cell growth through activating downstream signaling [4]. Since the first IF staining for an ErbB receptor (EGFR) was reported about 40 years ago [5], IF staining has been used to investigate expression, phosphorylation, dimerization, and endocytosis of ErbB receptors [6].

Before running an IF experiment for ErbB receptors, confirmation of receptor expression by an appropriate assay such as cytometry and immunoblotting (e.g., western blotting and ELISA) is highly recommended. We also recommend checking the cells under a light microscope after fixation and before immunostaining to ensure that there are enough cells to be stained.

Permeabilization with organic solvents means dissolving lipids from cell membranes and allowing antibody pass through the membrane bilayer into the cells, thus providing access intracellular or intraorganellar antigens [7]. Thus, permeabilization is needed to stain intracellular antigens. Since ErbB receptors are intrinsically anchored in the cell membrane, we do NOT recommend permeabilization of cell membrane unless you are staining intracellular ErbB receptors. On the other hand, since the membranes of methanol-fixed cells are compatible to allow antibodies pass through, we exclude the permeabilization step in the case of using cold methanol as a fixative even for intracellular staining. Enhanced background signal is a common undesirable consequence of IF staining for ErbB in overpermeabilized cells. We also avoid organic solvents (Tween 20, Triton X, etc.) in wash buffer for the same reason. Sometimes, fluorophore signal may be confused with autofluorescence, which is the natural emission of fluorescent light by biological samples [8]. Long-time exposure can enhance autofluorescence especially when using band-pass filter. To minimize autofluorescence, avoid long-time exposures. Generally, a 10–100 ms exposure in DAPI channel and a 1–3 s exposure for routine fluorophore-labeled antibodies are enough depending on the antibodies and the channels used. Avoiding aldehyde as a fixative is another way to reduce autofluorescence.

Cross-reaction of primary antibody and false positives may accrue particularly in cells that coexpress ErbB receptors. To prevent this, it is highly recommended to use a primary antibody with high specificity for the target receptor. The authors of research reports mention the manufacturer and the catalog number of the antibodies used in their studies. An appropriate antibody may be chosen from research reports in case of using commercial antibodies. It is also important to run an IF staining for target positive and negative cells as controls using the same antibody. Mostly, primary antibodies for IF can also be used for immunoblotting. Thus, validation of antibody specificity by immunoblotting may be useful for troubleshooting. It is always recommended to run positive and negative controls to ensure that the primary or secondary antibody is working properly.

The current guideline, which is the optimized protocol in our laboratory, describes a step-by-step procedure of IF staining for ErbB receptors, although it may also be applied for any other membrane targets.

2 Materials

2.1 Basic Materials

1. Sterile 24-well plates.
2. 12 mm diameter sterile coverslips.
3. Microscope slides.
4. Deionized water.
5. Methanol.
6. 1× phosphate-buffered saline (PBS): 8 g NaCl (final concentration: 137 mM), 0.2 g KCl (final concentration: 2.7 mM), 1.44 g Na_2HPO_4 (final concentration: 10 mM), 0.24 g KH_2PO_4 (final concentration: 1.8 mM), 800 ml distilled H_2O. Adjust the pH to 7.4 with 1 N HCl, and then add distilled H_2O to 1 l. Store at room temperature.
7. 1× Tris-buffered saline (TBS): 6.05 g Tris (final concentration: 50 mM), 8.76 g NaCl (final concentration: 150 mM), 800 ml distilled H_2O. Adjust pH to 7.6 with 1 N HCl and then add distilled H_2O to 1 l. Store at +4 °C.
8. Blocking solution: 1% bovine serum albumin (BSA) in TBS (alternatively 5% goat or donkey serum solution).
9. Primary antibody solution: 1 μg/ml primary antibody in blocking solution.
10. Secondary antibody solution: 1 μg/ml fluorescent dye-conjugated secondary antibody in blocking solution (*see* **Note 1**).
11. Permeabilization solution (0.2% Triton X in PBS; *see* **Note 2**).
12. DAPI solution: 300 nM DAPI solution in PBS.
13. Mounting media: 10 ml of 0.1 M Tris pH 8.0, 0.5 g N-propyl gallate, 90 ml glycerol. Store at −20 °C in darkness for a long time. Keep at +4 °C for maximum 2 weeks (*see* **Note 3**).
14. Hydrophobic barrier pen or nail polish.
15. Fluorescence microscope.

2.2 Additional Materials for Suspended Cells

1. Paraformaldehyde.
2. Centrifuge.
3. Hot plate.

2.3 Additional Materials for Paraffin-Embedded Tissue Sections

1. 100% ethanol.
2. 90% ethanol.
3. 70% ethanol.
4. Xylene.
5. Antigen retrieval solution (10 mM sodium citrate buffer): dissolve 2.94 g of Tri-sodium citrate in 1 l of distilled H_2O. Adjust

the pH to 6.0 with 1 N HCL and then add 0.5 ml of Tween 20. Store at +4 °C.

6. Oven.

3 Methods

3.1 For Adherent Cells

1. Place a sterile coverslip in per well of 24-well plate and seed approximately 5×10^4 cells in and culture until reach 50–80% confluency.
2. Treat the cells with desired agents.
3. Wash the coverslip by adding PBS into well twice for 5 min each at room temperature.
4. Fixation: Aspirate PBS and add −20 °C methanol into the well and incubate in −20 °C for 5 min (*see* **Note 4**).
5. Aspirate methanol and wash the coverslip with TBS twice for 5 min each at room temperature (*see* **Note 2**).
6. Blocking: Aspirate TBS and block nonspecific staining by adding blocking solution into the well, and incubate at room temperature for 30 min.
7. Aspirate blocking solution and wash the coverslip with TBS twice for 5 min each at room temperature.
8. (Immunostaining) Aspirate TBS and add primary antibody solution into the well. Incubate either overnight at +4 °C or for 1 h at room temperature.
9. Aspirate the primary antibody solution and wash the coverslip twice with TBS for 5 min each at room temperature.
10. Aspirate TBS and add fluorescence-conjugated secondary antibody solution into the well and incubate at room temperature in the dark for 1 h (*see* **Note 5**).
11. Aspirate the secondary antibody solution and wash the coverslip twice with TBS for 5 min each at room temperature.
12. Aspirate TBS and add DAPI solution into the well (*see* **Note 4**) and incubate at room temperature in the dark for 5 min.
13. Aspirate DAPI solution in dark and add 1 ml TBS (*see* **Note 6**) into the well. Washing is not required.
14. Mounting: Drop 5–10 μl of mounting media on a clean glass slide. Place the coverslip on the mounting media drop with the cells facing down (*see* **Note 3**).
15. Seal the coverslip using a hydrophobic barrier pen or nail polish and observe the stained cells under a fluorescence microscope. An example of convenient IF staining is shown in Fig. 1b.

3.2 For Suspended Cells

1. Transfer 10^6 cells into a new tube and pellet the cells by centrifugation at 200 × g for 3 min.
2. Aspirate the supernatant medium and resuspend the cell pellet in 960 μl PBS.
3. Fixation: Add 40 μl absolute paraformaldehyde to the cell suspension and pipette up and down several times. Incubate at room temperature for 15 min.
4. Spin down the cells at 200 × g for 3 min.
5. Aspirate the supernatant and wash the cells by adding 1 ml PBS, and spin down at 200 × g for 3 min.
6. Aspirate the supernatant and resuspend the cell pellet in 200 μl of deionized water. Drop 20 μl of the cell suspension on a slide and smear with the side of a pipette tip.
7. Incubate the slide on a 50 °C hot plate until the liquid is evaporated.
8. Check the cells under a microscope, and make sure that there are no crystals. If crystals are observed, wash the slide with deionized water.
9. Wipe the area around the cell spot with cotton swabs and surround the cell spot by drawing a circle with a hydrophobic barrier pen or nail polish and air-dry. Make sure that a coverslip can easily be placed in the circle.
10. Blocking: Cover the cell spot by blocking solution and incubate at room temperature for 30 min.
11. Immunostaining: Discard the blocking solution and cover the cell spot with primary antibody solution and incubate either overnight at +4 °C or for 1 h at room temperature.
12. Discard the primary antibody solution and wash the cell spot with TBS three times for 3 min each at room temperature.
13. Discard TBS and cover the cell spot with secondary antibody solution and incubate at room temperature in the dark for 1 h.
14. Discard the secondary antibody solution and wash the cell spot with TBS three times for 3 min each at room temperature.
15. Nuclear staining: Discard TBS and cover the cell spot with DAPI solution and incubate at room temperature in the dark for 5 min.
16. Mounting: Discard DAPI solution and drop 5 μl of mounting media on the cell spot and place a coverslip on it gently.
17. Seal the coverslip using the hydrophobic barrier pen or nail polish and observe the stained cells under fluorescence microscope.

3.3 For Paraffin-Embedded Tissue Sections

1. Sectioning: Cut formalin-fixed paraffin-embedded section at 1–20 μm thick (routinely 5 μm thick) using a rotary microtome and place the section on precleaned and positively charged microscope slide.
2. Deparaffinization: Heat the slide at 60 °C in a dry oven for 20 min to remove attached tissues and to soften the paraffin.
3. Wash the slide in xylene three times for 10 min each at room temperature.
4. Wash the slide in 100% ethanol three times for 10 min each at room temperature.
5. Wash the slide in 90% ethanol three times for 10 min each at room temperature.
6. Wash the slide in 70% ethanol three times for 10 min each at room temperature.
7. Wash the slide with deionized water for 5 min at room temperature (do NOT allow tissue to dry any time).
8. Antigen retrieval: Heat the slide in antigen retrieval solution at 95–100 °C in an oven for 20 min.
9. Remove the slide from oven and immerse it into room temperature antigen retrieval solution for 30 min or until the slide is cooled to room temperature.
10. Wash the slide with TBS three times for 3 min each at room temperature.
11. Wipe the area around the tissue section with cotton swab and surround the tissue by drawing a circle with a hydrophobic barrier pen or nail polish and air-dry. Make sure that the circle is wide enough for a coverslip to be easily placed in the circle.
12. Blocking: Cover the tissue section by blocking solution and incubate at room temperature for 30 min.
13. Immunostaining: Discard the blocking solution and cover the tissue section with primary antibody solution and incubate either overnight at +4 °C or for 1 h at room temperature.
14. Discard the primary antibody solution and wash the slide with TBS three times for 3 min each at room temperature.
15. Discard TBS and cover the tissue section with secondary antibody solution and incubate at room temperature in the dark for 1 h.
16. Discard the secondary antibody solution and wash the slide with TBS three times for 3 min each at room temperature.
17. Nuclear staining: Discard TBS and cover the tissue section with DAPI solution and incubate at room temperature in the dark for 5 min.

18. Mounting: Discard DAPI solution and drop 5 μl of mounting media on the tissue section and place a coverslip on it gently.
19. Seal the coverslip using the hydrophobic barrier pen or nail polish and observe the stained cells under fluorescence microscope.

4 Notes

1. This protocol is optimized for “indirect immunostaining.” In “direct immunostaining” the primary antibody must be conjugated with a fluorescent day. In this case, the treatment with secondary antibody is NOT required. In “double-immunostaining,” different antibodies can be mixed in one tube. Make sure that the final concentration of **each** antibody is 1 μg/ml.
2. In IF staining for extracellular targets membrane permeabilization is NOT required. For intracellular targets (e.g., receptor endocytosis) cells should be permeabilized after methanol or paraformaldehyde fixation, by incubating the cells in permeabilization solution for 10 min at room temperature.
3. Do NOT use any mounting media with DAPI in it, as brighter background and lower image contrast may result.
4. Paraformaldehyde can be used instead of methanol for fixation. When using paraformaldehyde, the slides may be fixed by incubation in 4% paraformaldehyde at room temperature for 15 min.
5. It is recommended to wrap the plate in aluminum foil to prevent light exposure.
6. Adding TBS into the well can facilitate coverslip removal from the well.

References

1. Noyes WF, Watson BK (1955) Studies on the increase of vaccine virus in cultured human cells by means of the fluorescent antibody technique. J Exp Med 102:237–242
2. Coons AH (1961) The beginnings of immunofluorescence. J Immunol 87:499–503
3. Beutner EH (1961) Immunofluorescent staining: the fluorescent antibody method. Bacteriol Rev 25:49–76
4. Wieduwilt MJ, Moasser MM (2008) The epidermal growth factor receptor family: biology driving targeted therapeutics. Cell Mol Life Sci 65:1566–1584
5. Shechter Y, Schlessinger J, Jacobs S, Chang KJ, Cuatrecasas P (1978) Fluorescent labeling of hormone receptors in viable cells: preparation and properties of highly fluorescent derivatives of epidermal growth factor and insulin. Proc Natl Acad Sci U S A 75:2135–2139
6. Wang Z, Zhang L, Yeung TK, Chen X (1999) Endocytosis deficiency of epidermal growth factor (EGF) receptor–ErbB2 heterodimers in response to EGF stimulation. Mol Biol Cell 10:1621–1636
7. Jamur MC, Oliver C (2010) Permeabilization of cell membranes. Methods Mol Biol 588:63–66
8. Minici M (2005) Cell and tissue autofluorescence research and diagnostic applications. Biotechnol Annu Rev 11:227–256

Chapter 8

Activation of Endosome-Associated Inert EGF Receptor Following Internalization

Yi Wang, Sukhmani Billing, and Zhixiang Wang

Abstract

Our current understanding of endosomal signaling is incomplete given our inability to generate endosomal signals in isolation from plasma membrane signals. We present a novel method to specifically activate epidermal growth factor (EGF) receptor (EGFR) following its internalization into endosomes. This method will allow us to address questions such as whether endosomal signaling is sufficient to activate signal transduction pathways and produce biological responses. In this method, the cells are treated with EGF in the presence of AG1478, a specific EGFR tyrosine kinase inhibitor, and monensin, which blocks recycling of EGFR. The treatment results in the internalization of nonactivated EGF–EGFR complex into endosomes. Next, the removal of AG1478 and monensin causes endosome-associated EGFR activation. Phosphorylated surface EGFR levels are undetectable during the treatment. This is a unique and novel approach to study endosomal signaling and its physiological relevance. The protocol is transferable to study endosomal signaling of other receptor tyrosine kinases.

Key words Endosome, Epidermal growth factor receptor (EGFR), Receptor tyrosine kinase (RTK), Endocytosis, Phosphorylation

1 Introduction

Epidermal growth factor (EGF) stimulates cell growth, differentiation, and proliferation by binding to its receptor (EGFR). Binding of EGF to EGFR stimulates dimerization, activating the tyrosine kinase activity of the receptor. As a result, multiple tyrosine residue sites on the receptor undergo transphosphorylation and/or autophosphorylation. The phosphorylated tyrosine residues serve as docking sites for the downstream signaling proteins [1, 2]. In addition to the activation of various signaling pathways, EGF binding results in the clustering of EGF receptor complexes and its internalization via coated pit and coated vesicles [1, 3]. The primary fate of internalized EGF–EGFR complex is its rapid degradation in lysosomes [1]. However, it is yet to be elucidated whether endocytosis primarily results in the inactivation of the ligand–receptor

Zhixiang Wang (ed.), *ErbB Receptor Signaling: Methods and Protocols*, Methods in Molecular Biology, vol. 1652, DOI 10.1007/978-1-4939-7219-7_8,

complex or if it also plays a crucial role in regulating signaling pathways [4]. It is well established that endocytosis of cell-surface EGFR into lysosomes results in its degradation, attenuating its signaling and that this process can function as a tumor suppressor pathway [4]. On the contrary, there is significant evidence indicating an involvement of endosome-associated EGF–EGFR complexes in cell signaling. Internalized EGFR is reported to be autophosphorylated and catalytically active [5]. EGFR has been reported to be endocytosed bound with various signaling molecules involved in Ras signaling, such as growth receptor binding protein 2 (Grb2), small guanosine triphosphatase (GTPase) activating protein (GAP), Shc adaptor protein, and Cbl [6]. Moreover, these signaling molecules are observed to remain bound with EGFR following its internalization [7, 8]. Further evidence for endosome-associated signaling is supported by endocytosis blocking experiments. Dominant-negative dynamin mutant inhibits EGFR endocytosis, resulting in enhanced EGF-dependent cell proliferation and decreased extracellular signal-regulated kinase (ERK) activation [9]. This suggests that endosomal signaling has an important role in initiating and controlling specific signaling pathways. Furthermore, inhibition of EGFR endocytosis by mutant dynamin or β-arrestin, following its transactivation by heterotrimeric guanine nucleotide-binding protein (G protein)-coupled receptors, eliminates ERK activation [10]. Phospholipase D mediated inhibition of EGFR endocytosis also abolishes EGF-stimulated ERK activation [11]. Thus, the physiological relevance of endosomal signaling still remains controversial.

Present approaches to study endosomal signaling are unable to distinguish between endosomal vs plasma membrane derived signals as well as are not suitable for studying the dynamics of endosomal signaling. Endocytosis inhibition remains a popular tool to study endosomal signaling; however, this approach has some major limitations. Blocking EGFR endocytosis eliminates endosomal EGFR signaling, resulting in its subsequent retention at the plasma membrane, thereby enhancing plasma membrane-derived EGFR signaling. Moreover, mutant dynamin or β-arrestin mediated inhibition of EGFR endocytosis has wide ranging effects on the endocytosis of other receptors that depend on these factors. This makes it challenging to conclude whether the observed signaling effects are the result of inhibiting the process of endocytosis in general or due to increased accumulation of EGFR at the cell surface.

These limitations in studying endosomal signaling can be resolved by specifically activating endosome-associated receptors without prior activation at the plasma membrane [12–15]. This protocol allows specific activation of endosome-linked EGFR without its prior activation at the plasma membrane, and without disrupting endocytic events. The protocol involves endosomal internalization of inactive (de-phosphorylated) EGF–EGFR complex and subsequent specific activation of the endosome-associated EGFR (Fig. 1). The protocol provides a powerful method to study endosomal signaling of EGFR. In addition, the protocol is transferable to study other tyrosine kinase receptors.

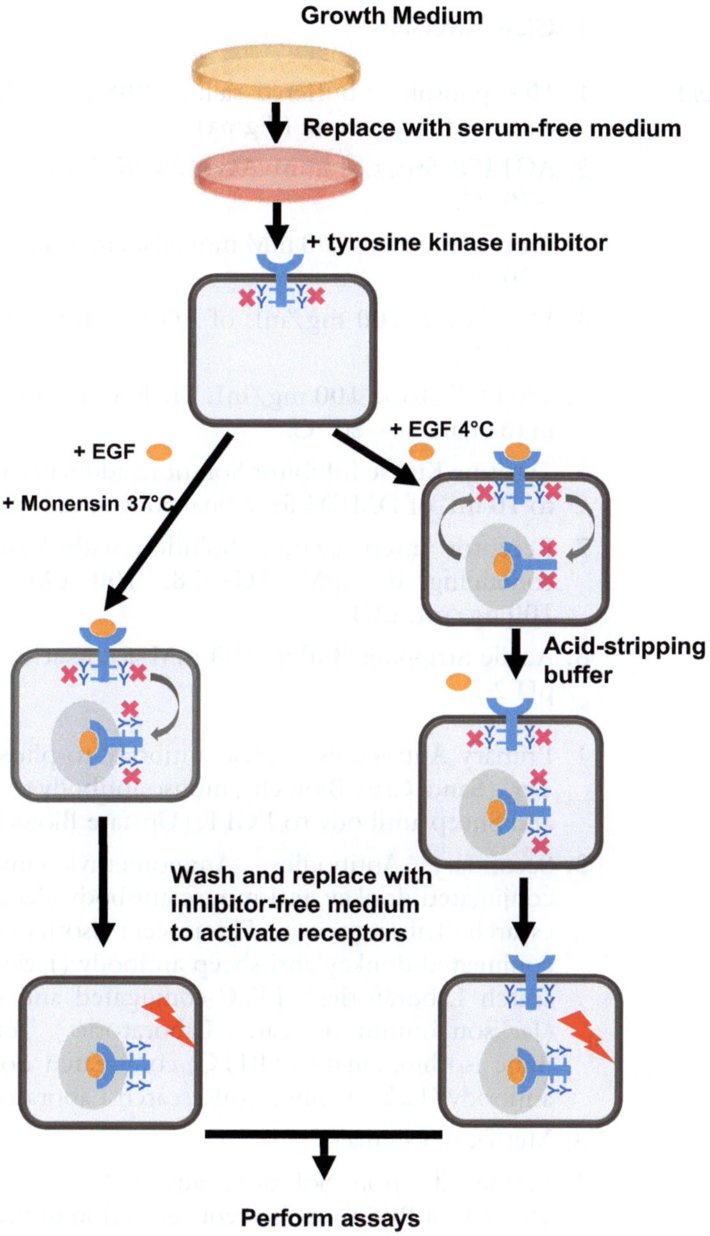

Fig. 1 Flow sheet summarizing the procedure for specific activation of endosome-associated EGFR. This figure is reproduced from the figure previously published (Mol. Cell. Biol. 22: 7279–7290, 2002) with permission

2 Materials

2.1 Cell Culture

1. Cell lines: Madin-Darby Canine Kidney Epithelial (MDCK) cells (ATCC) and BT-20 cells (ATCC).
2. Culture medium: Dulbecco's modified Eagle's medium (DMEM), 10% FBS, Penicillin and streptomycin (100 U/mL).
3. 24-well cell culture plate and 100-mm culture dishes.
4. Glass coverslips.

2.2 Cell Treatment

1. 10× phosphate-buffered saline (PBS): pH 7.2, without calcium and magnesium (Sigma).
2. AG1478 Stock: 1 mM AG1478 in distilled water stored at −20 °C.
3. Monensin Stock: 100 mM monensin in distilled water stored at −20 °C.
4. EGF Stock: 100 mg/mL of EGF in distilled water stored at −20 °C.
5. TR-EGF Stock: 100 mg/mL TR-EGF in distilled water stored in the dark at −20 °C.
6. Tyrosine Kinase Inhibitor Solution: add 5 μL of AG1478 Stock to 10 mL of DMEM for a final concentration of 0.5 μM.
7. Receptor Internalization Solution with Monensin: DMEM containing 0.5 μM AG1478, 100 μM monensin, and 100 ng/mL EGF.
8. Acidic Stripping Buffer: 100 mM acetic acid, 150 mM NaCl, pH 2.7.

2.3 Immunofluorescence

1. Primary Antibodies: mouse antibody to phosphotyrosine (p-Tyr) (Santa Cruz Biotech), mouse antibody to phospho-EGFR and Sheep antibody to EGFR (Upstate Biotechnology).
2. Secondary Antibodies: Aminomethylcoumarin (AMCA)-conjugated donkey anti-mouse antibody (Jackson ImmunoResearch Laboratories), Fluorescein isothiocyanate (FITC)-conjugated donkey anti-sheep antibody (Jackson ImmunoResearch Laboratories), FITC-conjugated anti-mouse antibody (Jackson ImmunoResearch Laboratories), Tetramethylrhodamine isothiocyanate (TRITC)-conjugated donkey anti-sheep antibody (Jackson ImmunoResearch Laboratories).
3. Methanol (Sigma).
4. Permeabilization Solution: add 0.2 mL Triton X-100 to 100 mL of PBS for a final concentration of 0.2%.
5. 0.1% Triton X-100 Solution: add 0.1 mL Triton X-100 to 100 mL PBS for a final concentration of 0.1%.

6. Primary Labeling Solution: 0.1% Triton X-100 containing 1:50 Sheep antibodies to EGFR and 1:100 Mouse antibodies to pTyr.
7. Secondary Labeling Solution: 0.1% Triton X-100 containing 1:200 TRITC-conjugated donkey anti-sheep IgG and FITC-conjugated donkey anti-mouse IgG.
8. Mounting Medium: dissolve 0.4 g of n-propyl gallate in 10 mL of 50% glycerol by heating.
9. Monensin-Free Receptor Internalization Solution: DMEM containing 0.5 μM AG1478 and 100 ng/mL TR-EGF.
10. Parafilm (Sigma).
11. Glycerol (Sigma).
12. Propidium iodide: 50 μg/mL (Sigma).

2.4 Equipment

1. Fluorescence microscope with FITC, TRITC, and DAPI (4′-6-diamidino-2-phenylindole) filters.
2. Standard equipment for cell culture: Laminar flow hood, CO_2 humidified incubator.

3 Methods

We describe two related systems to specifically activate endosome-associated EGFRs: one using monensin to inhibit EGFR recycling, and one without monensin. Monensin protocol gives better accumulation of nonactivated EGFR in endosomes and better activation of endosome-associated EGFR. The advantage of omitting monensin is that it avoids any possible side effects produced by monensin, such as changes in endosomal morphology.

3.1 Specific Activation of Endosome-Associated EGFR with Monensin Treatment

3.1.1 Immunofluorescent Applications

1. Plate MDCK or BT20 cells at 2×10^4 cells per well (40% confluence) onto coverslips in a 24-well culture dish with 1 mL/well Culture Medium and incubate at 37 °C in a humidified 5% CO_2 incubator.
2. Grow until the cells reach 60% confluence. Change the medium with 1 mL/well of Serum-Free Culture Medium. Incubate MDCK cells for 24 h or BT20 cells for 16 h at 37 °C, 5% CO_2 (*see* **Note 1**).
3. Aspirate the medium and add 300 μL/well Tyrosine Kinase Inhibitor Solution. Incubate for 15 min at 37 °C, 5% CO_2.
4. Aspirate the inhibitor solution and add 300 μL/well Receptor Internalization Solution with Monensin. Incubate for 30 min at 37 °C, 5% CO_2 (*see* **Note 2**).

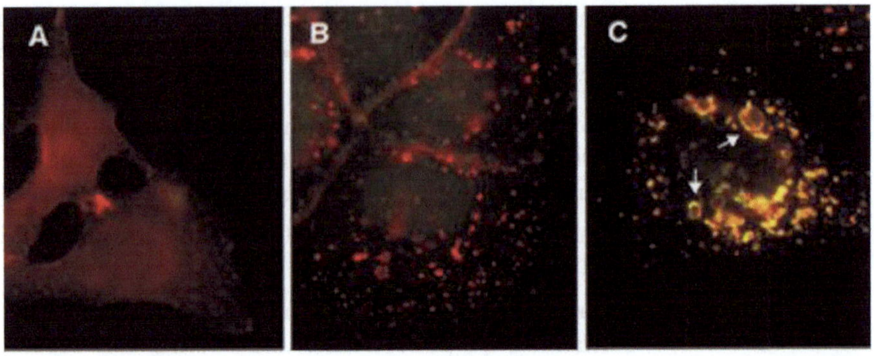

Fig. 2 Immunofluorescent localization of selectively activated endosome-associated EGFRs. BT20 cells are incubated with AG1478 for 15 min (**a**) then activated with EGF in the presence of monensin and AG1478 at 37 °C for 30 min (**b**). Some of the cells are washed with PBS three times and incubated with serum-free medium for 30 min (**c**). EGFR (*red*) and pTyr (*green*) localization is analyzed via indirect immunofluorescence. *White arrows* represent EGFR/pTyr (*yellow*) colocalization. This figure is reproduced from the figure previously published (Mol. Cell. Biol. 22: 7279–7290, 2002) with permission

5. Aspirate the solution. Wash the cells 3× with PBS (~ 0.5 mL/well, 1 min/wash). Add 1 mL/well of Serum-Free Culture Medium. Incubate for 30 min at 37 °C, 5% CO_2 (*see* **Note 3**).
6. Aspirate the medium and fix the cells in 300 μL of −20 °C methanol at RT for 5 min.
7. Wash the cells 3× with PBS (~ 0.5 mL/well, 1 min/wash).
8. Incubate the cells in 300 μL/well Permeabilization Solution at RT for 10 min.
9. Pipette 10 μL droplet of Primary Labeling Solution on a small piece of Parafilm. Incubate the coverslip face down on the droplet at RT for 1 h. Transfer the coverslip back to the 24-well dish.
10. Wash the cells 3× with PBS (~ 0.5 mL/well, 1 min/wash).
11. Pipette 10 μL droplet of Secondary Labeling Solution on a piece of Parafilm. Incubate the coverslip face down on the droplet at RT for 1 h. Transfer the coverslip back to the 24-well dish.
12. Wash the cells 3× with PBS (~ 0.5 mL/well, 1 min/wash).
13. Pipette 5 μL of Mounting Medium on a glass slide. Press the coverslip gently face down on the droplet and wipe off the excess medium. The cells can now be subjected to immunofluorescent analysis (Fig. 2).

3.1.2 Biochemical Applications

1. Plate MDCK or BT20 cells at 5×10^5 cells/dish (40% confluence) in 100-mm culture dishes within 8 mL/dish of Culture Medium and incubate at 37 °C in a humidified 5% CO_2 atmosphere.

2. Grow until the cells reach 60% confluence. Change the medium with 8 mL/dish of Serum-Free Culture Medium. Incubate MDCK cells for 24 h or BT20 cells for 16 h at 37 °C, 5% CO_2 (*see* **Note 1**).
3. Aspirate the medium and add 3 mL/dish Tyrosine Kinase Inhibitor Solution. Incubate for 15 min at 37 °C, 5% CO_2.
4. Aspirate the inhibitor solution and add 3 mL/dish of Receptor Internalization Solution with Monensin. Incubate for 30 min at 37 °C, 5% CO_2 (*see* **Note 2**).
5. Aspirate the solution. Wash the cells 3× with PBS (~ 10 mL/well, 1 min/wash). Add 8 mL/dish of DMEM and incubate for 30 min at 37 °C, 5% CO_2 (*see* **Note 3**).
6. The cells can now be subjected to biochemical analyses, such as immunoblotting, coimmunoprecipitation, and subcellular fractionation.

3.2 Specific Activation of Endosome-Associated EGFR Without Monensin Treatment

3.2.1 Immunofluorescent Applications

1. Plate MDCK or BT20 cells at 2 × 10^4 cells per well (40% confluence) onto coverslips in a 24-well culture dish with 1 mL/well Culture Medium and incubate at 37 °C in a humidified 5% CO_2 incubator.
2. Grow until the cells reach 60% confluence. Change the medium with 1 mL/well of Serum-Free Culture Medium. Incubate MDCK cells for 24 h or BT20 cells for 16 h at 37 °C, 5% CO_2 (*see* **Note 1**).
3. Aspirate the medium and add 300 μL/well Tyrosine Kinase Inhibitor Solution at 300 μL/well concentration. Incubate for 15 min at 37 °C, 5% CO_2.
4. Aspirate the inhibitor solution and add 300 μL/well Receptor Internalization Solution without Monensin at 300 μL/well concentration. Incubate for 30 min at 37 °C, 5% CO_2 (*see* **Note 4**).
5. Cool the cells to 4 °C.
6. Wash the cells 1× with 300 μL/well of ice-cold Acidic Stripping Buffer for 1 min.
7. Aspirate the stripping buffer. Wash the cells 3× with ice-cold PBS (~1 mL/well, 1 min/wash) (*see* **Note 5**).
8. Incubate the cells with 1 mL/well DMEM at 37 °C, 5% CO_2 for 30 min (*see* **Note 3**).
9. Remove the medium and fix the cells with 300 μL of −20 °C methanol for 5 min.
10. Wash the cells 3× with PBS (~0.5 mL/well, 1 min/wash).
11. Incubate the cells in 300 μL/well Permeabilization Solution for 10 min at RT.

12. Pipette 10 μL droplet of Primary Labeling Solution on a Parafilm strip. Incubate the coverslip face down on the droplet at RT for 1 h. Transfer the coverslip back to the 24-well dish.
13. Wash once with about 0.5 mL/well of PBS.
14. Pipette 10 μL droplet of Secondary Labeling Solution for Monensin-Free Application on a Parafilm piece. Incubate the coverslip face down on the droplet at RT for 1 h. Transfer the coverslip back to the 24-well dish.
15. Wash the cells 3× with PBS (~ 0.5 mL/well, 1 min/wash).
16. Pipette 5 μL of Mounting Medium on a glass slide. Press the coverslip gently face down on the droplet and wipe off excess medium. The cells can now be subjected to immunofluorescent analysis (Fig. 2).

3.2.2 Biochemical Applications

1. Plate MDCK or BT20 cells at 5×10^5 cells/dish (40% confluence) in 100-mm culture dishes within 8 mL/dish of Culture Medium and incubate at 37 °C in a humidified 5% CO_2 atmosphere.
2. Grow until the cells reach 60% confluence. Change the medium with 8 mL/dish of Serum-Free Culture Medium. Incubate MDCK cells for 24 h or BT20 cells for 16 h at 37 °C, 5% CO_2 (*see* **Note 1**).
3. Aspirate the medium and add 3 mL/dish Tyrosine Kinase Inhibitor Solution. Incubate for 15 min at 37 °C, 5% CO_2.
4. Aspirate the inhibitor solution and add 3 mL/dish of Receptor Internalization Solution without Monensin. Incubate for 30 min at 37 °C, 5% CO_2 (*see* **Note 4**).
5. Cool the cells to 4 °C.
6. Aspirate the solution. Wash the cells for 1 min with 3 mL/dish of Acidic Stripping Buffer.
7. Aspirate the buffer and wash the cells 3× with ice-cold PBS (~10 mL/dish, 1 min/wash) (*see* **Note 5**).
8. Add 8 mL/dish of DMEM and incubate for 30 min at 37 °C, 5% CO_2 (*see* **Note 3**).
9. The cells can now be subjected to biochemical analyses, such as immunoblotting, coimmunoprecipitation, and subcellular fractionation.

   Using these approaches, EGFR or other receptor tyrosine kinase will be successfully activated in endosomes. For troubleshooting, please refer to (*see* **Notes 6–10**).

4 Notes

1. At the end of this step, EGFRs are localized at the plasma membrane, in a nonphosphorylated state.
2. By the end of the incubation period, EGFRs will be internalized into endosomes, and stay nonphosphorylated.
3. The end of this step should ensure specific activation of endosome-associated EGFR, i.e., EGFRs are now localized to endosomes in a heavily phosphorylated form.
4. By the end of the incubation period, most TR-EGFs and EGFRs will be internalized into endosomes, but stay nonphosphorylated. The purpose of using TR-EGF is to verify stripping of the membrane-bound TR-EGF.
5. The stripping buffer is used to remove membrane-bound TR-EGF.
6. The two described protocols for the specific activation of endosome internalized EGFR were found to work well. Besides blocking EGFR recycling, monensin can have possible nonspecific side effects on cells. However, these nonspecific effects are minimized by washing monensin solution prior to EGFR stimulation. The protocol without monensin eliminates possible monensin associated nonspecific side effects. We did not detect any nonphysiological effects of monensin on endosome-associated EGFR signaling. Moreover, we prefer the monensin protocol, because the treatment with AG1478, monensin, and EGF resulted in more plasma membrane EGFRs to be internalized into endosomes, giving higher activation levels of endosome-associated EGFR.
7. EGFR phosphorylation occurs prior to EGF addition or in the unstimulated control: The probable cause is the leftover serum traces in the culture medium. To solve this issue, wash the cells with PBS or serum-free medium two or three times prior to incubation in the serum-free medium. In addition, increase the incubation period for the serum free medium, for example, up to 36 to 48 h for MDCK cells and 24 h for BT20 cells, while still maintaining cells below 80% confluence. Cellular overgrowth makes it harder to synchronize the cells.
8. EGFR fails to accumulate in endosomes following AG1478, EGF, and monensin treatment: The possible causes are: (1) slow endosomal accumulation of EGFR and (2) recycling of endosome-accumulated EGFR back to the plasma membrane. This can be resolved by (1) increasing the EGF, AG1478, and monensin incubation period and (2) increasing monensin concentration to block EGFR recycling.

9. Inadequate phosphorylation of endosome-associated EGFR following EGF, AG1478, and monensin removal: This can occur due to an inadequate removal of AG1478. To mitigate this problem: (1) determine the minimum AG1478 concentration required to inhibit EGFR activation via dose–response curve, (2) increase the duration and number of washes following AG1478 treatment, or (3) increase the activation time by increasing the incubation duration following the wash.
10. Phosphorylated plasma membrane associated EGFRs (Monensin-Free Protocol): The possible cause is an incomplete removal of plasma membrane associated EGF–EGFR complexes. This can be overcome by increasing the wash duration of acidic stripping buffer or lowering the pH of the acidic stripping buffer to pH 2.4.

References

1. Carpenter G (1987) Receptors for epidermal growth factor and other polypeptide mitogens. Annu Rev Biochem 56(1):881–914
2. Schlessinger J, Ullrich A (1992) Growth factor signaling by receptor tyrosine kinases. Neuron 9(3):383–391
3. Schlessinger J (1986) Allosteric regulation of the epidermal growth factor receptor kinase. J Cell Biol 103(6 Pt 1):2067–2072
4. Di Fiore PP, Gill GN (1999) Endocytosis and mitogenic signaling. Curr Opin Cell Biol 11 (4):483–488
5. Cohen S, Fava RA (1985) Internalization of functional epidermal growth factor:receptor/kinase complexes in A-431 cells. J Biol Chem 260(22):12351–12358
6. Wang Z, Tung PS, Moran MF (1996) Association of p120 ras GAP with endocytic components and colocalization with epidermal growth factor (EGF) receptor in response to EGF stimulation. Cell Growth Differ 7(1):123–133
7. Haugh JM, Huang AC, Wiley HS, Wells A, Lauffenburger DA (1999) Internalized epidermal growth factor receptors participate in the activation of p21(ras) in fibroblasts. J Biol Chem 274(48):34350–34360
8. Sorkin A (2001) Internalization of the epidermal growth factor receptor: role in signalling. Biochem Soc Trans 29(Pt 4):480–484
9. Vieira AV, Lamaze C, Schmid SL (1996) Control of EGF receptor signaling by clathrin-mediated endocytosis. Science 274 (5295):2086–2089
10. Ahn S, Maudsley S, Luttrell LM, Lefkowitz RJ, Daaka Y (1999) Src-mediated tyrosine phosphorylation of dynamin is required for beta2-adrenergic receptor internalization and mitogen-activated protein kinase signaling. J Biol Chem 274(3):1185–1188
11. Shen Y, Xu L, Foster DA (2001) Role for phospholipase D in receptor-mediated endocytosis. Mol Cell Biol 21(2):595–602
12. Wang Y, Pennock SD, Chen X, Kazlauskas A, Wang Z (2004) Platelet-derived growth factor receptor-mediated signal transduction from endosomes. J Biol Chem 279(9):8038–8046
13. Wang Y, Pennock S, Chen X, Wang Z (2002) Endosomal signaling of epidermal growth factor receptor stimulates signal transduction pathways leading to cell survival. Mol Cell Biol 22(20):7279–7290
14. Wang Y, Pennock S, Chen X, Wang Z (2002) Internalization of inactive EGF receptor into endosomes and the subsequent activation of endosome-associated EGF receptors. Epidermal growth factor. Sci STKE 2002(161): pl17
15. Kourouniotis G, Wang Y, Pennock S, Chen X, Wang Z (2016) Non-ligand-induced dimerization is sufficient to initiate the Signalling and endocytosis of EGF receptor. Int J Mol Sci 17 (8):1200

Chapter 9

Two-Pulse Endosomal Stimulation of Receptor Tyrosine Kinases Induces Cell Proliferation

Steven Pennock, Sukhmani Billing, Zhixiang Wang, and Yi Wang

Abstract

Signals transduced from ligand-activated receptor tyrosine kinases (RTKs) lead to a diverse array of biological outcomes, such as cell proliferation. Strict regulation of RTK activity is therefore necessary to prevent aberrancies in cell signaling that can lead to diseases such as cancer. RTKs are activated at the plasma membrane (PM) upon ligand binding. Contrary to the initial belief, RTK activity does not terminate immediately following endocytosis, instead RTKs remain active while being trafficked in endosomes. Here we describe a two-pulse endosomal stimulation approach which can specifically activate endosome-accumulated EGFR and drive cell proliferation.

Key words Endosome, Epidermal growth factor receptor (EGFR), Receptor tyrosine kinase (RTK), Two pulse, Cell proliferation

1 Introduction

Epidermal growth factor receptor (EGFR) belongs to the tyrosine kinase family of transmembrane cell surface receptors, or RTKs. Ligand binding and consequent dimerization of these receptors activates their intrinsic tyrosine kinase activity, leading to the tyrosine phosphorylation of target substrates. This further leads to the recruitment of signaling complexes, and ultimately, downstream transduction of signaling cascades leading to biological outcomes such as cell growth and survival [1, 2]. Just minutes following growth factor stimulation at the plasma membrane, activated RTKs rapidly cluster into clathrin-coated pits (CCP), further invaginate, and eventually pinch off as internalized clathrin coated vesicles (CCVs). These vesicles go on to shed their clathrin coats and enter the trafficking route as early endosomes, late endosomes/multivesicular bodies (MVBs), and finally degrade in lysosomes [3].

Endosomes are an important location not only for RTK degradation, but also for RTK signaling. Receptors undergo clathrin-mediated endocytosis only minutes after ligand stimulation, and yet

Zhixiang Wang (ed.), *ErbB Receptor Signaling: Methods and Protocols*, Methods in Molecular Biology, vol. 1652, DOI 10.1007/978-1-4939-7219-7_9, © Springer Science+Business Media LLC 2017

can spend between 1 and 2 h trafficking along the endocytic route in a kinase-active, phosphorylated state before being degraded. This would also appear to indicate that RTKs transduce a large proportion of their signals intracellularly. However, experimental systems that differentiate endosomal signals from PM-generated signals are not available [4]. We developed a novel approach which can specifically activate RTKs, such as epidermal growth factor (EGF) receptor (EGFR), following their endocytosis into endosomes [5–10].

The classical concept of how growth factors drive cell proliferation has been substantially modified recently (reviewed in [11–13]). Quiescent cells can be stimulated with a prolonged treatment (16 h) or two short pulses (1 h each, separated by 8 h apart) of ligands such as PDGF to stimulate cell proliferation. By using a system that can specifically activate endosome-associated RTKs, we demonstrated that endosome-derived RTK signals trigger cell proliferation via two-pulse stimulation.

Detailed here is an assay for mitogenic RTK signaling from endosomes. First, we activated RTKs specifically from endosomes in two short pulses spaced by 8 h, then we measured cell proliferation using BrdU incorporation method [9].

2 Materials

2.1 Cell Culture

1. Cell lines: Madin-Darby Canine Kidney (MDCK) epithelial cells (ATCC) and BT-20 cells (ATCC).
2. Culture medium: Dulbecco's Modified Eagle's Medium (DMEM), 10% FBS, penicillin and streptomycin (100 U/mL).
3. 24-well cell culture plate and 100-mm culture dishes.
4. Glass coverslips (*see* **Note 1**).

2.2 Cell Treatment

1. Phosphate-buffered Saline (PBS): 137 mM NaCl, 2.7 mM KCl, 10 mM Na_2HPO_4.
2. Serum free culture medium: DMEM (Gibco).
3. AG1478 Stock: 1 mM AG1478 in distilled water stored at −20 °C.
4. Monensin Stock: 100 mM monensin in distilled water stored at −20 °C.
5. EGF Stock: 100 mg/mL of EGF in distilled water stored at −20 °C.
6. Tyrosine kinase inhibitor solution: add 5 μL of AG1478 Stock to 10 mL of DMEM for a final concentration of 0.5 μM.
7. Receptor Internalization Solution with Monensin: DMEM containing 0.5 μM AG1478, 100 μM Monensin, and 100 ng/mL EGF.

2.3 DNA Synthesis

1. Bromodeoxyuridine (BrdU), Amersham (GE Healthcare, Mississauga, ON)
2. Methanol, −20 °C, Sigma (Oakville, ON).
3. HCl, 2 M, Sigma (Oakville, ON).
4. Parafilm, Sigma (Oakville, ON).
5. Mouse anti-BrdU antibody, Amersham (GE Healthcare, Mississauga, ON).
6. FITC-conjugated anti-mouse IgG antibody (Jackson ImmunoResearch Laboratories, West Grove, PA).
7. Propidium iodide, 50 μg/mL (Sigma).
8. Mounting Medium: dissolve 0.4 g of n-propyl gallate in 10 mL of 50% glycerol by heating.

2.4 Equipment

1. Fluorescence microscope with FITC, TRITC, and DAPI (4′-6-diamidino-2-phenylindole) filters.
2. Standard equipment for cell culture: Laminar flow hood, CO_2 humidified incubator.

3 Methods

Previous studies have shown that serum-arrested cells can be driven into S-phase using two 30 min pulses of ligand, spaced 8 h apart [13]. Moreover, the kinetics of proliferation were equivalent to the case where cells were exposed to mitogens in a continuous fashion. We demonstrate that two pulses of 1 h EGF stimulation spaced 8 h apart can stimulate cell proliferation in MDCK and BT20 cells. The strategy is depicted in Fig. 1.

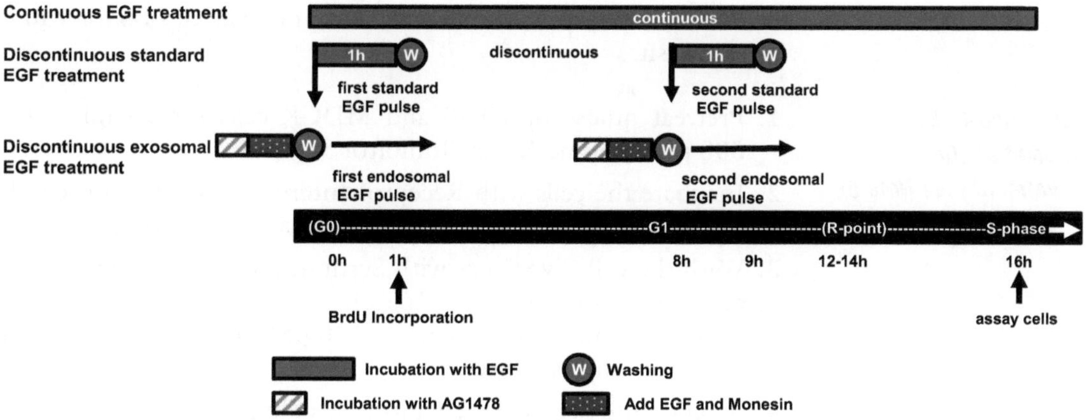

Fig. 1 Schematic of the continuous, discontinuous standard, and discontinuous endosome-associated EGFR stimulation assay. This figure is reproduced from the figure previously published (Mol Cell Biol 23: 5803–5815, 2003) with permission

3.1 Cell Culture and Serum Starvation

1. Plate MDCK or BT20 cells at 1×10^4 cells (40% confluent density) onto coverslips in a 24-well culture dish and grow at 37 °C in 1 mL/well of Culture Medium in a humidified 5% CO_2 atmosphere (*see* **Note 2**).
2. Grow the cells to 60% confluence (*see* **Note 3**).
3. Replace the Culture Medium with 1 mL/well of Serum-Free Culture Medium and incubate for 24 h (MDCK cells) or 16 h (BT20 cells) at 37 °C, 5% CO_2 (*see* **Note 4**).

3.2 Continuous Standard EGF Treatment

1. Remove the culture medium and add 300 μL EGF solution at 100 ng/mL.
2. Leave the EGF treatment for 16 h.
3. Add 25 μM bromodeoxyuridine (BrdU) in the solution after 1 h EGF treatment.

3.3 Two-Pulse Standard EGF Treatment

1. Remove the culture medium and add 300 μL EGF solution at 100 ng/mL (First Pulse).
2. Leave the EGF treatment in the cells at 37 °C, 5% CO_2 for 60 min.
3. Wash the cells three times with 1 mL/well of PBS for 1 min per wash.
4. Incubate the cells at 37 °C, 5% CO_2 in 1 mL per well of Serum-Free Culture Medium for 7 h.
5. Remove the Serum-Free Culture Medium and add the second pulse of EGF and BrdU solution for 1 h.
6. Wash the cells three times with 1 mL per well of PBS for 1 min per wash.
7. Incubate the cells at 37 °C, 5% CO_2 in 1 mL per well of Serum-Free Culture Medium with Brdu for 7 h.
8. Wash the cells three times with 1 mL per well of PBS for 1 min per wash.

3.4 Two-Pulse Exosomal EGF Treatment (See Note 5)

1. Pretreat quiescent BT20 and MDCK cells for 15 min with 300 μL Tyrosine Kinase Inhibitor Solution.
2. Incubate the cells with Receptor Internalization Solution with Monensin for 30 min at room temperature (*see* **Note 6**).
3. Wash the cells five times with Serum-Free Culture Medium to activate the internalized receptors. Then, maintain the cells in serum-free medium containing 100 μM monensin and 25 μM Brdu for 7 h (*see* **Note** 7).
4. Aspirate the Serum-Free Culture Medium.
5. Pretreat the cells with 0.5 μM AG1478 for 15 min.

6. Incubate the cells with 100 μM monensin and 100 ng/mL EGF for 30 min at room temperature.
7. Wash the cells five times with Serum-Free Culture Medium in order to activate internalized receptors. Then, maintain the cells in serum-free medium containing 100 μM monensin and 25 μM Brdu for 7 h (*see* **Note 8**).
8. Wash the cells three times with 1 mL per well of PBS for 1 min per wash.

3.5 Assay DNA Synthesis Using Bromodeoxyuridine (BrdU) Incorporation

1. Fix the cells with methanol (−20 °C) for 5 min.
2. Wash the cells three times with 1 mL per well of PBS for 1 min per wash.
3. Denature the DNA with 2 M HCl for 30 min at room temperature.
4. Wash the cells three times with 1 mL per well of PBS for 1 min per wash (*see* **Note 9**).
5. Place 10 μL of mouse anti-BrdU antibody on a Parafilm piece and float the coverslip face down for 1 h at room temperature (*see* **Note 10**).
6. Wash the cells three times with 1 mL per well of PBS for 1 min per wash.
7. Place 10 μL FITC-conjugated anti-mouse IgG antibody on a Parafilm strip and float the coverslip face down for 1 h at room temperature.
8. Wash the cells three times with 1 mL per well of PBS for 1 min per wash.
9. Incubate the cells with 50 μg/mL propidium iodide (to stain for total DNA).
10. Wash the cells three times with 1 mL per well of PBS for 1 min per wash.
11. Place 5 mL of Mounting Medium on a glass slide and put the coverslip face down on top of the Mounting Medium.
12. The stained cells can be analyzed and photographed using a Zeiss Axiovert 200 microscope (Thornwood, NY) and an AttoArc2 HBO 100 W light source (Atto Instruments, Rockville, MD). We visualized cell nuclei in the red and green channels and quantified digital images for BrdU incorporation. Percent DNA synthesis can be calculated as (Number of BrdU positive cells/Total number of cells analyzed) × 100. For each experimental treatment, we counted a minimum of 500 × 3 cells. Fig. 2 presents an example of the results that can be obtained using this approach (*see* **Note 11**).

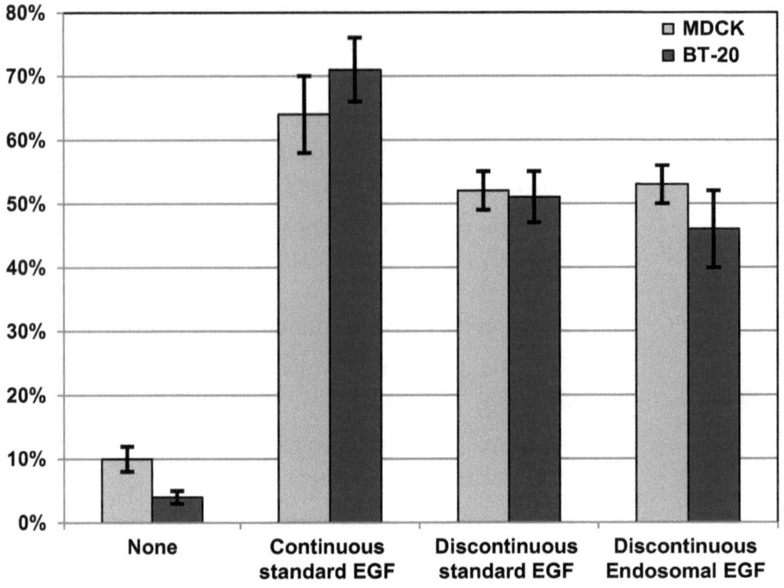

Fig. 2 Two pulses of standard, or endosomal EGFR signaling are sufficient to stimulate cell proliferation. DNA synthesis (S-phase entry) induced from a continuous EGF pulse, two pulses of standard EGF stimulation, or two pulses of exosomal EGF stimulation. This figure is reproduced from the figure previously published (Mol Cell Biol 23: 5803–5815, 2003) with permission

4 Notes

1. Do not use plastic coverslips. The high fluorescence background of plastic coverslip lowers the imaging quality for publication.
2. Both MDCK and BT-20 cells can be grown in similar media (DMEM) and do not require specialized culture techniques. Both cell lines tend to adhere well on coverslips.
3. Do not overgrow cells. Overgrowth of cells will make it harder to synchronize the cells. Confluence of more than 80% results in excessive autophosphorylation, which is difficult to silence using serum-starved medium.
4. At this stage, EGFRs are inactivated and localized to the plasma membrane.
5. Endosome activation treatment (i.e., endosomal accumulation of inactive ligand–receptor complexes followed by their activation) has been previous published by our group [6, 7].
6. By the end of the incubation period, EGFRs will be internalized into endosomes, and remain nonphosphorylated.
7. This step specifically activates endosome-associated EGFR. This is the first pulse of endosomal signaling.

8. This is the second pulse of endosomal signaling.
9. Make sure to wash the coverslips thoroughly. The residue HCL affects the following incubation with antibodies.
10. Optional incubation conditions for mouse anti-BrdU antibody is overnight at 4 °C. Ensure that the coverslips do not dry out by properly sealing the plate and by adding a sufficient amount of antibody solution.
11. As depicted in Fig. 2, continual exposure to EGF leads to a marked increase in proliferating cells over unstimulated controls. In addition, two 1-h pulses of standard EGF, or endosomal-EGF, spaced 8 h apart, can drive cells into S-Phase.

References

1. Jiang G, Hunter T (1999) Receptor signaling: when dimerization is not enough. Curr Biol 9 (15):R568–R571
2. Schlessinger J, Ullrich A (1992) Growth factor signaling by receptor tyrosine kinases. Neuron 9(3):383–391
3. Schlessinger J (1986) Allosteric regulation of the epidermal growth factor receptor kinase. J Cell Biol 103(6 Pt 1):2067–2072
4. Burke P, Schooler K, Wiley HS (2001) Regulation of epidermal growth factor receptor signaling by endocytosis and intracellular trafficking. Mol Biol Cell 12(6):1897–1910
5. Wiley HS, Burke PM (2001) Regulation of receptor tyrosine kinase signaling by endocytic trafficking. Traffic 2(1):12–18
6. Wang Y, Pennock SD, Chen X, Kazlauskas A, Wang Z (2004) Platelet-derived growth factor receptor-mediated signal transduction from endosomes. J Biol Chem 279(9):8038–8046
7. Wang Y, Pennock S, Chen X, Wang Z (2002) Internalization of inactive EGF receptor into endosomes and the subsequent activation of endosome-associated EGF receptors. Epidermal growth factor. Sci STKE 2002(161):pl17
8. Wang Y, Pennock S, Chen X, Wang Z (2002) Endosomal signaling of epidermal growth factor receptor stimulates signal transduction pathways leading to cell survival. Mol Cell Biol 22(20):7279–7290
9. Pennock S, Wang Z (2008) A tale of two Cbls: interplay of c-Cbl and Cbl-b in epidermal growth factor receptor downregulation. Mol Cell Biol 28(9):3020–3037
10. Pennock S, Wang Z (2003) Stimulation of cell proliferation by endosomal epidermal growth factor receptor as revealed through two distinct phases of signaling. Mol Cell Biol 23 (16):5803–5815
11. Kourouniotis G, Wang Y, Pennock S, Chen X, Wang Z (2016) Non-ligand-induced dimerization is sufficient to initiate the signalling and endocytosis of EGF receptor. Int J Mol Sci 17 (8):1200
12. Jones SM, Kazlauskas A (2000) Connecting signaling and cell cycle progression in growth factor-stimulated cells. Oncogene 19 (49):5558–5567
13. Jones SM, Kazlauskas A (2001) Growth factor-dependent signaling and cell cycle progression. FEBS Lett 490(3):110–116
14. Jones SM, Kazlauskas A (2001) Growth-factor-dependent mitogenesis requires two distinct phases of signalling. Nat Cell Biol 3 (2):165–172

Chapter 10

Study of EGFR Signaling/Endocytosis by Site-Directed Mutagenesis

Qian Wang and Zhixiang Wang

Abstract

As an essential tool in molecular biology and biochemistry, in vitro mutagenesis is important for mechanistic and functional studies of protein at the cellular level. Through site-directed substitution, insertion, or deletion of potential function residues, we may evaluate the specific biological role of proteins and understand the underlying mechanism. EGF ligand binding will activate EGFR and initiate signaling transduction for cell proliferation, differentiation, and survival; meanwhile, it will also induce the EGFR endocytosis for degradation to maintain cellular homeostasis. During this complicated process, some key residue(s) or sequences of EGFR molecule play essential regulatory roles. We describe in this chapter the inverse PCR-based site-directed mutagenesis protocol used in our studies on EGFR signaling and endocytosis.

Key words Epidermal growth factor receptor (EGFR), Polymerase chain reaction (PCR), Site-directed mutagenesis (SDM), Substitution, Deletion, Insertion

1 Introduction

Binding of ligands to EGFR induces the receptor dimerization, which results in the activation of intrinsic receptor tyrosine kinase and receptor transautophosphorylation of tyrosine residues within the C-terminal domain [1–4]. These phosphorylated residues serve as docking sites to recruit downstream signaling proteins and other important regulatory molecules like Grb2 and Cbl, for the initiation of both signal transduction and endocytosis of EGFR into the cell [5–9]. Through mutagenesis studies, several residues and motifs have been shown to play potentially important regulatory roles on EGFR signaling/endocytosis. Addition to the tyrosine residues (Y992, Y1045, Y1068, Y1086, Y1114, and Y1173) of receptor c-terminus, there are also several putative sequences that have been suggested to mediate EGFR endocytosis, like sequence 992–1022, 996QQGFF1000, and tyrosine-based sorting signals YXXΦ and NPXY such as 974YRAL [10–12]. Structure studies on ErbB receptors indicate that receptor dimerization, a crucial step in

Zhixiang Wang (ed.), *ErbB Receptor Signaling: Methods and Protocols*, Methods in Molecular Biology, vol. 1652, DOI 10.1007/978-1-4939-7219-7_10, © Springer Science+Business Media LLC 2017

EGFR activation and signaling/endocytosis, is mediated through the dimerization loop of CR1 domain within EGFR extracellular part [13–15]. Interestingly, by using kinase-dead mutant K721A, it is also shown that EGFR endocytosis is probably independent of kinase activity and tyrosine autophosphorylation [16]. In summary, our current understanding of EGFR signaling and endocytosis regulation is largely derived from various mutagenesis studies.

As the cornerstone of modern molecular biology, mutagenesis is a very important tool for studying protein structure–function relationship and regulating protein activities. Over the past few decades, many mutagenesis techniques have been developed, and of which the site-directed mutagenesis (SDM) has been widely used to studying EGFR signaling and endocytosis pathway, including the work in the author's laboratory [17–23]. Site-directed mutagenesis, as the name implies, allows researchers to specifically modify the gene at a selected location, and generally includes substitution (single or multiple amino acids), insertion, and deletion. Site-directed mutagenesis is typically performed using PCR-based methods, like traditional PCR, overlap extension PCR, and inverse PCR, by incorporating the desired modification (s) in PCR primers [24]. Here, in this chapter, we focus on the application of inverse PCR method in mutagenesis study of EGFR, including the three major types of site-directed mutagenesis we performed in our laboratory.

2 Materials

1. QuikChange® XL site-directed mutagenesis kit (Stratagene).
2. Pfu DNA polymerase (ThermoFisher Scientific).
3. Oligonucleotide DNA primers (IDT).
4. PCR master mix (ThermoFisher Scientific).
5. T4 DNA ligase (ThermoFisher Scientific).
6. pEYFP-N1/wtEGFR (constructed in our laboratory).
7. *E. coli* competent cell (made in our laboratory).
8. Thermal cycler (MJ Research PTC-100).
9. Agarose gel apparatus (Bio-Rad).

3 Methods

3.1 Overview of the General Principle (See Note 1)

1. Using two reverse-oriented synthetic oligonucleotide primers containing the desired mutations, the inverse PCR based EGFR site-directed mutagenesis utilizes the *Pfu* DNA polymerases to amplify the double strand of entire circular plasmid vector with EGFR (Fig. 1).

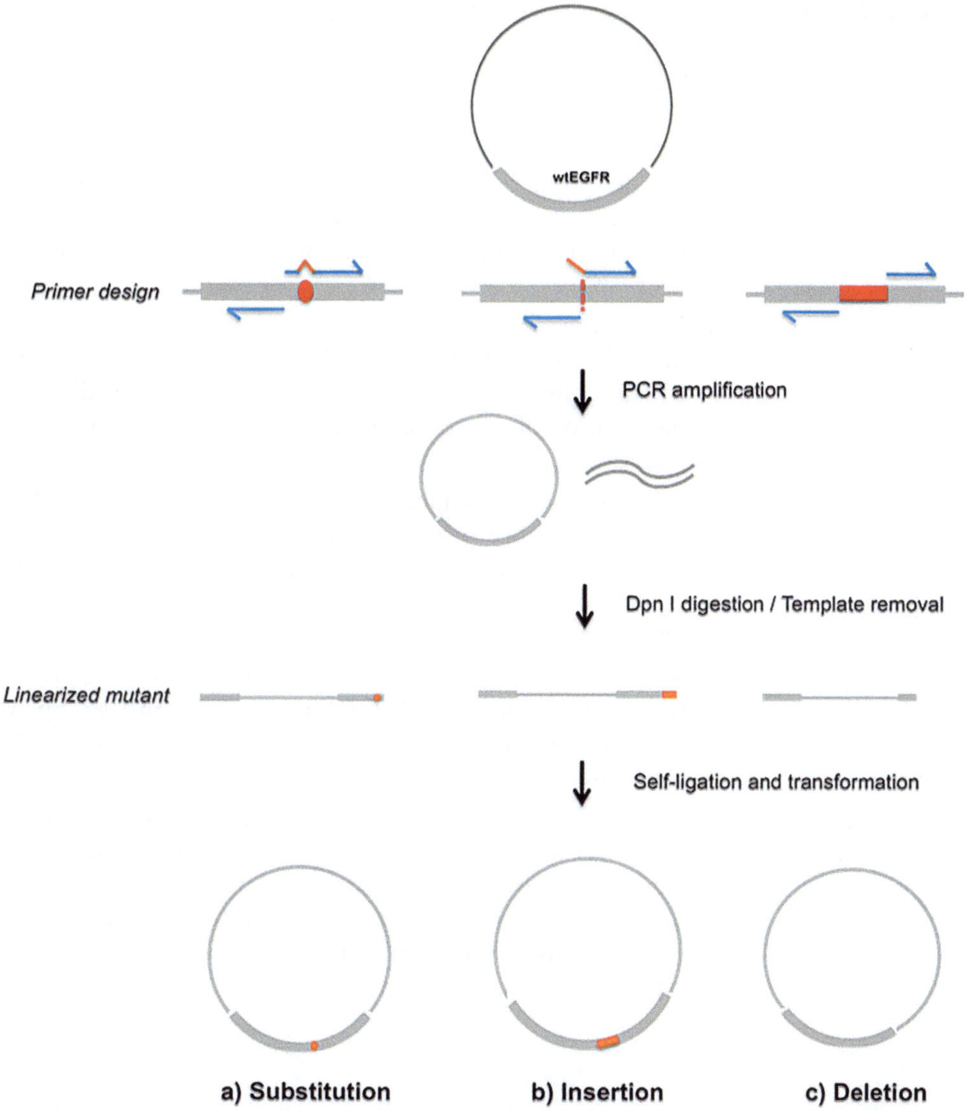

Fig. 1 Workflow of inverse PCR-based site-directed mutagenesis

2. With the superior proofreading ability, *Pfu* DNA polymerases can generate high-fidelity replication fragments with fewer errors to minimize the unexpected mutation, and the blunt-ended linearized amplicon is easy to self-ligate and recirculate when primers are synthesized 5′-phosphorylated.
3. Following amplification, the PCR product is treated with *Dpn* I, an endonuclease specific for methylated DNA, to digest the parent plasmid template. The selected undigested amplicon mutants will be further self-ligated and transformed into competent cells for analysis.

3.2 Primers Design

1. The specifically designed oligonucleotide primer is the key point for inverse PCR-based site-directed mutagenesis. The primer pair can be designed in either "back-to-back" orientation or overlapping way.
2. For "back-to-back" orientation, the oligonucleotide primers are designed to incorporate the point/multiple substitution (Fig. 1a) or inserted sequence (Fig. 1b) at the 5′-end, or miss the fragment to be deleted between the 5′-termini of primers (Fig. 1c).
3. In the case of insertion of large sequence, the insert can be divided and introduced at the 5′-termini of both primers. Because the yield PCR products are blunt-ended, the phosphorylation of primers at 5′-end is required for subsequent self-ligation.
4. When using QuikChange® site-directed mutagenesis kit from Stratagene, both primers should contain the desired changes, ideally in the middle of oligonucleotide, and primers should be complementary to the same sequence of opposite strands to generate nicked circular plasmid (Fig. 2).
5. An additional consideration made when designing primers for this kit is keeping the length between 25 and 45 bp, with a melting temperature ≥ 78 °C to prevent primer dimer formation (check the manufacturer's instructions).
6. Figure 3 shows the specific primers used in the author's lab to construct various EGFR mutants for EGFR endocytosis study, like EGFR1010LL/AA substitution, EGFR Δ1005–1017 deletion mutant, and ErbB2 mutant ($ErbB2_{EGFR1005-1017}$) with the insertion of the EGFR 1005–1017 fragment in its c-terminus [22, 23].

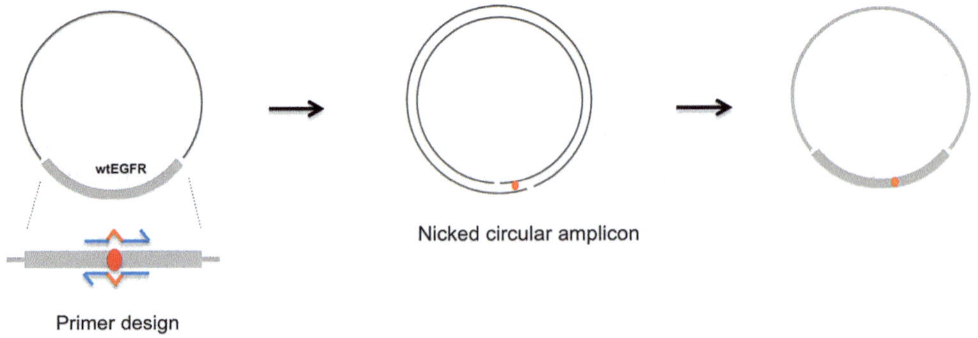

Fig. 2 Outline of QuickChange site-directed mutagenesis

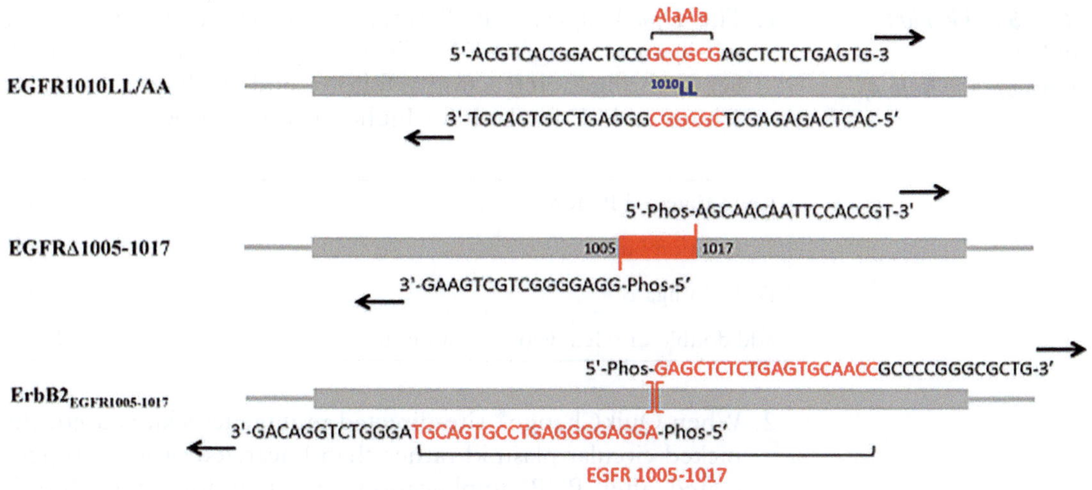

Fig. 3 Example of primer design for EGFR mutation using inverse PCR-based site-directed mutagenesis

3.3 Amplification Reaction

1. Set up reaction system as follows:

10× Reaction buffer	5 μl
10 mM dNTP	1 μl
Primer 1 (10 pmol/μl)	1 μl
Primer 2 (10 pmol/μl)	1 μl
Template plasmid	10–50 ng
Pfu DNA polymerase (2.5 U/μl)	1 μl
Add double-distilled deionized water to	50 μl

2. Perform PCR amplification using following conditions (*see* **Note 2**):

95°C	2 min	
95°C	30 sec	18 cycles
Tm -5°C	30 sec	18 cycles
68°C	1 min/kb of plasmid length	18 cycles
68°C	10 min	
4°C	Hold	

3.4 DpnI *Digestion of Template Plasmid (See Note 3)*

1. To digest the methylated template plasmid, add 1 μl of *Dpn* I endonuclease (10 U/μl) to the 50-μl PCR reaction and mix thoroughly, and incubate at 37 °C for 1 h.

2. *Dpn* I can be then heat-inactivated by incubation at 68 °C for 20 min. Optionally, the reaction can be further purified by agarose gel electrophoresis for subsequent processing.

3.5 Self-Ligation of PCR Product (See Note 4)

1. The *Dpn* I digested PCR reaction can be directly applied for self-ligation using T4 DNA ligase to generate circular plasmid. Set up the ligation reaction as follows, and incubate at 16 °C for 1 h, or at 4 °C for 16 h for higher ligation efficiency.

Dpn I-digested PCR reaction	2 μl
5× Ligation buffer	4 μl
T4 DNA ligase	1 μl
Add double-distilled deionized water to	20 μl

2. When QuikChange® site-directed mutagenesis kit is used, the nicked circular plasmid rather than linearized product is generated after PCR amplification, and therefore, the Dpn I-digested reaction can be directly transformed into competent cells, in which the nicked DNA will be repaired.

3.6 Transformation

1. Following self-ligation, 10% of the ligation product is transformed into *E. coli* DH5α competent cells using the standard method.
2. Briefly, add 2 μl of ligation reaction into competent cells thawed on ice, and swirl gently to mix thoroughly.
3. After incubation on ice for 30 min, the plasmid–cell mixture is heat-shocked at 42 °C for 45–90 s (depending on the type and volume of competent cells used, generally, we use 90 s for 100 μl of homemade *E. coli* competent cells), and put back on ice immediately for 2 min.
4. Add SOC medium (900 μl) and incubate with shaking at 37 °C for 1 h.
5. Spread 100 μl of the resulting culture on the LB plate with correct selection antibiotics (e.g., Kanamycin for pEYFP-N1 vector) and incubate at 37 °C overnight.

3.7 Verification of Transformants

1. Typically, 4–6 colonies are picked for cultural propagation, and plasmid DNA is isolated for sequencing analysis to verify the mutation. In the author's experience, the efficiency of inverse PCR-based site-directed mutagenesis on EGFR is over 80%.

4 Notes

1. Inverse PCR-based site-directed mutagenesis provides a useful mutagenesis method for protein function study, including EGFR signaling and endocytosis. By designing mutagenic primers with site-specific mutations, inverse PCR-based site-directed mutagenesis can easily introduce point mutation

(substitution), and large deletion and insertion mutations into EGFR. Conventional PCR-based mutagenesis methods involve multiple subcloning steps. Inverse PCR enables the amplification of entire circular plasmid using primers oriented in reverse direction, and therefore does not require the multiple restriction enzymatic treatment and gel purification before ligation [25]. Therefore, the major advantage for this one-step technique is the ability to generate mutated products relatively efficiently and cost-effectively. The main limitation for this approach lies in: (1) although the high-fidelity *Pfu* polymerase is used, PCR error with base mutation remains a big concern when amplifying full-length plasmid, and considering the processivity and fidelity of *Pfu* polymerases, the size of plasmid construct should be restricted to less than 10 kb, which limits the application for large gene targets; (2) relatively low efficiency of self-ligation of blunt-ended PCR products. Although the QuickChange SDM system does not need the self-ligation step, the primer dimer formation and limitation of primer melting temperature are common disadvantages. Modification on traditional QuickChange SDM protocol can efficiently overcome the abovesaid problems [26].

2. Considerations for PCR condition: (1) When setting up the PCR amplification conditions, if the primer has a melting temperature (Tm) $\geq$ 73 °C, which means that the annealing temperature for the PCR (generally = Tm −5 °C) is $\geq$68 °C, a two-step PCR condition with annealing/extension step at 68 °C is recommended. (2) The efficiency of the polymerization reaction by Pfu DNA polymerase is comparably lower than Taq polymerase; therefore, if no PCR band has been observed in gel electrophoresis or no colonies have been yielded after transformation, the extension time (generally 1 min/kb of plasmid size) could have been too short. In practice, the extension time and reaction cycle number can be increased for large template plasmid.
3. *Dpn* I digestion of parental plasmid is an important step for inverse PCR-base site-directed mutagenesis. Since *Dpn* I recognizes the m6 methylated sites, the template plasmid should be isolated from Dam methylase-positive host cells such as *E. coli* DH5α and JM109, and Dam methylase-negative strains like JM110 and SCS110 are not suitable for this mutagenesis protocol.
4. To self-ligate the blunt end of Pfu-yielded PCR product, the phosphorylation of DNA fragment at 5′-end is required. To do this, the phosphate group can be added to primers before amplification, or treat the PCR product directly with polynucleotide kinase (PNK), which can work with T4 DNA ligase in

the same reaction buffer. Because the association of 5′-end phosphate group with 3′-end hydroxyl group is more transient and lack the hydrogen bond stabilization of cohesive ends, more incubation time may be needed for the self-ligation, although 1 h at 37 °C is good enough in the author's experience, ligation at 4 °C overnight might be more efficient.

References

1. Schlessinger J (2002) Ligand-induced, receptor-mediated dimerization and activation of EGF receptor. Cell 110:669–672
2. Bublil EM, Yarden Y (2007) The EGF receptor family: spearheading a merger of signaling and therapeutics. Curr Opin Cell Biol 19:124–134
3. Carpenter G (1987) Receptors for epidermal growth factor and other polypeptide mitogens. Annu Rev Biochem 56:881–914
4. Schlessinger J, Ullrich A (1992) Growth factor signaling by receptor tyrosine kinases. Neuron 9:383–391
5. Huang F, Sorkin A (2005) Growth factor receptor binding protein 2-mediated recruitment of the RING domain of Cbl to the epidermal growth factor receptor is essential and sufficient to support receptor endocytosis. Mol Biol Cell 16:1268–1281
6. Sorkin A, Helin K, Waters CM et al (1992) Multiple autophosphorylation sites of the epidermal growth factor receptor are essential for receptor kinase activity and internalization. Contrasting significance of tyrosine 992 in the native and truncated receptors. J Biol Chem 267:8672–8678
7. Wang Z, Moran MF (1996) Requirement for the adapter protein GRB2 in EGF receptor endocytosis. Science 272:1935–1939
8. Okutani T, Okabayashi Y, Kido Y et al (1994) Grb2/Ash binds directly to tyrosines 1068 and 1086 and indirectly to tyrosine 1148 of activated human epidermal growth factor receptors in intact cells. J Biol Chem 269:31310–31314
9. Okabayashi Y, Kido Y, Okutani T et al (1994) Tyrosines 1148 and 1173 of activated human epidermal growth factor receptors are binding sites of Shc in intact cells. J Biol Chem 269:18674–18678
10. Chang CP et al (1993) Ligand-induced internalization of the epidermal growth factor receptor is mediated by multiple endocytic codes analogous to the tyrosine motif found in constitutively internalized receptors. J Biol Chem 268:19312–19320
11. Sorkin A, Mazzotti M, Sorkina T et al (1996) Epidermal growth factor receptor interaction with clathrin adaptors is mediated by the Tyr974-containing internalization motif. J Biol Chem 271:13377–13384
12. Bonifacino JS, Dell'Angelica EC (1999) Molecular bases for the recognition of tyrosine-based sorting signals. J Cell Biol 145:923–926
13. Ferguson KM, Berger MB, Mendrola JM et al (2003) EGF activates its receptor by removing interactions that autoinhibit ectodomain dimerization. Mol Cell 11:507–517
14. Ogiso H et al (2002) Crystal structure of the complex of human epidermal growth factor and receptor extracellular domains. Cell 110:775–787
15. Garrett TP et al (2003) The crystal structure of a truncated ErbB2 ectodomain reveals an active conformation, poised to interact with other ErbB receptors. Mol Cell 11:495–505
16. Wang Q, Villeneuve G, Wang Z (2005) Control of epidermal growth factor receptor endocytosis by receptor dimerization, rather than receptor kinase activation. EMBO Rep 6:942–948
17. Saxon ML, Lee DC (1999) Mutagenesis reveals a role for epidermal growth factor receptor extracellular subdomain IV in ligand binding. J Biol Chem 274(40):28356–28362
18. Honegger AM, Dull TJ, Felder S et al (1987) Point mutation at the ATP binding site of EGF receptor abolishes protein-tyrosine kinase activity and alters cellular routing. Cell 51:199–209
19. Dawson JP, Berger MB, Lin CC et al (2005) Epidermal growth factor receptor dimerization and activation require ligand-induced conformational changes in the dimer Interface. Mol Cell Biol 25(17):7734–7742
20. Jiang X, Huang F, Marusyk A, Sorkin A (2003) Grb2 regulates internalization of EGF receptors through clathrin-coated pits. Mol Biol Cell 14(3):858–870
21. Sigismund S, Argenzio E, Tosoni D et al (2008) Clathrin-mediated internalization is essential for sustained EGFR signaling but

dispensable for degradation. Dev Cell 15 (2):209–219

22. Wang Q, Zhu F, Wang Z (2007) Identification of EGF receptor C-terminal sequences 1005-1017 and di-leucine motif 1010LL1011 as essential in EGF receptor endocytosis. Exp Cell Res 313(15):3349–3363
23. Wang Q, Chen X, Wang Z (2015) Dimerization drives EGFR endocytosis through two sets of compatible endocytic codes. J Cell Sci 128 (5):935–950
24. Reikofski J, Tao BY (1992) Polymerase chain reaction (PCR) techniques for site-directed mutagenesis. Biotechnol Adv 10(4):535–547
25. Ochman H, Gerber AS, Hartl DL (1988) Genetic applications of an inverse polymerase chain reaction. Genetics 120 (3):621–623
26. Zheng L, Baumann U, Reymond JL (2004) An efficient one-step site-directed and site-saturation mutagenesis protocol. Nucleic Acids Res 32(14):e115

Chapter 11

Using Percoll Gradient Fractionation to Study the Endocytic Trafficking of the EGFR

Julie A. Gosney and Brian P. Ceresa

Abstract

Scientists have separated subcellular compartments based on differences in their densities by gradient centrifugation for decades. With the proper equipment and thoughtful experimental design, density gradients are a reliable method for enriching various intracellular compartments to assess their protein composition, morphology, or biochemical activity. While variations of this basic technique have been used for separating whole cells, endosomes, synaptic vesicles, and viruses, we have found it especially useful for resolving the compartments of the endocytic pathway. In particular, this technique has been valuable for studying the regulation and signaling of the Epidermal Growth Factor Receptor (EGFR) while undergoing endocytic trafficking. In this article we detail the technical aspects of utilizing Percoll gradients to study the endocytic trafficking of the EGFR and associated proteins.

Key words Percoll gradient, Endocytosis, Epidermal growth factor receptor, Late endosome, Early endosome, Isopycnic gradient

1 Introduction

Many questions in cell biology revolve around understanding the composition and structure of subcellular compartments and the biochemical reactions that occur within those organelles. Being able to resolve compartments is a critical first step for these studies. Once isolated, these individual compartments can be probed by immunoblot for the presence of various proteins, examined with high-resolution microscopy for structural features, or monitored for enzymatic activity.

Density gradients are an efficient and reliable way to separate subcellular compartments. Briefly, these experiments mix a density media (such as Percoll) with the material of interest (cytosol) and the contents are separated via isopycnic centrifugation. As a density gradient forms, individual compartments separate within the gradient based on their buoyant density. Gradients are made from a variety of media such as Percoll, sucrose, and Optiprep; however,

Zhixiang Wang (ed.), *ErbB Receptor Signaling: Methods and Protocols*, Methods in Molecular Biology, vol. 1652, DOI 10.1007/978-1-4939-7219-7_11, © Springer Science+Business Media LLC 2017

Percoll remains the matrix of choice for many scientists. Percoll is a commercially available (GE Healthcare), stable, and inert matrix made from a heterogeneous mixture of colloidal silica coated with polyvinylpyrrolidone. During ultracentrifugation, the media creates a spontaneous, self-forming gradient throughout the sample tube with the least dense materials at the top of the tube and the densest materials at the bottom. When Percoll is mixed with post-nuclear supernatant (cytosol) and subjected to ultracentrifugation, organelles migrate to their unique isopycnic point within the gradient. The gradients can then be fractionated to separate the contents of differing densities.

An advantage of using a continuous gradient is the ability to resolve compartments with minute differences in density. This is an important consideration since most organelles have a range of densities due to organelle biosynthesis, fusion with other organelles, and/or organelle maturation. Density centrifugation overcomes this limitation. Early and late endosomes exhibit two separate ranges of densities (i.e., 1.035–1.042 g/mL and 1.048–1.060 g/mL, respectively) which can be easily resolved under the proper conditions [1]. Although this increases the range of fractions within the gradient that will contain these vesicles, their densities are distinct enough to still separate both, with minimal overlap. However, investigators should be aware that this separation procedure results in dilution of the samples in the Percoll, decreasing their concentration.

One of the most productive and established uses of Percoll gradients is to separate the compartments of the endocytic pathway [2, 3]. Our studies have used Percoll gradients to monitor the trafficking of the EGFR, a receptor tyrosine kinase (RTK) with important biological roles in developmental biology, wound healing, and cancer biology. When EGFR is ligand-bound at the plasma membrane, it dimerizes with another EGFR (or another ErbB family member) that causes kinase domain activation and transphosphorylation of intracellular tyrosine residues. Under physiologic conditions (low ligand concentration, physiologic expression of WT EGFRs, etc.), the active EGFR dimer undergoes clathrin-mediated endocytosis and is trafficked into an early endosome. Once internalized, the receptor can be targeted for lysosomal degradation, recycled to the plasma membrane, or trafficked to the nucleus or ER [4–6]. Using Percoll gradients in a variety of EGFR-expressing cell lines, we have been able to assess the kinetics of EGFR trafficking, proteins that regulate ligand-mediated and constitutive EGFR trafficking, and spontaneously arising defects in EGFR trafficking [7–10].

2 Materials

2.1 Cells and Cell Culture

1. HeLa cells are a human cervical adenocarcinoma cell line from ATCC (American Type Culture Collection) (*see* **Note 1**).
2. Dulbecco's Minimal Essential Media (DMEM).
3. Growth Media: DMEM supplemented with 5% fetal bovine serum (FBS), 100 units/mL penicillin, 100 units/mL streptomycin, and 2 mM glutamine.

2.2 Percoll Gradient

1. 2.5 M sucrose.
2. TES Buffer: 10 mM triethanolamine/1 mM EDTA/0.25 M sucrose, pH 7.2.(*see* **Note 2**)
3. Percoll (GE Healthcare): At least 24 h prior to experimentation, stock Percoll should be equilibrated with 2.5 M sucrose at a ratio of 9:1. The final 90% Percoll/0.25 M sucrose solution can be stored at 4 °C for up to 3 months.
4. Density Marker Beads (Cospheric): Cospheric offers a number of colored microspheres that migrate at distinct densities and are visible within the density gradient (Fig. 1).
5. Rubber policeman to scrape cells.
6. Protease and phosphatase inhibitors: We utilize a cocktail of 2 mM phenylmethylsulfonyl fluoride (PMSF) (Calbiochem), 1 mM sodium orthovanadate (Na_3VO_4) (Sigma), 10 μM pepstatin (Sigma), and 1 μM aprotinin (Roche) (*see* **Note 3**).
7. Peristaltic pump (Pharmacia Biotech, Pump P-1), 0.8 mm internal diameter (i.d.) Silicone tubing (Bio-Rad), 50 μL disposable micropipette (Accu-glass) for fractionating resolved gradients.
8. EGF ligand (available from multiple sources including Invitrogen, ProSpec, and R&D Systems): stock ligand is reconstituted according to manufacturer's directions at a concentration of 1 μg/μL and stored in 2 μL aliquots at −20 °C. Aliquots are thawed on ice for ~15 min immediately prior to dilution in media, followed by incubation with cells. The remainder of the aliquot is discarded to avoid loss of ligand efficacy from freeze/thawing.
9. ^{125}I-EGF in Binding Buffer: DMEM supplemented with 20 mM HEPES, 0.1% bovine serum albumin, pH 7.3.
10. 15 mL conical tubes and a low speed tabletop centrifuge (Beckman Coulter Allegra 25R).
11. Ultracentrifugation equipment: 16 mm × 67 mm OptiSeal™ polypropylene tubes, VTi 65.1 vertical rotor (Beckman Coulter), Beckman Coulter Optima L-100 XP Ultracentrifuge (*see* **Note 4**).

A

B

$$R_f \text{ of density beads} = \frac{\text{distance from beads to top of gradient tube}}{\text{distance from top to bottom of gradient tube}}$$

$$R_f \text{ of fractions} = 1 - \left(\frac{\text{number of drops collected from top of tube}}{\text{number of drops collected from entire gradient tube}} \right)$$

Fig. 1 (**a**) The colored Cospheric density beads will migrate to their buoyant density within the gradient during centrifugation. *Arrows* indicate location of colored beads in a 17% Percoll gradient. (**b**) Equations used to calculate R_f values of the density beads and individual fractions collected from Percoll gradients. Because the density of each bead is known, the R_f value of each bead can be correlated to the R_f values of fractions in the sample tubes to estimate the density of the vesicles in each fraction of the gradient

12. 1.5 mL Eppendorf tubes and Eppendorf 5424 Microfuge.
13. 0.1 M sodium acetate buffer, pH 4.4 containing 0.1% Triton X-100.
14. 100 μM fluorescein-β-D-galactopyranoside (Sigma Aldrich).
15. Fluorescence plate reader (BioTek).

2.3 SDS-PAGE and Immunoblot

1. Laemmli 6× sample buffer.
2. MW markers (e.g., Bio-Rad SDS-PAGE Standards).
3. SDS-PAGE supplies and transfer equipment: Large format gels (18 cm × 16 cm glass plates) (SE400 Vertical Unit, GE Healthcare) and Electrophoretic Transfer tank (Trans-Blot Cell, Bio-Rad) (*see* **Note 5**).
4. Nitrocellulose (available from Amersham etc.)

5. Ponceau S (available from Sigma etc.) for staining transferred proteins.
6. ECL reagents.
7. Digital dark room (Fotodyne imaging system).

2.4 Antibodies

1. In order to determine the location of subcellular compartments within the gradient, it is essential to select highly specific marker proteins for each organelle to detect via immunoblot. For example, early endosomes can be identified using an Early Endosome Autoantigen-1 (EEA1) antibody (available from Cell Signaling, BD Biosciences, etc.), and late endosomes/lysosomes can be identified using a Lysosomal-Associated Membrane Protein 1 or 2 (LAMP1 or LAMP2) antibody (available from Cell Signaling, BD Biosciences, The University of Iowa Hybridoma Bank, etc.) (*see* **Note 6**).
2. EGFR (available from Santa Cruz, Cell Signaling, etc.) and phosphorylated EGFR (pEGFR) antibodies (available from Cell Signaling; pY1045, pY1068, etc.) are useful to determine the localization of the receptor within the Percoll gradient. Localization of EGFR or pEGFR with other organelle markers can be used as an indication of the type of endosomes EGFR is localized to at that time point.

2.5 Radioligands

1. ^{125}I-EGF (NEX-428, PerkinElmer).
2. Binding Buffer: DMEM supplemented with 20 mM HEPES, 0.1% bovine serum albumin, pH 7.3.
3. 12 mm × 75 mm plastic test tubes.
4. Beckman gamma counter (Wizard-2, PerkinElmer).

3 Methods

3.1 Ligand Stimulation, Cell Lysis, and Postnuclear Supernatant (PNS) Preparation

1. Grow HeLa cells in 15 cm tissue culture dishes in growth media.
2. When cells have reached confluency (~80–90%), serum starve the cells for approximately 2 h by washing two times with PBS pH 7.4 and adding 10 mL DMEM (37 °C) to each dish and returning to the incubator.
3. Incubate serum starved cells with EGF (10 ng/mL) that has been freshly diluted from a 1 μg/μL aliquot for 15 min at 37 °C (*see* **Note 7**).
4. Wash cells two times with PBS pH 7.4 at room temperature and equilibrate the dishes on ice (4 °C). Aspirate PBS and quickly wash cells in ice-cold TES buffer, always keeping the dishes on ice to prevent further EGFR trafficking.

5. Aspirate the TES and add 2 mL fresh TES supplemented with protease and phosphatase inhibitors. Monitor the cells with a light microscope (10× objective) until they begin to swell, but before they burst, and harvest by scraping the cells with a rubber policeman. Pipet the harvested cells up and down 40 times with a P1000 pipet to disperse and lyse the cells.
6. Transfer the lysed cells to a pre-chilled 15 mL conical tube and prepare the post-nuclear supernatant (PNS) by centrifuging at ~200 × *g* for 10 min at 4 °C in a low-speed tabletop centrifuge. Transfer the PNS to a fresh tube and resuspend the pellet in 2 mL TES supplemented with fresh protease and phosphatase inhibitors. Centrifuge the resuspended pellet at ~200 × *g* for 10 min at 4 °C. Collect the supernatant and pool with the first supernatant, yielding a final PNS of ~4 mL per sample.

3.2 Percoll Gradient Fractionation of Early and Late Endosomes

1. Combine the PNS with TES and the 90% Percoll solution to yield a final Percoll concentration of 17% in a total volume of 11.5 mL.
2. Mix the final gradient solutions, load into OptiSeal ultracentrifuge tubes using a P1000 pipet, and seal the caps to the tube using an OptiSeal sealer. The tubes are then loaded into the pre-chilled VTi65.1 vertical rotor, and a spacer is placed on top of each tube (*see* **Note 8**). A cap is then screwed into each tube-holder in the rotor that contains a sample tube (*see* **Note 9**).
3. Density marker beads are used to determine the density of migrated materials in the Percoll gradient and allow comparisons between experiments. The beads are added to a separate tube containing 17% Percoll in TES buffer, centrifuged, and photographed (Fig. 1). The photograph can be printed, and the migration of each bead throughout the tube can be measured and used to calculate an R_f value for each density bead. The R_f values are plotted against the density of the beads to generate a standard curve that can be used to approximate the density of each fraction based on R_f value. Immunoblotting for organelle markers can be used as a substitute for density beads to verify the integrity of the gradient.
4. Load the VTi65.1 rotor into an ultracentrifuge and spin at 50,000 × *g* for 25 min with maximum acceleration and maximum brake. Be careful when removing the samples from the rotor so as not to disturb the gradients. Some slight movement is acceptable and inevitable, but any major disruptions or tilting of the tubes will cause the gradients to mix. If this occurs, the samples can be remixed thoroughly (i.e., by carefully turning the tube upside down, making sure that the lid remains sealed) and re-centrifuged.

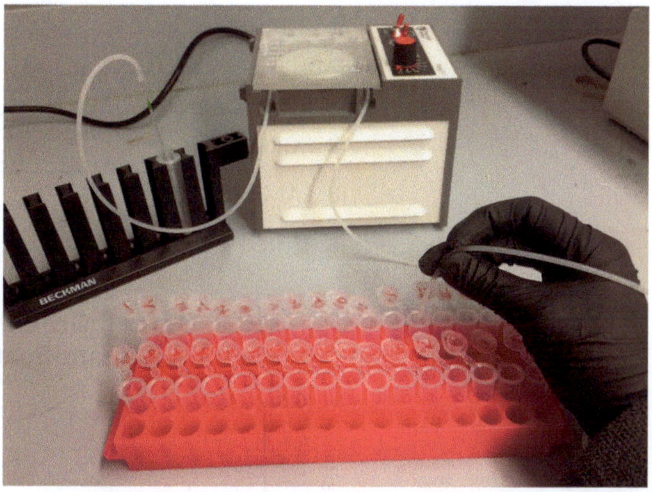

Fig. 2 A glass capillary tube connected to the peristaltic pump tubing is carefully placed into the Percoll gradient OptiSeal tube until the bottom of the pipet touches the bottom of the gradient tube. The pump is then turned on, and the gradient is collected in ten-drop fractions into pre-chilled and labeled Eppendorf tubes

5. If a separate gradient was prepared using the density beads, take a picture of the beads tube as soon as possible after centrifugation, as warming to room temperature may cause diffusion of the beads.
6. The Percoll gradients containing PNS should be immediately fractionated so as to reduce sample diffusion in the gradient. To minimize proteolysis and dephosphorylation, fractionation is performed in a cold room to ensure the samples remain at 4 °C. Carefully, take the lid off of the 16 × 67 mm OptiSeal centrifuge tube and place a 50 μL glass micropipette connected to the silicone tubing to the bottom of the gradient tube (Fig. 2). Run the silicone tubing through the peristaltic pump, turn on the pump and collect the gradient into prechilled, labeled Eppendorf tubes in ten-drop fractions (~330 μL each). This generally yields about 30 fractions per gradient. Keep track of how many drops are collected in the last fraction, as the total number of drops is used to calculate the Rf value of each fraction (Fig. 1). Between gradients flush any remaining sample out of the peristaltic pump tubing with 1–2 mL of ice cold ddH_2O.
7. Each fraction is diluted in 6× SDS sample buffer containing 10% β-mercaptoethanol (βME), boiled at 100 °C for 3′, and centrifuged at 21,000 × *g for 10 min* to pellet Percoll (*see* **Note 10**). The fractions are now prepared for resolution on an SDS-PAGE gel and subsequent immunoblot analysis [8]

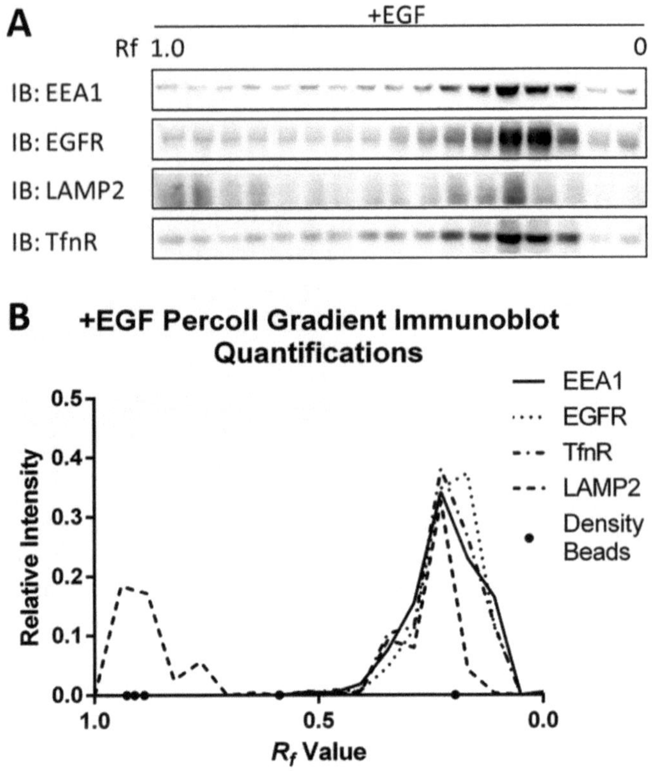

Fig. 3 Organelle marker proteins have distinct migrations within a Percoll gradient. HeLa cells were treated with 10 ng/mL EGF for 15′, harvested, and loaded onto a 17% Percoll gradient. (**a**) Every other fraction was resolved via SDS-PAGE and immunoblotted for markers of various organelles (EEA1 = early endosomes, LAMP2 = late endosomes/lysosomes, TfnR = plasma membrane/recycling/early endosomes) or the Epidermal Growth Factor Receptor (EGFR). Note that an R_f value of 1.0 represents the bottom of the gradient tube, while an R_f value of 0 represents the top of the gradient. (**b**) Immunoblots from (**a**) were quantified using NIH Image J and the relative density of each protein was plotted versus their R_f value. *Circles* on the *x*-axis represent density bead migration (R_f 0.93 = 1.109 g/mL, R_f 0.91 = 1.070 g/mL, R_f 0.89 = 1.057 g/mL, R_f 0.59 = 1.049 g/mL, R_f 0.20 = 1.042 g/mL)

(Fig. 3). For these studies, we typically run only the even fractions. The odd fractions can be probed later, if necessary. Band intensity from the immunoblots is quantified by NIH Image J software and the relative intensity can be plotted as a function of the R_f value.

8. The R_f values of the density beads can be calculated from the image, and the known bead densities can be added to the graph of the protein distribution (Figs. 1 and 3).

3.3 Separation of Compartments of the Late Endocytic Pathway

For some lines of investigation, further separation of endocytic compartments is desired. Separation of the various components of the late endocytic pathway can be achieved by pooling the fractions corresponding to the late endosomes from a 17% Percoll and centrifuging in a higher percentage of Percoll (i.e., 28%). This protocol allows resolution of the late endosome–lysosome hybrid compartments from late endosomes and lysosomes (Fig. 4) [10].

1. Resolve the early and late endosomes using a 17% Percoll gradient as described in Subheading 3.2, **steps 1–7**. Keep the samples on ice. Do not put in sample buffer.
2. Pool fractions corresponding to the late endosomal fractions (densities >1.060 g/mL or R_f = 1.0–0.7 on a 17% Percoll gradient—typically the first nine tubes if separated into ten-drop fractions). Adjust the volume of the combined fractions to 11.5 mL and 28% Percoll using the 90% Percoll solution and TES solution, taking into account that the pooled fractions have 17% Percoll.
3. Transfer samples to an OptiSeal tube and seal. Using the same Beckman VTi65 rotor and Beckman ultracentrifuge described above, centrifuge samples at 4 °C for 25 min at 50,000 × *g* with maximum acceleration and maximum brake.
4. Fractionate samples at 4 °C as described in Subheading 3.2 **step 6**.

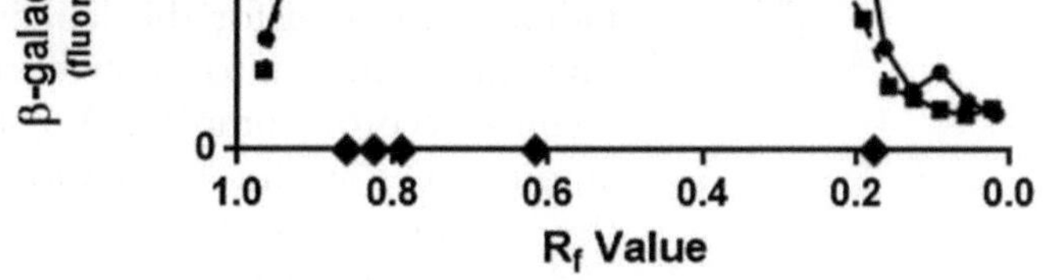

Fig. 4 siCON-transfected (control) and RAB7 siRNA-transfected HeLa cells were stimulated for 60 min with EGF. PNS was separated over a 17% Percoll gradient, and the late endosomal fractions (R_f 1.0-0.7) were then further resolved over a 28% Percoll gradient. Fractions from the 28% gradient were assayed for β-galactosidase activity, and each fraction was plotted as a percentage of total activity. Data are representative of four independent experiments. *Diamonds* on the *x*-axis represent migration of density beads (R_f 0.20 = 1.040 g/mL; R_f 0.48 = 1.055 g/mL; R_f 0.74 = 1.069 g/mL; R_f 0.79 = 1.080 g/mL; R_f 0.85 = 1.109 g/mL)

5. The distribution of lysosomes can be measured as a function of β-galactosidase activity. Mix 100 μL of each fraction and incubate with 250 μL of 0.1 M sodium acetate buffer, pH 4.4 containing 0.1% Triton X-100 and 100 μM fluorescein-β-D-galactopyranoside in individual wells of a 96-well plate. Incubate the samples at 37 °C for 3 h. Enzymatic activity is measured in a fluorescent plate reader at 480 nm excitation and 520 nm emission. Data are plotted as the relative β-galactosidase activity versus R_f value (Fig. 4).
6. The portion of each sample not used for enzymatic assays can be prepared in 6× SDS sample buffer as described in Subheading 3.2 **step** 7 and subjected to immunoblot analysis.

3.4 Isolation of Endosomes Containing ^{125}I-EGF–EGFR Complexes

An alternative method for tracking the EGF–EGFR complex through the endocytic pathway is to use EGF labeled with radioactive iodine (^{125}I-EGF). The advantages of this strategy are that radioactivity analysis is much faster than immunoblotting, the radiolabeled signal is very sensitive, and the detection procedure is noninvasive so the sample can be used for biochemical and immunoblot assays if desired (although one needs to keep in mind that there is radioactivity associated with the sample). The limitations of this strategy are that the use of radioactivity requires authorization (usually from the institution), monitoring the equipment and workspace for radioactivity, and the fact that many scientists do not like working with radioactivity.

1. A confluent 10 cm dish of HeLa cells are serum starved by washing twice with PBS and incubated with 37 °C DMEM for 2 h to decrease basal EGFR activity and increase the cell surface levels of EGFR.
2. DMEM is removed and cells are incubated at 37 °C for 7 min with 3 mL of prewarmed 1 ng/mL ^{125}I-EGF in Binding Buffer. Due to the expense and waste issues, the volume of radioactivity used for this labeling step should be minimized, but still sufficient to allow formation and detection of the ligand–receptor complex. A 7 min incubation is a sufficient amount of time for the radioligand to bind the EGFR and the ^{125}I-EGF–EGFR complex to enter the endocytic pathway and synchronize the wave of ligand–receptor endocytosis.
3. The radioligand containing media is collected (*see* **Note 11**). Cells are washed four times with room temperature PBS to remove unbound, extracellular radioligand. Pre-warmed growth media is added back to the cells for another 8 min. This is sufficient time to traffic most of the ligand–receptor complexes to the early endosomes. A longer incubation time is required to accumulate the ^{125}I-EGF–EGFR complexes into the late endosomes (*see* **Note 12**).

4. At the end of the total incubation, cells are washed twice with room temperature PBS and placed on ice to equilibrate to 4 °C. Cells are subjected to a brief wash with ice-cold TES, and then harvested in 2 mL TES supplemented with protease and phosphatase inhibitors. PNS supernatant is prepared as described in Subheading 3.1 **steps 5** and **6**.
5. The radioligand containing PNS is incorporated into a 17% Percoll solution, transferred to an OptiSeal tube and centrifuged as described in Subheading 3.2 **steps 1–4**.
6. Radiolabeled gradients should be fractionated at 4 °C into prechilled, labeled 12 × 75 mm plastic test tubes and kept on ice. These tubes fit in Beckman gamma counters, but the tube size may need to be adjusted for other gamma counters.
7. To identify the fractions that contain radioactivity, samples are counted in a gamma counter. Care should be taken to minimize the length of time that the samples are not maintained at 4 °C. This can easily be done by shortening the count time to 0.1 min and only counting ten samples at a time (*see* **Note 13**).
8. Once the counting is completed, the fractions can be assayed for enzymatic activity (such as β-galactosidase activity) or put in sample buffer for immunoblot analysis.

4 Notes

1. HeLa cells are used because they express physiologically relevant levels of EGFR (~50,000 receptors per cell) [11]. Other cell lines may have varying levels of receptor expression that could change the kinetics of receptor trafficking, dependent upon multiple factors (*see* **Note** 7).
2. TES buffer is used to lyse cells and maintain isotonic sucrose concentrations within the Percoll gradient. Some organelles are sensitive to osmotic lysis, while others (e.g., early endosomes) are not, so it is important to determine if your target compartment might be compromised during prolonged exposure to such buffers [12].
3. Lysis of the cell and its subcellular compartments can liberate both proteases and phosphatases. In order to get reproducible data, it is important to use various protease and phosphatase inhibitors during cell lysis and fractionation to minimize sample degradation and protein dephosphorylation. Phosphatase inhibitors are of particular importance when studying signaling molecules.
4. Similar ultracentrifuge equipment is offered through other manufacturers including Sorvoll and NuAire. It is imperative

that tubes, rotors and centrifuges are designed to work together.

5. Many labs are equipped with the supplies and equipment to perform immunoblotting using a mini-gel system. For Percoll gradient samples, large acrylamide gels (18 × 16 cm glass plates) allow the investigator to immunoblot using a larger sample size and run more samples on the same gel. This allows a direct comparison of protein immunodetection. Further, the larger size of the nitrocellulose facilitates cutting the nitrocellulose horizontally, allowing the investigator to probe the same samples for a variety of organelle markers of varying molecular weights.
6. When resolving early and late endosomes, EEA1 and LAMP1/2 are the typical organelle markers used to show proper resolution. However, the fractions can be probed for the presence of other marker proteins such as plasma membrane, endoplasmic reticulum (ER), and Golgi. For example, Na/K-ATPase is a common marker used for plasma membrane (antibodies available from Cell Signaling), and Calnexin is a protein that is localized to ER (antibodies available from BD Transduction, Assay Designs, etc.).
7. When studying EGFR trafficking, the specific ligand and its concentration as well as incubation time are crucial factors to consider. Certain ligands will trigger the receptor to be rapidly recycled back to the plasma membrane, while others will target the receptor for lysosomal degradation [13]. Further, very high concentrations of ligand will induce internalization via non-clathrin-mediated mechanisms. While this induces rapid internalization of the receptors, it can oversaturate the endocytic pathway and trigger trafficking to other compartments. Generally, less than 10 ng/mL of EGF is recommended to target the receptor for clathrin-mediated endocytosis into early endosomes [14]. Lastly, incubation time of cells with ligand should be carefully considered, as the temporal relationship of the receptor with various endosomes is discreet. If the receptor is to be studied within a specific compartment, such as early endosomes, sufficient time must be allowed for the receptor to be internalized, but limited so that trafficking does not progress into late endosomes or lysosomes. The temporal regulation of the receptor should be monitored carefully within a cell line to determine the optimal incubation time to study the desired compartment. We recommend performing preliminary studies with immunofluorescence to monitor EGFR trafficking and kinetics within each cell line.
8. A vertical rotor must be used to create a continuous density gradient in a short period of time. If a vertical ultracentrifuge

rotor is not available, a near-vertical rotor or a swinging bucket rotor may be substituted. However, the use of these rotors will require optimization of the duration and rate of ultracentrifugation.

9. When loading the VTi65.1 rotor, be sure to place the spacer on top of the tube, followed by the screw-top cap. The cap can be screwed into place by hand, but must be sealed with a torque wrench, turning to 100 lbs of pressure (but do not exceed 100 lbs).
10. When preparing Percoll gradient fractions for immunoblot, it is crucial that the Percoll is removed from the sample so that it runs properly on the gel. In order to pellet Percoll, the samples must be boiled in sample buffer prior to centrifugation; the supernatant is resolved by SDS-PAGE. Centrifugation prior to the addition of sample buffer will not pellet the Percoll.
11. All waste at this point should be considered radioactive and placed in the appropriate waste container. In addition, if proteins are resolved by SDS-PAGE the electrode buffer should be considered radioactive as well.
12. The presence of the radioactivity only indicates which fractions contain ^{125}I; investigators may need to verify that the ^{125}I-EGF is intact and/or bound to the EGFR.
13. It has been our experience that one can get a sufficient signal with 0.1 min count time. While the data may have less variability with longer counts, having the samples at room temperature for a longer period of time is not justified.

References

1. Kornilova E, Sorkina T, Beguinot L, Sorkin A (1996) Lysosomal targeting of epidermal growth factor receptors via a kinase-dependent pathway is mediated by the receptor carboxyl-terminal residues 1022-1123. J Biol Chem 271 (48):30340–30346
2. Marsh M, Schmid S, Kern H, Harms E, Male P, Mellman I, Helenius A (1987) Rapid analytical and preparative isolation of functional endosomes by free flow electrophoresis. J Cell Biol 104(4):875–886
3. van der Goot FG (1997) Separation of early steps in endocytic membrane transport. Electrophoresis 18(14):2689–2693. doi:10.1002/elps.1150181426
4. Felder S, LaVin J, Ullrich A, Schlessinger J (1992) Kinetics of binding, endocytosis, and recycling of EGF receptor mutants. J Cell Biol 117(1):203–212
5. Stamos J, Sliwkowski MX, Eigenbrot C (2002) Structure of the epidermal growth factor receptor kinase domain alone and in complex with a 4-anilinoquinazoline inhibitor. J Biol Chem 277(48):46265–46272. doi:10.1074/jbc.M207135200
6. Vieira AV, Lamaze C, Schmid SL (1996) Control of EGF receptor signaling by Clathrin-mediated endocytosis. Science 274:2086–2089
7. Ceresa BP, Bahr SJ (2006) Rab7 activity affects epidermal growth factor: epidermal growth factor receptor degradation by regulating Endocytic trafficking from the late endosome. J Biol Chem 281(2):1099–1106
8. Rush JS, Ceresa BP (2013) RAB7 and TSG101 are required for the constitutive recycling of unliganded EGFRs via distinct mechanisms. Mol Cell Endocrinol 381(1–2):188–197. doi:10.1016/j.mce.2013.07.029

9. Rush JS, Quinalty LM, Engelman L, Sherry DM, Ceresa BP (2012) Endosomal accumulation of the activated epidermal growth factor receptor (EGFR) induces apoptosis. J Biol Chem 287(1):712–722. doi:10.1074/jbc.M111.294470
10. Vanlandingham PA, Ceresa BP (2009) Rab7 regulates late endocytic trafficking downstream of multivesicular body biogenesis and cargo sequestration. J Biol Chem 284(18):12110–12124
11. Berkers JA, van Bergen en Henegouwen PM, Boonstra J (1991) Three classes of epidermal growth factor receptors in HeLa cells. J Biol Chem 266(2):922–927
12. Schroter CJ, Braun M, Englert J, Beck H, Schmid H, Kalbacher H (1999) A rapid method to separate endosomes from lysosomal contents using differential centrifugation and hypotonic lysis of lysosomes. J Immunol Methods 227(1–2):161–168
13. Ouyang X, Gulliford T, Huang G, Epstein RJ (1999) Transforming growth factor-alpha short-circuits downregulation of the epidermal growth factor receptor. J Cell Physiol 179(1):52–57. doi:10.1002/(SICI)1097-4652(199904)179:1<52::AID-JCP7>3.0.CO;2-M
14. Sigismund S, Woelk T, Puri C, Maspero E, Tacchetti C, Transidico P, DiFiore PP, Polo S (2005) Clathrin-independent endocytosis of ubiquinated cargos. Proc Natl Acad Sci U S A 102(8):2760–2765

Chapter 12

Analysis of Epidermal Growth Factor Receptor-Induced Cell Motility by Wound Healing Assay

Junfeng Tong and Zhixiang Wang

Abstract

Wound healing assays are well-defined and low-cost assays to study cell proliferation and migration rates of different cells and culture conditions as well as cell polarity, tissue matrix remodeling, and actin cytoskeletal structure regulation. The assay procedure generally involves growing a confluent cell monolayer and then creating a wound by scratching a line through the monolayer to destroy or displace certain cells. The open gap created by this wound is healed as cells move in and fill the damaged area. This wound healing process can take several hours to days depending on the cell type, culture conditions, and the width of the wound. The healing process is investigated microscopically over certain time intervals as the cells move into the open gap and close the wound.

Key words Cell motility, Cell migration, Wound healing assay, Scratch assay, Epidermal growth factor receptor (EGFR)

1 Introduction

Cell migration plays important roles in the processes that control morphogenesis, inflammation, and metastasis. The study of cell migration is an essential part of the cell biology, embryology, immunology, and neuroscience fields [1]. Cell motility is the ability of cells to translocate onto a solid substratum by consumption of energy. Cancer cells that overexpress the receptor tyrosine kinase ErbB2 family members (EGFR, ErbB2, Her3, and Her4) result in extensive metastatic progression [2–5]. The invasive behavior of cancer cells depends on their motility. Over the years, a variety of methods and techniques have been developed to analyze the cell migratory behavior, including wound healing/scratch assay, Boyden chamber assay, transwell invasion assay, time-lapse live imaging of cells, and chemotaxis analysis in three-dimensional environment [1, 6–8]. This protocol describes a simple assay to analyze cell motility using wound healing method. A wound is created by scratching a line in a confluent cell monolayer. In response to the

Zhixiang Wang (ed.), *ErbB Receptor Signaling: Methods and Protocols*, Methods in Molecular Biology, vol. 1652, DOI 10.1007/978-1-4939-7219-7_12, © Springer Science+Business Media LLC 2017

wound, cells polarize and initiate protrusion toward the wound, and migrate perpendicular to the wound edge to fill the wound area. Time-lapse photography is used to record the closure of the wound by moving cells. Although this protocol is written specifically for EGFR, the assay can be adapted to characterize other growth factor-induced cell migration and proliferation rates of cells or under different culture conditions including activation and inhibition.

2 Materials

Dulbecco's modified Eagle's medium with supplements (10% fetal bovine serum, antibiotics); mouse epidermal growth factor, phosphate-buffered saline (PBS); cell line of choice; tissue culture dishes (60 mm); P20 and P200 pipet tips; phase-contrast microscope or time-lapse microscope; imaging software for analyzing wound closure (Image J), fine tip Sharpie marker.

3 Methods

3.1 Seeding Cells

1. Seed COS-7 cells into two 60 mm dishes at 1.0×10^6 cells per dish in 3.0 mL of complete DMEM media. Spread evenly (*see* **Note 1**).
2. Incubate cells in a 37 °C incubator with 5% CO_2.

3.2 Prepare Cells for Scratch

1. Allow cells to be grown to just more than 90% confluent with no gaps between cells (*see* **Note 2**). This usually takes approximately 24 h.
2. On the following day, aspirate and discard the media and wash the cells once time with room temperature sterile PBS.
3. Add 3.0 mL of serum deprivation media (DMEM with 0.1% FBS) to the dishes and incubate overnight.

3.3 Scratch Cells to Generate Wound

1. On the following day, aspirate and discard the media and wash cells once with room temperature sterile PBS.
2. Draw a straight reference line with a fine tip marker on the bottom of the dish. Moving perpendicular to the line, scratch two separate wounds through the cell monolayer with a sterile 20 μL or 200 μL pipet tip. The width of wounds should be about 0.5 mm. Try to maintain consistent width and straight edges for all the experiments (*see* **Note 3**).
3. Rinse the cells gently with room temperature sterile PBS to wash off floating cells, cell debris, and make sure no sheets of cells lift off the dish.
4. Visualize cells under a light microscope, repeat wash if necessary.

3.4 Treat the Cells with EGF

1. After washing off cell debris and unattached cells, for control dish, add 3.0 mL of serum-free DMEM media containing antibiotics.
2. For EGF treated dish, add 3.0 mL of serum-free DMEM antibiotics containing media supplemented with 50 ng/mL mouse EGF.

3.5 Imaging

1. Place the dish under a phase-contrast microscope. Use the line on the dish bottom as a reference to orient the measurements. Select field of view just above and below each line and adjust the dish so that the line just appears at the edge of the field of view.
2. Alternatively, reference points at the bottom of dish can be made by a fine tip marker to help find the same field of view. The captured image should have both sides of wounds visible, which allows tracking cell migration on both edges. Avoid imaging fields with uneven wound edges or wound edge with accumulation of layers of scratched cells.
3. Take a photo using 10× magnification.
4. Incubate the dishes for 6 h, place the dish under a phase-contrast microscope, find the same field of view matching the reference line or reference point, and take another photo.
5. Repeat this procedure at 12 h, 18 h, and 24 h. The time frame required for incubation depends on the cell type, culture conditions and the width of the wound and should be determined empirically. Fast moving cells will need shorter time intervals. Keep same time intervals between different dishes when taking images.
6. After each observation, replace the old media with 3 mL of fresh serum-free DMEM antibiotics containing media supplemented with 50 ng/mL mouse EGF. If time-lapse imaging technique is performed for wound closure observation, it can generate results that are more reliable.

3.6 Data Analysis

1. The acquired images are used to calculate the migration rate of cells by using image analyzing software such as Image J.
2. Length of cell migration is the distance between one side of the wound and the other side. Cells Migration Rate = length of cell migration (μm)/migration time (h).
3. Alternatively, percentage closure (percentage of the surface area of the migrated cells into the defined wound area) can be used to indirectly measure the cells migration rate. Percent closure (%) = Migrated cell surface area/Total surface area ×100. Migrated cell surface area = length of cell migration (μm) × length ×2; Total surface area = width × length of the initial wound area.
4. It is recommended to repeat each sample three times to generate reproducible results (*see* **Notes 4** and **5**).

4 Notes

1. Because cell clumps will generate an uneven cell density, ensure that cells are dispersed individually.
2. It is important to obtain monolayers of equal confluency to start with between different dishes.
3. The time required to close the wound depends on the cell type and wound size. The larger the wound, the longer time it takes to close the wound. In order to get consistent results, it is important to scratch wounds with a constant width and straight edges.
4. Wound healing assays are simple and low-cost assays that have been used by researchers for years to study cell proliferation and migration rates of different cells and culture conditions as well as cell polarity, tissue matrix remodeling, and actin cytoskeletal structure regulation. It does not require specific chemoattractants, gradient chambers, and special equipment. The disadvantage of this assay is that it is difficult to maintain consistency in terms of the width, size and straight edge of the wound. It requires skills to adjust speed and angle of pipette tips to scratch wound with minimum variations. In addition, the scratch can cause damages to the cells at the edge of the wound, which may affect cell migration and other morphological features. The lack of a defined gap between cells will likely have effect on obtaining consistent results. To overcome this inconsistency, some commercial products such as CytoSelect™ Wound Healing Assay Kit from Cell Biolab, Inc. or Ibidi provide a proprietary treated inserts that can create a defined wound gaps. After cells form a monolayer around the insert, the insert is removed, leaving a wound gap between the cells with increased consistency and less cell damage.
5. The wound healing assay is generally conducted on non-transfected homogeneous cell monolayers. It is also suggested to measure transfected cells migration to study the effect of specific protein overexpression on cell migration [9]. In this case, the cells are observed under fluorescence microscope. It is recommended to use stable cell line for this purpose to avoid variations by various transfection efficiencies. As cell migration may be affected by other factors such as cell crowding and strength of cell–cell adhesions, it should be cautious to interpret individual transfected cells migration rate if there are not enough transfected cells in the leading edge of the wound or the overall transfection efficiency is low.

References

1. Vicente-Manzanares M, Horwitz AR (2011) Cell migration: an overview. Methods Mol Biol 769:1–24
2. Chen P, Xie H, Sekar MC, Gupta K, Wells A (1994) Epidermal growth factor receptor-mediated cell motility: phospholipase C activity is required, but mitogen-activated protein kinase activity is not sufficient for induced cell movement. J Cell Biol 127(3):847–857
3. Garcia R, Franklin RA, McCubrey JA (2006) EGF induces cell motility and multi-drug resistance gene expression in breast cancer cells. Cell Cycle 5(23):2820–2826
4. Kiuchi T, Ortiz-Zapater E, Monypenny J, Matthews DR, Nguyen LK et al (2014) The ErbB4 CYT2 variant protects EGFR from ligand-induced degradation to enhance cancer cell motility. Sci Signal 7(339):ra78
5. Ueda Y, Wang S, Dumont N, Yi JY, Koh Y et al (2004) Overexpression of HER2 (erbB2) in human breast epithelial cells unmasks transforming growth factor beta-induced cell motility. J Biol Chem 279(23):24505–24513
6. Falasca M, Raimondi C, Maffucci T (2011) Boyden chamber. Methods Mol Biol 769:87–95
7. Marshall J (2011) Transwell((R)) invasion assays. Methods Mol Biol 769:97–110
8. Sixt M, Lammermann T (2011) In vitro analysis of chemotactic leukocyte migration in 3D environments. Methods Mol Biol 769:149–165
9. Liang CC, Park AY, Guan JL (2007) In vitro scratch assay: a convenient and inexpensive method for analysis of cell migration in vitro. Nat Protoc 2(2):329–333

Part III

Recently Developed Methods in the Studies of ErbB Receptor Signaling

Chapter 13

Cell Cycle Synchronization of HeLa Cells to Assay EGFR Pathway Activation

Ping Wee and Zhixiang Wang

Abstract

Progression through the cell cycle causes changes in the cell's signaling pathways that can alter EGFR signal transduction. Here, we describe drug-derived protocols to synchronize HeLa cells in various phases of the cell cycle, including G1 phase, S phase, G2 phase, and mitosis, specifically in the mitotic stages of prometaphase, metaphase, and anaphase/telophase. The synchronization procedures are designed to allow synchronized cells to be treated for EGF and collected for the purpose of Western blotting for EGFR signal transduction components.

S phase synchronization is performed by thymidine block, G2 phase with roscovitine, prometaphase with nocodazole, metaphase with MG132, and anaphase/telophase with blebbistatin. G1 phase synchronization is performed by culturing synchronized mitotic cells obtained by mitotic shake-off. We also provide methods to validate the synchronization methods. For validation by Western blotting, we provide the temporal expression of various cell cycle markers that are used to check the quality of the synchronization. For validation of mitotic synchronization by microscopy, we provide a guide that describes the physical properties of each mitotic stage, using their cellular morphology and DNA appearance. For validation by flow cytometry, we describe the use of imaging flow cytometry to distinguish between the phases of the cell cycle, including between each stage of mitosis.

Key words Cell cycle, Synchronization, Interphase, Mitosis, Prometaphase, Metaphase, Anaphase, Telophase, EGFR, Epidermal growth factor receptor

1 Introduction

Cell cycle progression brings about tremendous changes within the cell, complete with different needs and requirements for the cell at that moment. At each step of the cell cycle, kinases including cyclin-dependent kinases (CDKs) phosphorylate their specific but wide-ranging set of substrates, leading not only to changes to their substrate but also to changes in their pathways. Mass spectrometry studies of the phosphoproteome of cells synchronized to different cell cycle phases reveal massive waves of cell cycle-dependent phosphorylations, especially during mitosis [1–4]. Whether one's protein of interest or one of their important downstream effectors is

Zhixiang Wang (ed.), *ErbB Receptor Signaling: Methods and Protocols*, Methods in Molecular Biology, vol. 1652, DOI 10.1007/978-1-4939-7219-7_13, © Springer Science+Business Media LLC 2017

affected by cell cycle progression are important considerations. For example, this cell cycle regulation of a protein can affect the interpretation of results when looking at a total cell population. It can also affect the efficacy of certain targeted therapies, as their intended target may respond differently depending on cell cycle phase. Importantly, a hallmark of cancer, including those with epidermal growth factor receptor (EGFR) overactivation, is their ability to sustain chronic proliferation, implying that cell cycle effects are more poignant on these continuously dividing cells.

The EGFR has been heavily studied for its role in affecting the cell cycle. A pulse of EGF stimulation during early G1 and another at late G1 have been shown to be sufficient for driving cells past the Restriction Point of the cell cycle, the point at which the cell must proceed through the cell division program [5, 6]. During mitosis, we and others have observed significant changes to EGFR-mediated signaling pathways [1, 7–9], to EGFR endocytosis kinetics [10], and to EGFR function [11, 12]. However, much work remains to be done in characterizing the EGFR function during the entire cell cycle.

Here, we describe drug-derived protocols for synchronizing HeLa cells in various phases of the cell cycle: G1 phase, S phase, G2 phase, and during mitosis, including prometaphase, metaphase, and anaphase/telophase (Fig. 1). The synchronization protocols are given for the purpose of harvesting them for assaying EGFR-mediated pathways by Western blot. The HeLa cell line is the most commonly used mammalian model system for cell cycle research. Its entire cell cycle lasts for approximately 24 h, with mitosis lasting for 40–60 min. A HeLa cell contains 1.4×10^5 EGFR molecules, a number that is close to many transformed and untransformed cells [13].

Synchronization in S phase is performed using the widely used double thymidine block. Synchronization in G2 is performed using the CDK inhibitor roscovitine after S phase synchronization

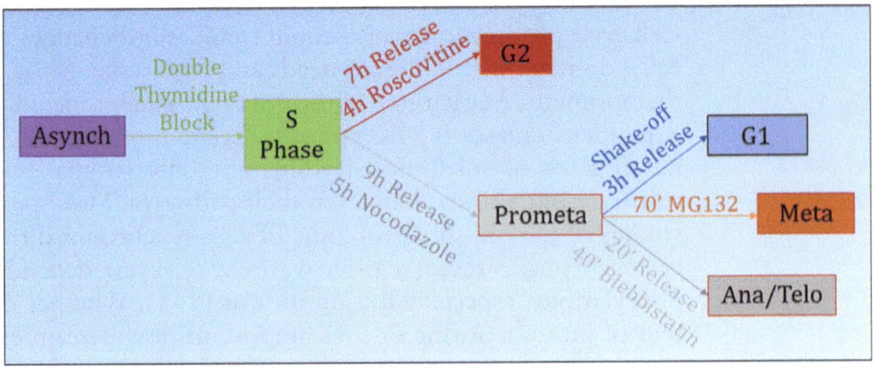

Fig. 1 Overview of cell synchronization protocols. Asynch = Asynchronous, Prometa = Prometaphase, Meta = Metaphase, Ana = Anaphase, Telo = Telophase

release. Mitotic synchronization relies on the use of the popular reversible microtubule depolymerizer nocodazole [14], which synchronizes cells in prometaphase. Metaphase cells are obtained by releasing cells from nocodazole block and treating with MG132, a proteasome inhibitor. Anaphase/telophase cells are obtained by prometaphase release followed by treatment with blebbistatin, a myosin II inhibitor [15]. Mitotic cells growing in cell culture can be separated from interphase cells using the mitotic shake-off method, which uses mechanical agitation to detach lowly adherent mitotic cells. Synchronization of HeLa cells in G1 is performed by collecting cells using the mitotic shake-off method, and replating them in fresh media for progression into G1.

It is necessary to obtain accompanying data for evaluating the purity and yield of the synchronization methods. Here, we include methods for validating cell synchronization methods. We provide the expression of cell cycle-specific markers that can be assayed by Western blotting. We also provide a definitive guide to distinguish between each stage of mitosis using the DNA and cell morphology characteristics of HeLa cells under microscopy. Lastly, we provide a protocol for analyzing cell cycle synchrony by flow cytometric measurement of DNA content.

EGFR activation by its ligand EGF can be performed at various concentrations. Low physiological EGF concentrations are considered to be 1–20 ng/mL, whereas high EGF concentrations are >20 ng/mL. Generally, 50 ng/mL EGF is used in our lab for high concentration treatments, using treatment times of 5 min, 15 min, or 30 min to strongly stimulate EGFR activation and to strongly activate the ERK and AKT pathways, as observed by Western blotting using phospho-specific antibodies. A list of the major proteins and their phosphorylation sites resulting from EGF stimulation of EGFR is included in Table 1.

Table 1
For Western blotting analysis of EGFR and EGFR-mediated signaling pathways, we assay for the following proteins and the following sites of phosphorylated proteins

Protein	Reference
EGFR/p-EGFR Y992, Y1045, Y1068 Y1086, Y1148, Y1173	[41]
Raf-1/p-Raf-1 S259, Y340, Y341	[42, 43]
MEK/p-MEK S218–220	[44]
ERK1/2/p-ERK1/2 T202/Y204	[45]
AKT/p-AKT T304, S473	[46]
S6 K/p-p70 S6 Kinase T389	[47, 48]
4E–BP1/p-4E-BP1 S65	[49]

2 Materials

2.1 Solutions, Reagents, and Chemicals

Cell lysis buffer A: 50 mM Tris–HCl (pH 7.5), 0.25% CHAPS. Add fresh: Protease Inhibitor Mix (0.03%), Na_3VO_4 (0.01%) (inhibits protein phosphotyrosyl phosphatases).

Cell lysis buffer B: 500 mM Tris–HCl (pH 7.5), 10% Nonidet P-40.

DAPI (4′,6-diamidine-2′-phenylindole dihydrochloride) stock: 30 μM in PBS. Working concentration: 300 nM.

DRAQ5 (Biostatus): 0.5 mM. Working concentration: 5 mM.

Flow Cytometry Blocking Buffer: TBS, 0.01% Triton X-100, 4% BSA.

Growth medium: Dulbecco's modified Eagle's medium (DMEM) (Sigma) in 10% FBS (Hyclone) and 1× Antibiotic–Antimycotic Solution (Sigma).

PBS: 1× Phosphate Buffered Saline sterilized by autoclaving.

Protease Inhibitor Mix: 0.1 mM AEBSF, 10 μg/mL aprotinin, 1 μM pepstatin in 100% EtOH.

Mounting Medium: *n*-propyl gallate in 10 mL boiled 50% glycerol.

SDS Loading Buffer (4×): 200 mM Tris–HCl (pH 6.8), 400 mM DTT, 8% SDS, 0.4% Bromophenol Blue, 40% glycerol.

TBS: Tris buffered saline sterilized by autoclaving.

Trypsin: 0.25% trypsin, 0.03% EDTA (ethylenediaminetetraacetic acid).

2.2 Drug Stock and Working Concentrations

Epidermal Growth Factor stock (Upstate): 50 μg/mL in DMSO. Store at −20 °C. Working concentration: 50 ng/mL.

Thymidine stock (Sigma): 200 mM in PBS. Dissolve completely and filter-sterilize. Store at 4 °C. Working concentration: 2 mM.

Roscovitine stock (Sigma): 100 mM roscovitine in DMSO. Store at −20 °C. Working concentration: 50 μM.

Nocodazole stock (Sigma): 1 mg/mL in DMSO. Store at −20 °C. Working concentration: 20 ng/mL.

MG132 stock (Calbiochem): 25 mM in DMSO. Store at −20 °C. Working concentration: 25 μM.

(S)-(−)-Blebbistatin stock (Sigma): 5 mM in DMSO. Store at −20 °C. Working concentration: 50 μM.

2.3 Cell Culture

Grow HeLa cells in cell culture growth medium in a humidifier incubator at 5% CO_2 and at 37 °C.

2.4 Equipment

Standard tissue culture equipment.

Cell culture dishes: 100-mm, 150-mm.

Cell culture plate: 24-well plate.

Centrifuge for 15 mL centrifuge tubes that can reach 1000 × *g* at 4 °C.

Fluorescence microscope.

Imaging flow cytometer equipped with 647 nm laser.

Microcentrifuge that can reach 14,000 × *g* at 4 °C.

3 Methods

3.1 G1 Synchronization

G1 synchronization is performed by obtaining nocodazole-arrested prometaphase cells by mitotic shake-off, and releasing them back into the cell cycle (*see* **Note 1**). Cells will adhere in 1–3 h and proceed toward G1.

1. Perform prometaphase synchronization with nocodazole in 150-mm dish (Subheading 3.4).
2. Aspirate media and add 6 mL of DMEM.
3. Perform mitotic shake-off. Knock dish against the sidewall of lab bench for 5 min. Care should be taken to keep consistency in the force used to shake the plates between different samples.
4. Transfer media containing detached mitotic cells to 15 mL centrifuge tube.
5. Centrifuge cells at 1000 × *g* for 5 min.
6. Aspirate media from centrifuge tube. Add 1 mL growth media to cells and resuspend well.
7. Add cells to 100-mm plate and add 4 mL growth media.
8. Put culture dish in incubator and incubate for 1–3 h.

3.2 S Phase Synchronization: Double Thymidine Block

The double thymidine block is a highly effective and widely used protocol for synchronization of cells in early S phase (*see* **Note 2**). Excess thymidine inhibits the formation of dCTP, an essential precursor of DNA, and thus halts DNA replication [16]. Importantly, it has been reported that excess thymidine does not block cells at the G1/S border, but that it instead considerably slows down the rate of DNA synthesis and the progression through S phase [17]. Care should be taken to keep thymidine incubation times consistent.

1. Seed 2.0×10^6 HeLa cells in 100-mm plates with 8 mL cell culture medium.
2. Grow HeLa cells overnight to ~40% confluency.
3. Aspirate the medium and add 5 mL growth media with 500 μL 200 mM thymidine (a final concentration of 2 mM).
4. Incubate for 16 h.
5. Aspirate the medium and release cells from S-phase block by washing cells with 15 mL PBS three times.
6. Add 8 mL growth medium.

7. Incubate for 8 h.
8. Aspirate the medium and add 5 mL growth media with 500 μL 200 mM thymidine (a final concentration of 2 μM).
9. Incubate for 16 h.

3.3 Late G2 Synchronization: Roscovitine

Roscovitine inhibits CDKs by competing with ATP at the ATP binding sites of various CDKs, including CDK1, CDK2, CDK5, and CDK7 [18]. Here, we release cells from the S phase block, allow them to grow into G2, and use roscovitine to prevent the CDK1/cyclin B1 complex (also known as maturation-promoting factor) from activating, which prevents cells from entering mitosis (*see* **Note 3**).

1. Perform S-phase synchronization by double thymidine block (*see* **Note 4**).
2. Release cells from S-phase block by washing cells with 15 mL PBS three times.
3. Add 8 mL growth medium.
4. Incubate for 7 h.
5. Aspirate and add 5 mL of warm DMEM with 2.5 μL 100 mM roscovitine (a final concentration of 50 μM).
6. Incubate for 4 h.

3.4 Prometaphase Synchronization: Nocodazole

The microtubule depolymerizing agent nocodazole may be the most commonly used agent for inducing mitotic prometaphase arrest. It has a high affinity to tubulin, and prevents tubulin-composed spindle microtubules from interacting properly with the kinetochores of chromosomes [19]. These chromosomes are therefore not brought to metaphase plate, and cannot proceed past the spindle assembly checkpoint (SAC) [14] (*see* **Note 5**). This protocol for mitotic synchronization is performed in 150-mm culture dishes, and is performed with two dishes at once to match the yield of interphase cells collected from one 100-mm dish. A cell confluency of ~70% is ideal for mitotic shake-off, as mitotic cells shake off more easily at lower confluences.

1. Seed 5.0×10^6 HeLa cells in each 150-mm plate with 20 mL cell culture medium.
2. Grow HeLa cells overnight to ~40% confluency.
3. Aspirate the medium and add 10 mL growth media with 100 μL 200 mM thymidine (a final concentration of 2 mM).
4. Incubate for 16 h.
5. Release cells from S-phase block by aspirating the medium and washing cells with 30 mL PBS three times.
6. Add 10 mL growth medium.
7. Incubate for 8 h.

8. Repeat thymidine block and release (**steps 3–5**; *see* **Note 4**).
9. Add 10 mL growth medium.
10. Incubate for 9 h.
11. Aspirate and add 10 mL warm DMEM with 2 μL of 100 μg/mL nocodazole (a final concentration of 20 ng/mL).
12. Incubate for 5 h.

3.5 Metaphase Synchronization: MG132

The spindle assembly checkpoint (SAC) ensures all chromosomes are properly aligned prior to commencing anaphase [20]. Upon satisfying the requirements of the SAC, the anaphase-promoting complex (APC) is turned on, and initiates the proteasomal degradation of cyclin B and securin [21–23]. This proteasomal activity allows sister chromatids to separate and transition from metaphase to anaphase. To synchronize cells in metaphase, the proteasome inhibitor MG132 is added to cells in prometaphase, allowing chromosomes to proceed past the SAC, but preventing cells from proceeding into anaphase [2].

1. Perform prometaphase synchronization.
2. Release cells from prometaphase arrest by washing cells with 30 mL PBS three times.
3. Add 10 mL warm DMEM with 10 μL of 25 mM MG132 (a final concentration of 25 μM) to each dish.
4. Incubate for 70 min.

3.6 Anaphase/Telophase Synchronization: Blebbistatin

Anaphase and telophase are the most difficult phases to synchronize due in part to their short duration (20–30 min). There are no drugs known to completely block cells in anaphase or telophase. However, the myosin II inhibitor blebbistatin may be used to extend the duration of anaphase and telophase [15] (*see* **Note 6**).

1. Perform prometaphase synchronization.
2. Release cells from prometaphase arrest by washing cells with 30 mL PBS three times.
3. Add 10 mL warm DMEM to each dish.
4. Incubate for 20 min.
5. Add 100 μL 5 mM blebbistatin (a final concentration of 50 μM).
6. Incubate for 40–55 min.

3.7 Interphase Cell Harvest After EGF Stimulation

Here, we describe the method for harvesting cells synchronized in G1 phase (Subheading 3.1), S phase (Subheading 3.2), and G2 phase (Subheading 3.3) for the purpose of Western blot analysis.

1. Add 5 μL 50 μg/mL EGF (a final concentration of 50 ng/mL) to each 100-mm dish containing 5 mL of media for required time.

2. Aspirate and wash cells with PBS. Aspirate as much PBS as possible.
3. Add 500 μL Cell Lysis Buffer A.
4. Put dish on ice and scrape cells using cell scraper.
5. Transfer scraped cells to microcentrifuge tube and incubate on ice for 5 min from the addition of Cell Lysis Buffer A.
6. Add 60 μL Cell Lysis Buffer B.
7. Incubate for 15 min on rotator at 4 °C.
8. Centrifuge at 14,000 × *g* at 4 °C for 15 min.
9. Collect supernatant immediately and store at −80 °C until ready for use.

3.8 Mitotic Shake-Off After EGF Stimulation

Here, we describe the mitotic shake-off method used to isolate cells synchronized in prometaphase (Subheading 3.4), metaphase (Subheading 3.5), and anaphase/telophase (Subheading 3.6) for the purpose of Western blot analysis. We describe the protocol as performed with two 150-mm dishes, as the yield of mitotic cells from two 150-mm dishes is roughly equivalent to the yield of interphase cells from one 100-mm dish, as measured by protein concentration.

1. Add 10 μL of 50 μg/mL EGF (a final concentration of 50 ng/mL) to each 150-mm dish containing 10 mL of media for required time (*see* **Note 7**).
2. Aspirate media and add 6 mL cold DMEM to each dish.
3. Perform mitotic shake-off. Place dishes on ice and knock the dishes to the sidewall of ice bucket for 5 min (*see* **Note 8**). Care should be taken to keep consistency in the force used to shake the plates between different samples.
4. Transfer cells to 15 mL centrifuge tube.
5. Centrifuge cells at 1000 × *g* for 5 min.
6. Add 6 mL ice-cold PBS to each dish and shake gently until centrifugation is complete.
7. Aspirate media from centrifuge tube. Add remaining cells from dishes to centrifuge tube.
8. Centrifuge cells at 1000 × *g* for 5 min.
9. Aspirate media from centrifuge tube.
10. Add 500 μL Cell Lysis Buffer A to cells. Transfer cells to microcentrifuge tube and incubate for 5 min on ice.
11. Add 60 μL Cell Lysis Buffer B.
12. Incubate for 15 min on rotator at 4 °C.

13. Centrifuge at 14,000 × *g* at 4 °C for 15 min.
14. Collect supernatant immediately and store at −80 °C until ready for use.

3.9 Validation of Synchronization: Western Blotting Markers

For Western blots, probing with cell cycle markers is necessary to provide confidence that samples are in the cell cycle stage they are claimed to be. Table 2 shows the expression of proteins throughout the cell cycle of HeLa cells that can be used as markers of synchronization. A popular family of cell cycle markers includes the cyclin proteins, since specific cyclins are produced and degraded at specific phases of the cell cycle to drive cell cycle progression [24, 25]. G1 cells will be negative for p-Histone 3 Ser10, Cyclin B1. Early G1 cells will be negative for Cyclin D, but late G1 cells will have maximal levels of Cyclin D (*see* **Note 9**). S phase cells will have the highest level of Cyclin E [26]. G2 phase cells will have the highest level of Cyclin A [26–28]. Mitotic cells are highly positive for p-Histone 3 Ser10. Prometaphase cells can be distinguished from metaphase cells by their positive staining for p-BubR1 Ser676 [29]. Anaphase/telophase cells can be distinguished by their positive p-Histone-3 Ser 10 staining, but negative Securin, Cyclin B1, and Cyclin A staining [29, 30].

1. Harvest cells as described in Subheadings 3.7 and 3.8.
2. Measure the protein concentration of the lysates.
3. Dilute protein sample with one part SDS Loading Buffer (4×) to three parts solubilized protein.
4. Vortex samples and denature proteins by boiling samples at 95 °C for 5 min.
5. Load samples for SDS-PAGE electrophoresis and perform Western blotting. Table 2 lays out the expression of cell cycle markers that can be used to validate the synchronization methods.

Table 2
Expression of proteins throughout the cell cycle in HeLa cells

	Cyclin D	Cyclin E	Cyclin A	Cyclin B	p-Histone 3 Ser10	Securin	p-BubR1 Ser676
G0–Early G1	–	–	–	–	–	–	–
Late G1	++	+	–	–	–	+	–
Early S-phase	–	++	+	–	–	+	–
Late G2	–	–	++	++	–	+	–
Prometaphase	–	–	++	++	++	++	–
Metaphase	–	–	+	++	++	++	++
Anaphase/Telophase	–	–	–	–	++	–	–

3.10 Validation of Synchronization: Microscopy

The synchronization treatments described earlier can be done concurrently with cells in tissue cultures dishes and cells seeded in 24-well plates. The cells in the 24-well plates can be fixed and the synchronization efficiency of mitotic cells can be determined by indirect fluorescence microscopy. The stages of mitosis can be easily distinguished by their DNA and cellular morphology under fluorescence microscopy. Although the stages of mitosis are well known, the observable physical demarcations that define each stage are less well described. Here, we describe the physical properties of each mitotic stage that distinguishes them when viewed by indirect fluorescence microscopy in combination with differential interference contrast (DIC) microscopy (*see* **Note 10**). A simple protocol for staining DNA of cells in 24-well plates is also provided.

1. Seed 5×10^4 HeLa cells on each coverslip in 24-well plate with 0.5 mL growth medium.
2. Perform cell synchronization method of choice. Adapt protocols from 150-mm or 100-mm dishes to 24-well plate.
3. Wash cells once with PBS. Aspirate PBS.
4. Fix cells by adding 0.5 mL ice-cold MeOH. Fix for 20 min at −20 °C.
5. Wash cells once with PBS. Aspirate PBS.
6. Add 500 μL PBS with 1:100 DAPI. Incubate for 10 min on ice, protected from light.
7. Aspirate and wash twice with PBS.
8. Mount cells onto glass slide with 2.8 μL mounting medium.
9. Seal edges with nail polish.
10. Visualize cells by indirect immunofluorescence for DAPI and by DIC for cell morphology (*see* **Note 11**). Count the number of cells in interphase, prophase, prometaphase, metaphase, anaphase, telophase, and any morphologically abnormal cells.

The following is a guide we use to identifying each mitotic stage, as well as the sub-stages often found in the literature. Prophase is the period when the chromosomes begin to condense, but a nuclear shape is still largely maintained since the nuclear envelope is still intact. Prometaphase is a long period of transition between prophase and metaphase, and begins after the nuclear envelope has broken down [31] and the nucleolus has disappeared [32]. During early prometaphase, the chromosomes appear disorganized and the cell is not yet round. During mid prometaphase, the cell begins to round. In late prometaphase, cells are rounded and proceeds until all sister chromatids are attached to spindle microtubules in a bioriented fashion at the metaphase plate, at which point cells are in metaphase. Early anaphase follows, where the cell is still round but the chromatids have started separating. Late anaphase sees the cell and sister chromatids further pulled apart, until the cell membrane begins to pinch slightly and the chromosomes almost reach their respective poles. Early telophase begins when the chromosomes have been pulled to opposite poles of the cell and when the ingression furrow has constricted significantly. Mid telophase is when the cell has essentially fully constricted and is awaiting cytokinetic abscission. At late telophase, the chromosomes despiralize, and the nuclear envelop has reformed. Here, the daughter cells also lose the characteristic round mitotic shape. Cytokinesis occurs when the contractile ring begins constricting the plasma membrane. An intracellular bridge remains, where the process of abscission completely separates the cells [33]. Cytokinesis occurs after late anaphase until late telophase (*see* **Note 12**).

3.11 Validation of Synchronization: Flow Cytometry

Flow cytometry analysis can be used to measure DNA content. G0–G1 cells have 2N DNA, S-phase cells have 2–4N DNA, and G2–M cells have 4N DNA (*see* **Note 13**). We use imaging flow cytometry to further distinguish between each stage of mitosis.

1a. For interphase cells, after completing cell cycle arrest, aspirate media and wash with PBS once. Add 0.8 mL of trypsin and incubate in incubator for 2.5 min. Add 4.2 mL PBS and detach cells by spraying cells off the plate. Collect cells in a precooled 15 mL tube. Centrifuge cells at 1000 × *g* for 5 min. Aspirate liquid and resuspend cells well in 0.5 mL PBS. Transfer cells to 4.5 mL of ice-cold 100% MeOH.

1b. For mitotic cells, complete mitotic shake-off (**steps 3–9** of Subheading 3.8). Resuspend cells well in 0.5 mL PBS. Transfer cells to 4.5 mL of ice-cold 100% MeOH.

2. Fix cells for 20 min at −20 °C.
3. Centrifuge cells at 1000 × *g* for 5 min. Aspirate liquid and resuspend cells in 1 mL Flow Cytometry Blocking Buffer. Incubate for 10 min on ice.
4. Centrifuge cells at 1000 × *g* for 5 min. Aspirate liquid and resuspend cells in 0.5 mL Flow Cytometry Blocking Buffer with p-Histone-3 Ser10 primary antibody. Incubate for 1 h on ice.
5. Centrifuge cells at 1000 × *g* for 5 min. Aspirate liquid and wash cells with 1 mL ice-cold PBS. Repeat washing step.
6. Centrifuge cells at 1000 × *g* for 5 min. Aspirate liquid and resuspend cells in 0.5 mL Flow Cytometry Blocking Buffer with proper fluorescent-conjugated secondary antibody. Incubate for 1 h on ice away from light.
7. Wash cells twice (**step 5**).

8. Centrifuge cells at 1000 × *g* for 5 min. Aspirate liquid and resuspend cells well in 200 μL PBS with 2 μL 0.5 mM DRAQ5 (a final concentration of 5 μM). Add cells to flow cytometry tube and incubate for 15 min at room temperature while protecting from light.
9. Run samples by imaging flow cytometry analysis. We use 647 nm for excitation (max = 642 nm) and 642–745 for emission. We obtain 100,000 cellular events per sample excluding doublets to achieve statistical significance.
10. Use DNA content to distinguish between 2N, 2–4N, and 4N, for G0–G1, S-phase, and G2–M respectively. Gate for p-Histone-3 staining to obtain the proportion of cells in mitosis. We distinguish between mitotic stages by tagging cells as their respective stage (*see* **Note 14**).

4 Notes

1. Gl synchronization can also be performed with lovastatin (20 μM) for early Gl [34], or mimosine (400 μM) [35], actinomycin D (0.1 μg/mL), or cycloheximide (10 μg/mL) [36] for late Gl. Other cell types may be synchronized to G0 by serum starvation or contact inhibition. HeLa cells do not properly undergo G0 synchronization by serum starvation, due to their p53 inactivation.
2. S phase synchronization can also be performed using DNA replication inhibitors, such as the DNA polymerase inhibitor aphidocolin (5 μg/mL) [37] and hydroxyurea (1.0–2.5 mM) [38].
3. G2 synchronization has also been performed with the reversible CDK inhibitor RO-3306 (9 μM), which specifically inhibits CDKl/Cyclin Bl activity [2, 39].
4. For G2 and mitotic synchronization, a prior single thymidine block may be sufficient, rather than a double thymidine block. This may also help minimize potential side effects.
5. Since it destabilizes spindle microtubules, nocodazole treatment leads to many unattached kinetochores, and thus synchronizes cells in early prometaphase [19]. Taxol treatment however stabilizes spindle microtubules, so that almost all kinetochores can become attached, yet the spindle microtubules are unable to move the chromosomes to the metaphase plate. Therefore, taxol treatment arrests cells in late prometaphase. Mitotic synchronization can also be achieved using vinca alkaloids, including colchicine and colcemid.
6. Blebbistatin inhibits the ingression of the cleavage furrow. The anaphase/telophase synchronization protocol can be further

tweaked to increase the number of anaphase cells or of telophase cells by increasing or decreasing the time of blebbistatin treatment.

7. We perform EGF stimulation prior to cell detachment from the dish in order to minimize potential changes in the cell's response to EGF. EGF stimulation of mitotic cells should be done with minimal shaking of plates in order to obtain high yields. Especially when dealing with mitotic cells on culture dishes, always pipette solutions to the side walls of the dishes.
8. Mitotic shake-off is performed on ice in order to slow intracellular signaling. In order to make a smoother shaking surface, we compress the ice down.
9. Cyclin D1 expression can vary highly in other tumor cell lines, especially if they have perturbations in the Ras or Rb pathways.
10. DIC may be replaced with any phase contrast microscope, or any fluorescent stain (e.g., phalloidin-rhodamine) that makes cell morphology distinguishable.
11. Validation of synchronization may also be performed by live imaging fluorescence microscopy using HeLa H2B-GFP cells. However, the user must be certain that the conditions in the microscope's growth chamber completely reflect the conditions of the incubator.
12. Other markers may be used to better differentiate between mitotic stages. Prophase may be more easily visualized by translocation of CDK1/Cyclin B1 to the nucleus. Tubulin staining can help differentiate early anaphase cells from prometaphase cells.
13. If using HeLa H2B-GFP cells, it is important to note that flow cytometric DNA content quantification cannot be performed by quantifying GFP, as H2B-GFP levels do not reflect DNA content.
14. We use the Amnis ImageStream imaging flow cytometer with the IDEAS software. We have also used the protocol by Filby et al. [40] in order to automate the subdivision of mitotic stages, in prometaphase, metaphase, anaphase, or telophase.

References

1. Dephoure N, Zhou C, Villén J et al (2008) A quantitative atlas of mitotic phosphorylation. Proc Natl Acad Sci U S A 105:10762–10767. doi:10.1073/pnas.0805139105
2. Dulla K, Daub H, Hornberger R et al (2010) Quantitative site-specific phosphorylation dynamics of human protein kinases during mitotic progression. Mol Cell Proteom 9:1167–1181. doi:10.1074/mcp.M900335-MCP200
3. Holt LJ, Tuch BB, Villén J et al (2009) Global analysis of Cdk1 substrate phosphorylation sites provides insights into evolution. Science 325:1682–1686. doi:10.1126/science.1172867
4. Olsen JV, Vermeulen M, Santamaria A et al (2010) Quantitative phosphoproteomics reveals widespread full phosphorylation site occupancy during mitosis. Sci Signal 3:ra3. doi:10.1126/scisignal.2000475

5. Jones SM, Kazlauskas A (2001) Growth-factor-dependent mitogenesis requires two distinct phases of signalling. Nat Cell Biol 3:165–172. doi:10.1038/35055073
6. Pennock S, Wang ZX (2003) Stimulation of cell proliferation by endosomal epidermal growth factor receptor as revealed through two distinct phases of signaling. Mol Cell Biol 23:5803–5815. doi:10.1128/Mcb.23.16.5803-5815.2003
7. Wee P, Shi H, Jiang J et al (2015) EGF stimulates the activation of EGF receptors and the selective activation of major signaling pathways during mitosis. Cell Signal 27:638–651. doi:10.1016/j.cellsig.2014.11.030
8. Dangi S, Shapiro P (2005) Cdc2-mediated inhibition of epidermal growth factor activation of the extracellular signal-regulated kinase pathway during mitosis. J Biol Chem 280:24524–24531. doi:10.1074/jbc.M414079200
9. Kiyokawa N (1997) Mitosis-specific negative regulation of epidermal growth factor receptor, triggered by a decrease in ligand binding and dimerization, can be overcome by overexpression of receptor. J Biol Chem 272:18656–18665. doi:10.1074/jbc.272.30.18656
10. Liu L, Shi H, Chen X, Wang Z (2011) Regulation of EGF-stimulated EGF receptor endocytosis during M phase. Traffic 12:201–217. doi:10.1111/j.1600-0854.2010.01141.x
11. Mardin B, Isokane M, Cosenza M et al (2013) EGF-induced centrosome separation promotes mitotic progression and cell survival. Dev Cell 25:229–240. doi:10.1016/j.devcel.2013.03.012
12. Astuti P, Pike T, Widberg C et al (2009) MAPK pathway activation delays G2/M progression by destabilizing Cdc25B. J Biol Chem 284:33781–33788. doi:10.1074/jbc.M109.027516
13. Klein S, Kaszkin M, Barth H, Kinzel V (1997) Signal transduction through epidermal growth factor receptor is altered in HeLa monolayer cells during mitosis. Biochem J 322(Pt 3):937–946
14. Zeive G, Turnbull D, Mullins J, McIntosh J (1980) Production of large numbers of mitotic mammalian cells by use of the reversible microtubule inhibitor nocodazole. Exp Cell Res 126:397–405
15. Matsui Y, Nakayama Y, Okamoto M et al (2012) Enrichment of cell populations in metaphase, anaphase, and telophase by synchronization using nocodazole and blebbistatin: a novel method suitable for examining dynamic changes in proteins during mitotic progression. Eur J Cell Biol 91:413–419. doi:10.1016/j.ejcb.2011.12.008
16. Bjursell G, Reichard P (1973) Effects of thymidine on deoxyribonucleoside triphosphate pools and deoxyribonucleic acid synthesis in Chinese hamster ovary cells. J Biol Chem 248:3904–3909
17. Bostock CJ, Prescott DM, Kirkpatrick JB (1971) An evaluation of the double thymidine block for synchronizing mammalian cells at the G1-S border. Exp Cell Res 68:163–168. doi:10.1016/0014-4827(71)90599-4
18. Whittaker SR, Walton MI, Garrett MD, Workman P (2004) The Cyclin-dependent kinase inhibitor CYC202 (R-Roscovitine) inhibits retinoblastoma protein phosphorylation, causes loss of Cyclin D1, and activates the mitogen-activated protein kinase pathway. Cancer Res 64:262–272. doi:10.1158/0008-5472.CAN-03-0110
19. Lara-Gonzalez P, Westhorpe FG, Taylor SS (2012) The spindle assembly checkpoint. Curr Biol 22:R966–R980. doi:10.1016/j.cub.2012.10.006
20. Li R, Murray AW (1991) Feedback control of mitosis in budding yeast. Cell 66:519–531. doi:10.1016/0092-8674(81)90015-5
21. Holloway SL, Glotzer M, King RW, Murray AW (1993) Anaphase is initiated by proteolysis rather than by the inactivation of maturation-promoting factor. Cell 73:1393–1402. doi:10.1016/0092-8674(93)90364-V
22. Hornig NCD, Knowles PP, McDonald NQ, Uhlmann F (2002) The dual mechanism of separase regulation by securin. Curr Biol 12:973–982. doi:10.1016/S0960-9822(02)00847-3
23. Waizenegger IC, Giménez-Abián JF, Wernic D, Peters JM (2002) Regulation of human separase by securin binding and autocleavage. Curr Biol 12:1368–1378. doi:10.1016/S0960-9822(02)01073-4
24. Darzynkiewicz Z, Gong J, Juan G et al (1996) Cytometry of cyclin proteins. Cytometry 25:1–13. doi:10.1002/(SICI)1097-0320(19960901)25:1<1::AID-CYTO1>3.0.CO;2-N
25. Pozarowski P, Darzynkiewicz Z (2004) Analysis of cell cycle by flow cytometry. Methods Mol Biol 281:301–311. doi:10.1385/1-59259-811-0:301
26. Dulić V, Lees E, Reed SI (1992) Association of human cyclin E with a periodic G1-S phase protein kinase. Science 257:1958–1961. (80-). doi:10.1126/science.1329201

27. Kraft C, Herzog F, Gieffers C et al (2003) Mitotic regulation of the human anaphase-promoting complex by phosphorylation. EMBO J 22:6598–6609. doi:10.1093/emboj/cdg627
28. Pines J, Hunter T (1990) Human cyclin a is adenovirus E1A-associated protein p60 and behaves differently from cyclin B. Nature 346:760–763. doi:10.1038/346760a0
29. Elowe S, Hümmer S, Uldschmid A et al (2007) Tension-sensitive Plk1 phosphorylation on BubR1 regulates the stability of kinetochore-microtubule interactions. Genes Dev 21:2205–2219. doi:10.1101/gad.436007
30. Widrow RJ, Laird CD (2000) Enrichment for submitotic cell populations using flow cytometry. Cytometry 39:126–130. doi:10.1002/(SICI)1097-0320(20000201)39:2<126::AID-CYTO5>3.0.CO;2-V
31. Güttinger S, Laurell E, Kutay U (2009) Orchestrating nuclear envelope disassembly and reassembly during mitosis. Nat Rev Mol Cell Biol 10:178–191. doi:10.1038/nrm2641
32. Gavet O, Pines J (2010) Progressive activation of CyclinB1-Cdk1 coordinates entry to mitosis. Dev Cell 18:533–543. doi:10.1016/j.devcel.2010.02.013
33. Eggert US, Mitchison TJ, Field CM (2006) Animal cytokinesis: from parts list to mechanisms. Annu Rev Biochem 75:543–566. doi:10.1146/annurev.biochem.74.082803.133425
34. Keyomarsi K, Sandoval L, Band V, Pardee AB (1991) Synchronization of tumor and normal cells from G1 to multiple cell cycles by lovastatin. Cancer Res 51:3602–3609
35. Park SY, Im JS, Park SR et al (2012) Mimosine arrests the cell cycle prior to the onset of DNA replication by preventing the binding of human Ctf4/and-1 to chromatin via Hif-1?? Activation in HeLa cells. Cell Cycle 11:761–766. doi:10.4161/cc.11.4.19209
36. Kim JH, Gelbard AS, Perez AG (1968) Inhibition of DNA synthesis by actinomycin D and cycloheximide in synchronized HeLa cells. Exp Cell Res 53:478–487. doi:10.1016/0014-4827(68)90317-0
37. Pedrali-Noy G, Spadari S, Miller-Faurès A et al (1980) Synchronization of HeLa cell cultures by inhibition of DNA polymerase alpha with aphidicolin. Nucleic Acids Res 8:377–387
38. Pfeiffer SE, Tolmach LJ (1967) Inhibition of DNA synthesis in HeLa cells by Hydroxyurea. Cancer Res 27:124–129
39. Vassilev LT, Tovar C, Chen S et al (2006) Selective small-molecule inhibitor reveals critical mitotic functions of human CDK1. Proc Natl Acad Sci U S A 103:10660–10665. doi:10.1073/pnas.0600447103
40. Filby A, Perucha E, Summers H et al (2011) An imaging flow cytometric method for measuring cell division history and molecular symmetry during mitosis. Cytometry A 79:496–506. doi:10.1002/cyto.a.21091
41. Wu P, Wee P, Jiang J et al (2012) Differential regulation of transcription factors by location-specific EGF receptor signaling via a spatio-temporal interplay of ERK activation. PLoS One 7:e41354. doi:10.1371/journal.pone.0041354
42. Marais R, Light Y, Paterson HF, Marshall CJ (1995) Ras recruits Raf-1 to the plasma membrane for activation by tyrosine phosphorylation. EMBO J 14:3136–3145
43. Dhillon AS, Kolch W (2002) Untying the regulation of the Raf-1 kinase. Arch Biochem Biophys 404:3–9
44. Zheng CF, Guan KL (1994) Activation of MEK family kinases requires phosphorylation of two conserved Ser/Thr residues. EMBO J 13:1123–1131
45. Crews CM, Alessandrini A, Erikson RL (1992) The primary structure of MEK, a protein kinase that phosphorylates the ERK gene product. Science 258:478–480. doi:10.1126/science.1411546
46. Woodgett JR (2005) Recent advances in the protein kinase B signaling pathway. Curr Opin Cell Biol 17:150–157. doi:10.1016/j.ceb.2005.02.010
47. Pullen N, Thomas G (1997) The modular phosphorylation and activation of p70(s6k). FEBS Lett 410(1):78–82
48. Romanelli A, Dreisbach VC, Blenis J (2002) Characterization of phosphatidylinositol 3-kinase-dependent phosphorylation of the hydrophobic motif site Thr389 in p70 S6 kinase 1. J Biol Chem 277:40281–40289. doi:10.1074/jbc.M205168200
49. Galbaugh T, Cerrito MG, Jose CC, Cutler ML (2006) EGF-induced activation of Akt results in mTOR-dependent p70S6 kinase phosphorylation and inhibition of HC11 cell lactogenic differentiation. BMC Cell Biol 7:34. doi:10.1186/1471-2121-7-34

Chapter 14

Analysis of Constitutive EGFR Signaling Regulating IRF3 Transcriptional Activity in Cancer Cells

Gao Guo, Ke Gong, and Amyn A. Habib

Abstract

Epidermal growth factor receptor (EGFR) plays an important role in various types of human cancers. Overexpression of EGFR leads to a constitutive tyrosine phosphorylation of multiple tyrosine residues in the EGFR. Recently, we have demonstrated that overexpressed EGFR oscillates between two distinct and mutually exclusive modes of signaling depending on the presence or absence of ligand. EGFR constitutively activates transcription factor IRF3, which results in transcription of its target genes. Addition of EGF causes a loss of IRF3 transcriptional activity and activation of canonical signaling pathways such as ERK. The mechanistic basis of this bimodal signaling appears to be the association of a distinct set of signaling proteins with EGFR in the absence or presence of ligand. In this chapter, we describe a detailed protocol for analyses of constitutive EGFR signaling with a focus on IRF3 target genes.

Key words Epidermal growth factor receptor (EGFR), IRF3, Bimodal signaling, Chromatin immunoprecipitation (ChIP)

1 Introduction

The epidermal growth factor receptor (EGFR) is a cell surface glycoprotein that mediates the actions of EGF [1]. Overexpression of EGFR has been documented in a variety of human cancers such as lung cancers, epithelial cancers, and glioblastoma [2–4]. EGFR gene amplification and overexpression are found in 40–50% of GBMs and about half of these tumors express the constitutively active oncogenic mutant EGFRvIII along with EGFR wild type (EGFRwt) [5, 6]. Both EGFRwt and EGFRvIII have been shown to be tumorigenic in glioma cells, and while signaling from EGFR-vIII is constitutive, signaling from overexpressed EGFRwt may be constitutive or ligand induced [7–12]. Activation) of the EGFR results in formation of specific signaling complexes that culminate in gene transcription and a biological response [1]. Constitutive EGFR signaling has been reported for constitutively active EGFR mutants. Signaling by EGFRwt is classically driven by the binding

Zhixiang Wang (ed.), *ErbB Receptor Signaling: Methods and Protocols*, Methods in Molecular Biology, vol. 1652, DOI 10.1007/978-1-4939-7219-7_14, © Springer Science+Business Media LLC 2017

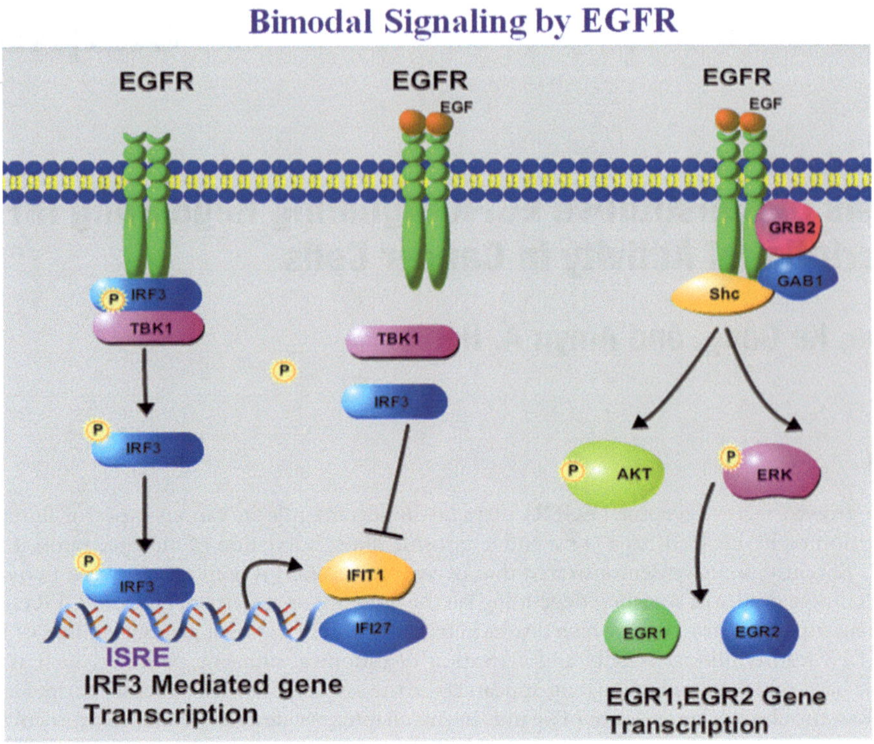

Fig. 1 Bimodal EGFR signaling by EGFR. Overexpression of EGFR results in constitutive tyrosine phosphorylation of EGFR and stimulates the formation of a IRF3/TBK1 complex, resulting in activation of IRF3. Addition of EGF induces dissociation of the complex, a loss of TBK1 and IRF3 phosphorylation, leading to a loss of IRF3 activation. The ligand-activated EGFR now becomes associated with Shc and activates ERK

of ligand to the EGFR. However, a noncanonical or constitutive EGFR signaling has also been reported, particularly in cancer cells that overexpress EGFRwt [11, 13] as shown in Fig. 1.

IRF3 is a transcription factor that plays a key role in antiviral innate immunity [14]. We have recently reported that overexpression of the EGFR in cancer cells results in a bimodal program of signal transduction and found that activation of IRF3 is regulated by constitutive EGFR signaling in cancer cells. Chromatin immunoprecipitation (ChIP) experiments were performed to investigate binding of IRF3 to the promoters of *IFI27* and *IFIT1* in cancer cells in the presence or absence of EGF. We find that IRF3 occupies the *IFIT1* and *IFI27* promoters in the absence of EGF in cancer cells. Addition of EGF decreases the binding of IRF3 to the promoter regions of both *IFIT1* and *IFI27*. These experiments support a model in which EGFR activates IRF3 in the absence of ligand and this activation is lost with addition of EGF.

2 Materials

2.1 Cells

1. U251 and U87 cells overexpressing wild-type EGFR (U251EGFR and U87EGFR) were generated as described in [8]; MDA-MB-468 cells were obtained from the American Type Culture Collection (ATCC).
2. Culture medium: Dulbecco's modified Eagle's medium (DMEM) supplemented with 10% fetal bovine serum, penicillin (100 U/mL), streptomycin (100 U/mL). Medium for U251EGFR and U87EGFR are supplemented with 200 μg/mL G418.
3. EGF protein (PeproTech).

2.2 Reagents and Buffers

1. Fixation buffer: 37% formaldehyde.
2. Glycine solution 1.25 M.
3. 200× Protease Inhibitor Cocktail II (MilliPore).
4. Protein A Agarose Beads (MilliPore).
5. 100× PMSF. Dissolve 34 mg PMSF in 1 mL DMSO and store aliquots at −20 °C.
6. SDS Lysis Buffer: 50 mM Tris–HCl (pH 7.5), 150 mM NaCl, 5 mM EDTA, 0.5% NP-40 (vol/vol), 1% Triton X-100, 1 × Protease Inhibitor Cocktail II.
7. ChIP Dilution Buffer: 20 mM Tris–HCl, pH 8.0, 150 mM NaCl, 2 mM EDTA, 1% Triton X-100, 0.01%SDS, 1× 1 × Protease Inhibitor Cocktail II.
8. Low-Salt Wash Buffer: 20 mM Tris–HCl, pH 8.0, 150 mM NaCl, 2 mM EDTA, 1% Triton X-100, 0.1% SDS, 1 mM PMSF.
9. High-Salt Wash Buffer: 20 mM Tris–HCl, pH 8.0, 500 mM NaCl, 2 mM EDTA, 1% Triton X-100, 0.1% SDS, 1 mM PMSF.
10. LiCl Wash Buffer: 10 mM Tris–HCl, pH 8.0, 1 mM EDTA, 250 mM LiCl, 1% deoxycholic acid, 0.5% NP-40, 1 mM PMSF.
11. TE Buffer: 10 mM Tris–HCl,pH 8.0, 1 mM EDTA.
12. Elution Buffer: 1% SDS, 100 mM $NaHCO_3$.
13. 25:24:1 phenol–chloroform–isoamyl alcohol: Mix phenol, chloroform, and isoamyl alcohol.
14. 100% ethanol.
15. 70% ethanol.

2.3 Antibodies

1. Anti-IRF3 (Santa Cruz).
2. Anti-Rabbit IgG (Cell Signaling).

3 Methods

3.1 Cross-Linking and Cell Harvesting (see Note 1)

1. Cells are cultured at 80% confluency on 150 cm^2 dishes (~1×10^7 cells).
2. Cells are serum-starved overnight.
3. EGF is added to medium at a final concentration of 50 ng/mL and incubates at 37 °C for 6 or 24 h.
4. Add formaldehyde dropwise to the media to a final concentration of 1%.
5. Mix gently and incubate at room temperature (RT) for 15 min (*see* **Note 2**).
6. Add glycine to a final concentration of 125 mM.
7. Incubate at RT for 5 min on a shaker.
8. Rinse cells twice with cold PBS.
9. Harvest cells using a cell scraper.
10. Centrifuge cells at 1000 × *g* for 5 min at 4 °C.
11. Carefully aspirate off supernatant and resuspend the pellet in 500 μL of ChIP lysis buffer (*see* **Note 3**).

3.2 Cell Sonication

1. Sonicate the cell lysate (4710 Series Ultrasonic Homogenizer 40% amplitude) for 5 cycles of 10 s (*see* **Note 4**).
2. Centrifuge at 12,000 × *g* for 5 min at 4 °C, remove supernatant to a fresh microfuge tube. Sonicated chromatin can be stored at −80 °C.

3.3 Immunoprecipitation

1. Prepare 100 μL of sonicated chromatin (Subheading 3.2, **step 2**).
2. For each ChIP, dilute sonicated chromatin 1:10 with ChIP dilution Buffer to a final volume of 1 mL.
3. Add 60 μL of Protein A Agarose for each IP.
4. Incubate for 1 h at 4 °C with rotation.
5. Centrifuge at 2000 × *g* for 1 min.
6. Remove 10 μL (1%) of chromatin to serve as your input sample and store at −20 °C until Subheading 3.4, **step 5**.
7. Add 2 μg primary antibody to the supernatant fractions (*see* **Note 5**).
8. Incubate overnight at 4 °C with rotation.
9. Add 60 μL of Protein A Agarose and rotate for 1 h at 4 °C.
10. Centrifuge at 2000 × *g* for 1 min and remove supernatant.
11. Add 1 mL of low-salt wash buffer to the Protein A Agarose-antibody/chromatin complex.
12. Rotate for 5 min at RT.
13. Centrifuge at 2000 × *g* for 1 min and remove supernatant.

14. Repeat **steps 11–13** with 1 mL high-salt and LiCl wash buffer, respectively.
15. Add 1 mL TE buffer.
16. Rotate for 1 min at RT.
17. Centrifuge at 2000 × *g* for 1 min and remove supernatant.
18. Wash twice with 1 mL TE buffer.

3.4 Elution and Reverse Cross-Links

1. Add 100 μL of elution buffer to the tube containing protein/antibody complex and the input prepared at Subheading 3.3, **step 6** (*see* **Note 6**).
2. Incubate for 15 min at RT with rotation.
3. Centrifuge at 2000 × *g* for 1 min and transfer supernatant into fresh tubes.
4. Repeat **steps 1–3** and combine eluates (total 200 μL).
5. Add 8 μL 5 M NaCl to the eluates.
6. Incubate at 65 °C for 4 h to overnight (*see* **Note 7**).
7. Add 4 μL 0.5 M EDTA, 8 μL 1 M Tris–HCl and 1 μL Proteinase K.
8. Incubate at 45 °C for 1–2 h.

3.5 DNA Purification

1. Add an equal volume of phenol–chloroform–isoamyl alcohol and vortex vigorously 10 s.
2. Centrifuge at 20,000 × *g* at 4 °C for 10 min.
3. Carefully transfer the top phase to a fresh tube.
4. Add 2 volume of ice-cold 100% ethanol and incubate for 1 h at −20 °C.
5. Centrifuge at 20,000 × *g* at 4 °C for 30 min.
6. Add 1 mL 70% ethanol, invert the tube several times.
7. Centrifuge at 20,000 × *g* at 4 °C for 5 min.
8. Remove supernatant and air-dry for 15 min.
9. Resuspend DNA pellet in 50 μL ultrapure water.

3.6 Quantitative ChIP-PCR

1. For PCR 2 μL of DNA prepared from Subheading 3.5, **step 9** are mixed with Platinum Taq DNA polymerase (Invitrogen). The following primers are used: For IFI27 promoter: 5′-CAT-GAGGGGAGAAAGATGTCTGCAGTT-3′ (Forward), 5′-CC TCCCTCCCAGTCT TACCCAAAGAAG-3′ (Reverse). For IFIT1 promoter: 5′-CCCCCGTCAGCAGGA ATTCCGC TAGCTTTA-3′ (Forward) and 5′-GCCAGGCTCCTCTGA GATCTGGCT-3′ (reverse). Primers are validated by using 1:100 dilutions of input DNA.
2. Remove 10 μL of each PCR product for analysis by 2% agarose gel.

4 Notes

1. A schematic for the procedure is shown in Fig. 2.
2. Cross-link with formaldehyde for no more than 15 min and wash well with PBS. Excessive cross-linking can reduce antigen accessibility and sonication efficiency.

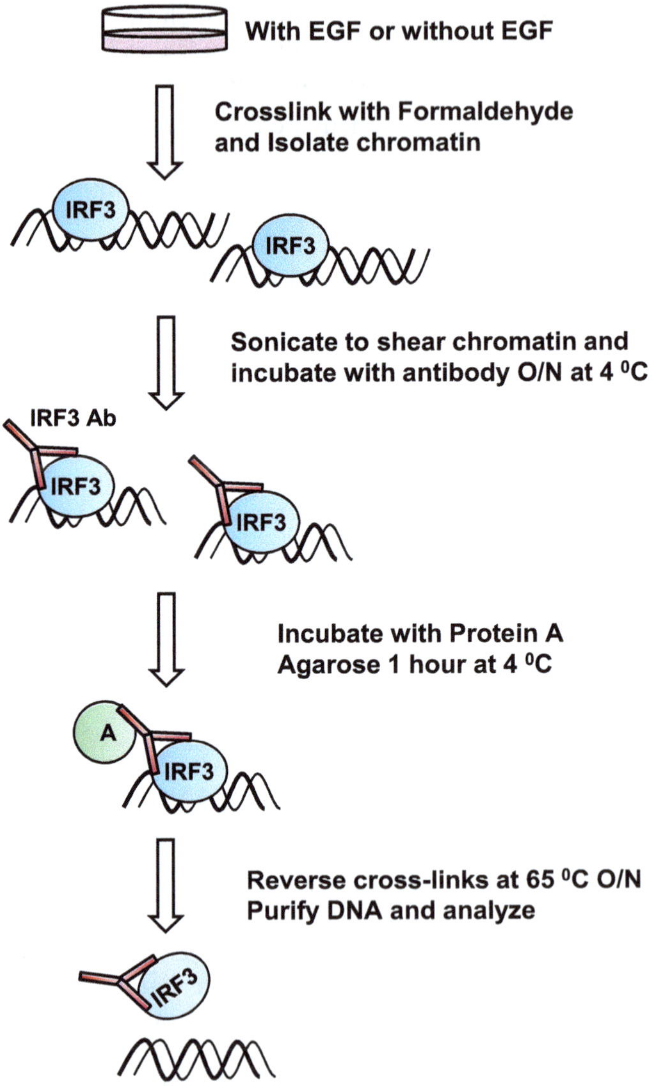

Fig. 2 Schematic representation of the steps involved in ChIP using IRF3 antibody. Cells are treated with EGF or control vehicle. Formaldehyde is added to cause cross-links between the DNA and protein. Whole cell extract is prepared and the cross-linked chromatin is sheared by sonication to reduce the average DNA fragment size. IRF3 antibody is added to capture the complex. Protein A/G Plus agarose beads are then added to collect the antibody–protein complex. The formaldehyde cross-linking was reversed by heating, followed by DNA purification

3. Add protease inhibitors fresh just before use.
4. Be sure to keep the samples on ice all the time. Sonication conditions should be optimized for different cell lines. The ideal DNA fragment size after sonication is 0.5–1 kb.
5. The amount of antibody to be added should be determined empirically. Use 2–5 μg of antibody in the first instance. This could be increased to 10 μg if no signal is observed.
6. Prepare the elution solution before use.
7. A minimum of 4 h is required.

Acknowledgment

This work was supported in part by NIH grants R01 NS062080 and by the Office of Medical Research, Departments of Veterans Affairs (AH).

References

1. Lemmon MA, Schlessinger J (2010) Cell signaling by receptor tyrosine kinases. Cell 141 (7):1117–1134
2. Veale D et al (1993) The relationship of quantitative epidermal growth factor receptor expression in non-small cell lung cancer to long term survival. Br J Cancer 68(1):162–165
3. Eckstein N et al (2008) Epidermal growth factor receptor pathway analysis identifies amphiregulin as a key factor for cisplatin resistance of human breast cancer cells. J Biol Chem 283 (2):739–750
4. Bredel M et al (1999) Epidermal growth factor receptor expression and gene amplification in high-grade non-brainstem gliomas of childhood. Clin Cancer Res 5(7):1786–1792
5. Hatanpaa KJ, Burma S, Zhao D, Habib AA (2010) Epidermal growth factor receptor (EGFR) in glioma: signal transduction, neuropathology, imaging and radioresistance. Neoplasia 12(9):675–684
6. Huang PH, Xu AM, White FM (2009) Oncogenic EGFR signaling networks in glioma. Sci Signal 2(87):re6
7. Nishikawa R et al (1994) A mutant epidermal growth factor receptor common in human glioma confers enhanced tumorigenicity. Proc Natl Acad Sci U S A 91(16):7727–7731
8. Chakraborty S et al (2014) Constitutive and ligand-induced EGFR signalling triggers distinct and mutually exclusive downstream signalling networks. Nat Commun 5:5811
9. Acquaviva J et al (2011) Chronic activation of wild-type epidermal growth factor receptor and loss of Cdkn2a cause mouse glioblastoma formation. Cancer Res 71(23):7198–7206
10. Wong AJ et al (1992) Structural alterations of the epidermal growth factor receptor gene in human gliomas. Proc Natl Acad Sci U S A 89 (7):2965–2969
11. Guo G et al (2015) Ligand-independent EGFR signaling. Cancer Res 75(17):3436–3441
12. Endres NF et al (2013) Conformational coupling across the plasma membrane in activation of the EGF receptor. Cell 152(3):543–556
13. Puliyappadamba VT et al (2013) Opposing effect of EGFRWT on EGFRvIII-mediated NF-kappaB activation with RIP1 as a cell death switch. Cell Rep 4(4):764–775
14. Fitzgerald KA et al (2003) IKKepsilon and TBK1 are essential components of the IRF3 signaling pathway. Nat Immunol 4(5):491–496

Chapter 15

Measurement of Epidermal Growth Factor Receptor-Derived Signals Within Plasma Membrane Clathrin Structures

Stefanie Lucarelli, Ralph Christian Delos Santos, and Costin N. Antonescu

Abstract

The epidermal growth factor (EGF) receptor (EGFR) is an important regulator of cell growth, proliferation, survival, migration, and metabolism. EGF binding to EGFR triggers the activation of the receptor's intrinsic kinase activity, in turn eliciting the recruitment of many secondary signaling proteins and activation of downstream signals, such as the activation of phosphatidylinositol-3-kinase (PI3K) and Akt, a process requiring the phosphorylation of Gab1. While the identity of many signals that can be activated by EGFR has been revealed, how the spatiotemporal organization of EGFR signaling within cells controls receptor outcome remains poorly understood. Upon EGF binding at the plasma membrane, EGFR is internalized by clathrin-mediated endocytosis following recruitment to clathrin-coated pits (CCPs). Further, plasma membrane CCPs, but not EGFR internalization, are required for EGF-stimulated Akt phosphorylation. Signaling intermediates such as phosphorylated Gab1, which lead to Akt phosphorylation, are enriched within CCPs upon EGF stimulation. These findings indicate that some plasma membrane CCPs also serve as signaling microdomains required for certain facets of EGFR signaling and are enriched in key EGFR signaling intermediates. Understanding how the spatiotemporal organization of EGFR signals within CCP microdomains controls receptor signaling outcome requires imaging methods that can systematically resolve and analyze the properties of CCPs, EGFR and key signaling intermediates. Here, we describe methods using total internal reflection fluorescence microscopy imaging and analysis to systematically study the enrichment of EGFR and key EGFR-derived signals within CCPs.

Key words Epidermal growth factor receptor, Total internal reflection fluorescence (TIRF) microscopy, Clathrin, Signaling microdomain, Akt, Gab1, Spatiotemporal signal organization, Image analysis

1 Introduction

The epidermal growth factor (EGF) receptor (EGFR or ErbB1) is one of the four members of the ErbB family of receptor tyrosine kinases. EGFR and ErbBs are key regulators of many aspects of cell physiology, including cell growth, proliferation, migration, adhesion, differentiation, and metabolism [1–4]. Several hormone

Stefanie Lucarelli and Ralph Christian Delos Santos contributed equally to this work.

Zhixiang Wang (ed.), *ErbB Receptor Signaling: Methods and Protocols*, Methods in Molecular Biology, vol. 1652, DOI 10.1007/978-1-4939-7219-7_15, © Springer Science+Business Media LLC 2017

ligands activate EGFR, including EGF, transforming growth factor α (TGFα), amphiregulin and HB-EGF, and these ligands have nonredundant physiological functions [5]. Upon binding EGF or other ligands at the plasma membrane (PM), EGFR forms asymmetric dimers, resulting in the activation of its cytosolic kinase domain, autophosphorylation of multiple tyrosine residues on the EGFR C-terminus, and initiation of several signaling cascades [6–8]. These phosphorylated residues form docking sites for proteins with Src homology 2 (SH2) or phospho-tyrosine binding (PTB) domains, including Grb2, Cbl, Shc, and phospholipase Cγ [9].

One of the first signaling intermediates recruited to active EGFR via Grb2 binding is growth factor receptor-associated binding protein 1 (Gab1), an adaptor protein required for EGF-stimulated phosphatidylinositol 3-kinase (PI3K) activation and Akt phosphorylation [10]. EGF stimulation elicits Gab1 phosphorylation on several tyrosine residues, allowing Gab1 to interact with the p85 regulatory subunit of PI3K [10, 11], which in turn results in the production of phosphatidylinositol-3,4,5-trisphosphate (PIP3) at the plasma membrane [12]. PIP3 binds and recruits the kinases PDK1 and Akt to the PM. In addition to PIP3 binding, Akt activation requires phosphorylation on T308 and S473 by PDK1 and mTORC2, respectively [13]. A large number of Akt substrates regulate specific aspects of cell metabolism and survival [14]. Hence, a critical control point leading to Akt activation and regulation of cell metabolism and survival by EGF stimulation is the binding of PI3K to Gab1, resulting in production of PIP3 at the PM.

Collectively, the proteins that initially bind to specific phosphotyrosine motifs on the EGFR C-terminus trigger a large number of signaling pathways that mediate the control of cell physiology by EGFR [9]. As a result, EGFR is a key regulator of several specific stages of organ development, evinced from the multiorgan failure leading to embryonic or perinatal death in mice with genetic knockout of the EGFR gene [15–21]. Importantly, dysregulation of EGFR resulting from genetic amplification, overexpression, aberrant activation, mutation or altered subcellular localization is commonly observed in many cancer types [2, 22–24], and is often associated with poor disease outcome, including in triple-negative breast, lung and colon cancers [23, 25, 26]. Thus, EGFR signaling must be tightly controlled to allow for sufficient signaling and ensure normal function in cell growth, development, and tissue homeostasis, and also to restrict amplification of EGFR signaling that contributes to tumor growth.

1.1 Ligand-Dependent Recruitment of EGFR to Clathrin-Coated Pits

Concomitant to the recruitment of cytosolic signaling proteins and the activation of signaling pathways following stimulation, EGF binding leads to receptor internalization by clathrin-mediated endocytosis [27]. Internalization of EGFR may lead to receptor degradation in the lysosome, receptor recycling back to the plasma membrane, or traffic to a number of other cellular compartments including the nucleus [27, 28]. In some contexts such as upon stimulation with elevated (>20 ng/mL) EGF levels, EGFR can also undergo clathrin-independent internalization [29].

In many cells, EGF stimulation elicits clathrin-mediated endocytosis of EGFR as a result of recruitment of the receptor to clathrin-coated pits (CCPs), dynamic assemblies of clathrin and other proteins at the cell surface (Fig. 1) [30]. Notably, many EGFR-proximal signals such as phosphorylated Gab1, Akt and PLCγ1 exhibit maximal activation within 1–3 min of EGF

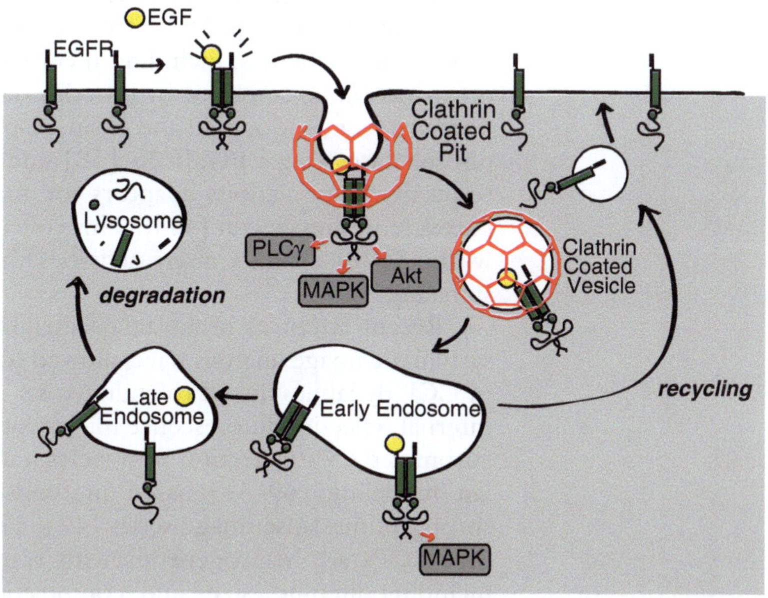

Fig. 1 EGF stimulation regulates EGFR membrane traffic and signaling. EGF binding to EGFR leads to the formation of asymmetric receptor dimers and triggers the activation of the EGFR kinase domain. As a result, ligand-bound EGFR at the plasma membrane elicits the activation of several signaling pathways including those leading to the activation of mitogen-activated protein kinase (MAPK), phospholipase Cγ1 (PLCγ1), and Akt, as well as recruitment to clathrin-coated pits (CCPs), eventually leading to receptor endocytosis. Following internalization, EGFR resident within clathrin-coated vesicles is routed to early endosomes, from which it may transit to late endosomes and lysosomes (leading to EGFR degradation and signal termination), undergo recycling (leading to continued EGFR signaling), or route to other cellular compartments such as the nucleus (not shown). The activation of specific EGFR signals can occur at the plasma membrane or in various endosomal compartments

stimulation [31, 32], a time that coincides with EGFR residence within CCPs [27]. As such, the microenvironment provided by CCPs may be particularly important for regulation of specific aspects of signaling by EGFR.

CCPs initiate by the recruitment of clathrin, the adaptor protein AP-2 and a number of other proteins to a small (50–100 nm), invaginating regions of the plasma membrane enriched in receptor cargo destined for internalization [30, 33]. Following a period of assembly of clathrin, the adaptor protein AP-2, and >50 specific proteins recruited from the cytosol, and a poorly understood maturation process, some CCPs undergo scission from the cell surface, mediated by dynamin GTPases, resulting in the production of internalized vesicles harboring EGFR or other cargo receptors. Proteins are recruited to CCPs as a result of interaction with the appendage domains of α- and β2-adaptin subunits of AP-2, by interaction with the N-terminal β-propeller domain of clathrin heavy chain or through an interaction with a protein recruited in this manner [34–37]. These CCP proteins have a wide range of biochemical activities, including membrane curvature stabilization or formation, exemplified by proteins harboring BAR and ENTH domains (e.g., endophilin and epsin, respectively), lipid kinases and phosphatases (e.g., PIK3C2α [38] and synaptojanin1 [39, 40], respectively) or various adaptors for formation of protein complexes (e.g., intersectin [41]). This collection of proteins enriched within CCPs exhibits many activities that may regulate receptor signaling.

Recent advances in advanced imaging techniques coupled to systematic image analysis have allowed for a deeper understanding of CCP dynamics linked to endocytosis. These studies utilize total internal reflection fluorescence microscopy (TIRF-M), followed by automated CCP detection by wavelet transform [42–48] or Gaussian modeling [49–51] based methods, and tracking of CCPs through time-lapse image series [47]. These studies have revealed that CCPs are heterogeneous with respect to several properties including lifetime, size, and composition [40, 42–46, 51–55]. Specifically, CCPs differ in the recruitment of specific cytosolic factors, as can be exemplified by the recruitment of specific lipid phosphatases to a subset of CCPs [40], the recruitment of the adaptor protein TTP to certain CCPs [56], and more broadly the recruitment of various cytosolic proteins, each to only a subset of CCPs [54]. In many cases, the profile of cytosolic proteins recruited to CCPs is regulated by the receptor cargo within these structures and govern CCP behavior and fate. For example, the recruitment of ARH and/or dab2 to CCPs harboring low-density lipoprotein receptors (LDLR) results in an increase in CCP size [43]. The internalization of δ-opioid receptors occurs through recruitment of these receptors to a functionally distinct subpopulation of CCPs with higher than average lifetimes [55].

Several lines of evidence indicate that the clathrin-mediated endocytosis of EGFR involves CCPs with distinct composition and/or functional profile than CCPs harboring other receptors. The clathrin-mediated endocytosis of transferrin receptor (TfR), but not EGFR is acutely sensitive to depletion of AP-2 [57, 58] and TTP [56], while that of EGFR (but not of TfR) requires clathrin phosphorylation [59] and is regulated by phosphatidic acid [53].

These studies indicate that cells harbor hundreds or thousands of CCPs, and that these structures differ in their composition of internalizing receptor cargo, composition of additional proteins recruited from the cytosol, fate (e.g., vesicle formation vs. disassembly at the plasma membrane), lifetimes, and likely also in their function. Thus, in order to resolve how CCPs function to control EGFR signaling, it is important to also employ techniques that can identify specific subpopulations of CCPs with specific features, such as the cohort of CCPs harboring EGFR.

1.2 Membrane Traffic and Compartmentalization of EGFR Signaling

Upon scission of CCPs harboring EGFR from the plasma membrane, the receptor is located within clathrin-coated vesicles that rapidly undergo disassembly of their clathrin coat in a process mediated by Hsc70 and auxilin [60], resulting in delivery of EGFR to Rab5-positive early endosomes [27]. From here, EGFR may undergo membrane traffic to late endosomes and eventually lysosomes for degradation, resulting in signal termination, or recycling back to the plasma membrane where it can participate in additional signaling events and thus sustain signaling (Fig. 1) [27].

The transit of EGFR through distinct compartments, including different microdomains within a single organelle (e.g., within a CCP) or different organelles within the endolysosomal network (e.g., early endosomes) provides EGFR with a series of distinct platforms and environments, each enriched in distinct proteins and lipids, within which the receptor may trigger activation of distinct signaling pathways. Indeed, a seminal study by Schmid et al. found that expression of a dominant-negative mutant of dynamin that blocked EGFR internalization also altered EGFR signaling [61]. Many of the signaling proteins that associate with active, phosphorylated EGFR at the plasma membrane such as Grb2, Shc, and Cbl are co-internalized with phosphorylated (active) EGFR into endosomes and remain associated with the receptor therein to maintain signaling, as detected by fractionation [62–66] and microscopy-based approaches [65, 67]. In addition to co-internalization with active receptors, the specific experimental activation of EGFR at endosomes and not the plasma membrane results in activation of Ras, Erk, and Akt, but not PLCγ signals [68].

1.3 Plasma Membrane Clathrin Microdomains Are Required for EGFR Signaling

Despite the ability of EGFR to signal from endosomes, cells lacking expression of all three isoforms of dynamin leading to a virtual absence of EGFR internalization nonetheless exhibited normal EGF-stimulated activation of Akt and Erk [69]. This indicates that the plasma membrane is the primary locale for EGFR signaling. Importantly, the plasma membrane is not a uniform compartment for activation of specific signals. We recently showed that EGF-stimulated Akt phosphorylation and activation requires receptor residence within CCPs at the plasma membrane, but not receptor endocytosis [70]. We also resolved that EGF stimulation results in the enrichment of certain signals, such as Gab1 and phosphorylated Gab1, within CCPs [70]. These findings indicate that some CCPs form signaling microdomains required for the activation of Gab1-PI3K-Akt signaling [70], and perhaps other signals as well. Interestingly, clathrin was not required for EGF-stimulated EGFR autophosphorylation, nor for the ligand-stimulated activation of MAPK signaling, demonstrating that the control of EGFR signaling within plasma membrane clathrin structures is specific to certain signals [70]. Further, we uncovered that clathrin was required in a similar capacity to regulate Met-receptor dependent activation of Gab1 and Akt [71]. As a result of this work, we proposed that a subset of CCPs function as plasma membrane signaling microdomains, and that specific receptor signaling processes, such as EGF-stimulated Akt activation, requires enrichment of key signaling intermediates (e.g., phosphorylated Gab1) within CCPs [70–72].

Consistently, stable, flat clathrin lattices that may rarely lead to endocytic events recruit G-protein coupled receptors such as CCR5 [73] and CCPs have function as signaling platforms for the activation of Gαi protein signaling by the delta opioid receptor [74]. Indeed a recent report demonstrated that β-arrestin-2 rapidly dissociated from β1-adrenergic receptor upon activation of the latter, resulting in rapid recruitment of β-arrestin-2 to plasma membrane CCPs [75]. Moreover, another recent study indicated that CCPs enhance EGFR signaling by enhancing EGFR clustering, thus potentiating receptor autophosphorylation [76]. However, we did not observe changes in EGFR phosphorylation upon clathrin perturbation [70]. Nonetheless, these studies collectively demonstrate that plasma membrane clathrin structures (CCPs) are unique signaling microdomains that are enriched in specific signals (phosphorylated Gab1, β-arrestin-2) and are required for the activation of specific pathways by a number of diverse signaling receptors (Fig. 2).

1.4 Other Plasma Membrane Signaling Microdomains

In addition to CCPs serving as signaling microdomains to facilitate EGFR signaling, the plasma membrane is comprised of a number of dynamic yet mechanistically and functionally unique nanoscale and microscale compartments that have key roles in regulation of signal transduction; we here refer to these collectively as *plasma*

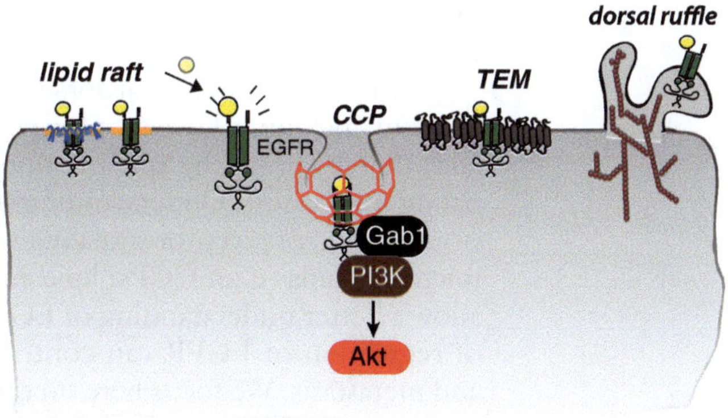

Fig. 2 EGFR signaling within plasma membrane microdomains. Ligand binding to EGFR triggers the activation of several receptor-proximal signaling intermediates and pathways. Many of the signals activated by EGFR exhibit spatiotemporal organization within specific plasma membrane microdomains. These plasma membrane microdomains include: (1) lipid rafts, a heterogeneous ensemble of membrane domains formed by lipid organization, some of which include specific proteins such as caveolin, (2) clathrin-coated membrane microdomains, which are a subset of clathrin-coated pits (CCPs) that are enriched in signaling proteins such as Gab1 and are required for EGF-stimulated activation of Gab1-PI3K-Akt signaling [70], (3) tetraspanin-enriched microdomains (TEMs) and (4) actin-dependent structures such as dorsal ruffles. Each of these microdomains has unique functions in the spatiotemporal organization and regulation of EGFR signaling in specific contexts [72]

membrane microdomains. These plasma membrane microdomains include structures defined by lipids (e.g., lipid rafts, glycosynapses), by proteins (e.g., clathrin-coated pits, tetraspanin domains) or by larger interactions with the cytoskeleton, as we discuss in a recent review [72] (Fig. 2).

Lipid rafts are distinct cholesterol- and sphingolipid-enriched membrane microdomains [77–80]. Lipid rafts are heterogeneous structures, yet are broadly classified as caveolae or noncaveolar, planar lipid rafts ranging in size from 10 to 200 nm [81, 82] and with lifetimes on the order of μs–ms to min [83, 84]. Indeed EGFR can localize to lipid rafts under some circumstances [85–91]. Paradoxically, disruption of lipid rafts may either enhance EGFR and MAPK phosphorylation [85, 89, 92], or may impair activation of MAPK and Akt [93–95], indicating that control of EGFR signaling by cholesterol-enriched membrane microdomains is context-specific.

The self-organization of tetraspanin proteins also leads to the formation of a class of ~200 nm microdomains termed tetraspanin-enriched microdomains (TEMs) [96]. TEMs also regulate EGFR signaling by sequestering the receptor or key signaling intermediates, and the nature of this regulation is dependent on the particular tetraspanin proteins expressed [72, 96]. The dynamic partitioning of EGFR between TEMs and CCPs regulates EGFR endocytosis and signaling [97].

Each microdomain harbors unique combinations of lipids and proteins including kinases, phosphatases and signaling intermediates, thereby imposing spatiotemporal organization on specific signaling processes. Understanding the control of spatiotemporal organization of receptor signaling within specific plasma membrane microdomains (e.g., CCPs, lipid rafts, and TEMs) is important to allow a better understanding of EGFR function and how disruption of regulation of EGFR can contribute to cancer growth, survival and metastasis. We focus here on the study of specific EGFR signals within CCPs. However, as many plasma membrane microdomains are also diffraction-limited structures, the methods described here can be readily adapted to study receptor-derived signals within other structures.

1.5 Measurement of EGFR Signaling Within Diffraction-Limited Structures

To understand the regulation and functional outcome of signaling by EGFR and other receptors, there is a need to examine not just signal activation but also to resolve the localization and/or enrichment of signals within specific microdomains (e.g., CCPs). Recent methods involving TIRF-M and automated image analysis have been described for the detection, tracking, and analysis of CCPs and CCP dynamics in time-lapse image series [40, 42–48, 51, 53, 98]. These approaches have revealed important new information about the dynamics of CCPs, including the recruitment of cytosolic proteins to these structures. Such approaches use live-cell time-lapse imaging and can indeed be used to study the recruitment and enrichment of EGFR-elicited signals to CCPs or other cell surface microdomains. However, there are two key limitations to these approaches involving time-lapse imaging of living cells, which can be overcome by approaches involving the quantitative study of CCPs in single image frames:

1. The activation of specific EGFR signals elicits a temporal profile in which certain signals exhibit peak activation within 1–3 min following EGF stimulation. The reliable tracking of the complete population of CCPs within an image series requires a time-lapse image series of ~3–5 min in length (with 1 s framerate, as used in [51]). Hence, EGFR signaling undergoes robust changes in profile at a faster timescale than that which is required for study of CCP dynamics. Study of single image frames, either by selecting specific frames corresponding to specific times of EGF stimulation, or sample fixation after specific times of EGF stimulation allows resolution of EGFR signaling with the required temporal resolution.
2. While the detection of protein localization can be done by controlled expression of fluorescently tagged fusion proteins, the detection of activated proteins, e.g., resulting from phosphorylation, is difficult to do in living cells. The latter requires

establishment of fluorescent biosensors, which have been described for a limited number of proteins, such as EGFR itself, c-Src, or AMPK [99–101]; however, these are not readily available for the majority of EGFR signals. Hence, methods that allow study of the spatiotemporal organization of receptor signaling within a series of fixed samples also allow the detection of phosphorylated protein by immunofluorescence staining with antibodies that specifically recognize phosphorylated proteins.

Here, we describe the use of methods for the systematic, quantitative measurement of the spatiotemporal properties of activated (ligand-bound) EGFR and key signals (phosphorylated Gab1) within plasma membrane CCPs using TIRF-M. This form of microscopy requires cells that exhibit close, uniform apposition of the plasma membrane to the coverslip for which the cultured human retinal pigment epithelial cell line ARPE-19 (henceforth, RPE cells) is ideally suited. This imaging is coupled to analysis of a series of single image frames, as we have recently described [70, 71, 102–104]. We first describe a method to selectively label activated and ligand-bound EGFR by treatment with fluorescently labeled EGF (Subheading 3.1). Next, we describe a method to label phosphorylated Gab1 by immunofluorescence staining (Subheading 3.2). Subsequently, we describe total internal reflection microscopy (TIRF-M) imaging (Subheading 3.3), and automated image analysis (Subheading 3.4) suitable for each of the labeling approaches used in Subheadings 3.1 and 3.2. Finally, we describe a method for simultaneous labeling of EGFR, phosphorylated Gab1 and clathrin, followed by analysis of these three-channel TIRF-M images, which allows resolution of the enrichment of specific signals (e.g., pGab1) within specific subsets of CCPs (e.g., positive for EGFR) (Subheading 3.5). These methods allow quantitative, systematic, and unbiased analysis of the enrichment and organization of EGFR signals within CCPs at the plasma membrane, allowing the dissection of how this spatiotemporal organization of receptor signaling intermediates by clathrin controls receptor function.

In order to facilitate the study of CCPs, a series of RPE cells expressing fluorescently tagged clathrin light chains (LC) have been previously described, including eGFP-tagged clathrin light chain a (eGFP-LCa) or Tag-RFP-T-tagged clathrin light chain a (RFP-LCa) [51] (*see* **Note 1**). Importantly, the expression of these LCa fusion proteins does not perturb CCP dynamics, assembly or clathrin-mediated endocytosis [51]. Other cell lines that may be useful for TIRF-M for imaging of EGFR signaling within membrane microdomains include BSC-1 [40, 42–46, 52, 53], NIH-3T3 fibroblasts [54], HeLa [105], HEK293 [55, 75], or COS-7 cells [106]. Here, we describe the use of RPE eGFP-LCa or RPE RFP-LCa cells for the study of the spatiotemporal organization of EGFR signaling.

2 Materials

2.1 Reagents and Materials

1. *For culture of RPE cells*: DMEM/F12 media (Gibco, Thermo Fisher Scientific, Waltham, MA), supplemented with 10% fetal bovine serum (Gibco) and 5% penicillin/streptomycin solution (Gibco). RPE cells are grown as described in [51, 70].
2. Dulbecco's Phosphate Buffered Saline (dPBS, Sigma-Aldrich).
3. Trypsin solution (Gibco).
4. *Serum-starvation medium*: Dulbecco's Modified Eagle's Medium (DMEM), with 1000 mg/L glucose, L-glutamine, and sodium bicarbonate, liquid, sterile-filtered, suitable for cell culture (Sigma-Aldrich), supplemented with 15 mM HEPES (BioShop Canada).
5. *Working stock fluorescent EGF solution* (for Subheading 3.1): rhodamine-EGF (Tetramethylrhodamine Conjugate, Thermo Fisher Scientific) diluted to 1 μg/mL in *serum starvation medium*.
6. *Working stock unlabeled EGF solution* (for Subheading 3.2): human EGF (Gibco), diluted to 1 μg/mL in *serum starvation medium*.
7. *Phosphate buffered saline (PBS):* 2.7 mM KCl, 1.5 mM KH_2PO_4, 136.9 mM NaCl, 8.9 mM Na_2HPO_4, pH 7.4.
8. *Fixation solution*: 4% paraformaldehyde (PFA) in PBS.
9. *Quenching solution:* 100 mM glycine in PBS.
10. *Permeabilization solution*: 100 mM glycine, 0.1% Triton X-100 (TX-100) (Bioshop Canada) in PBS.
11. Superblock Blocking Buffer (Thermo Fisher Scientific).
12. *Primary antibody solution* (to detect pGab1): 1/400 dilution of anti-pGab1 antibody (e.g., pY627 antibody, from Applied Biological Mat. Inc., Richmond, BC) in PBS supplemented with 1% Bovine Serum Albumin (BSA; BioShop Canada).
13. *Secondary antibody solution* (to detect antibody-bound pGab1): 2.5 μg/mL of A488-conjugated secondary goat anti-rabbit antibody (Jackson Immuno Research, West Grove, PA) in PBS supplemented with 1% BSA.
14. 4′,6-diamidino-2-phenylindole (DAPI) (Sigma-Aldrich): 2 mg/mL in H_2O.

2.2 Total Internal Reflection Fluorescence Microscopy (TIRF-M)

Access to a TIRF-M capable of independent illumination and imaging of two or three fluorescence channels, as per Subheadings 3.1–3.4 or Subheading 3.5, respectively, is required.

2.3 MATLAB and Relevant Custom Software Required

The analysis described herein requires the use of Matlab (Mathworks, Nantick, MA), version 2012 or newer. In addition to the successful installation of Matlab, there are several custom software components that must be successfully installed, as follows:

1. The cmeAnalysis package from the Danuser laboratory, available at: http://www.utsouthwestern.edu/labs/danuser/software/.
2. The EGFR signal analysis package from our group, available at: http://www.ryerson.ca/cellsurfacebio/software.html.

As described in their respective instruction documents, the packages consist of a series of .m MATLAB function files. Installation requires download of all .m files, copy of these .m files into specific folders, and designation of these folders within Matlab using the "Set Path" function.

3 Methods

3.1 Labeling for Measurement of the Recruitment of Activated EGFR to Clathrin-Coated Pits

Upon ligand binding, EGFR forms asymmetric dimers, leading to activation of the receptor tyrosine kinase domain, receptor autophosphorylation and activation of numerous downstream signaling pathways. Concomitantly, EGFR is recruited to CCPs as a result of several partly redundant mechanisms including EGFR binding to Grb2, EGFR ubiquitination, EGFR acetylation and EGFR binding to the clathrin adaptor protein AP-2 [107], and receptor dimerization [108, 109]. Activation of receptor-proximal signals occur concomitantly to EGFR residence within CCPs, and we previously uncovered that clathrin (but not receptor endocytosis) is required for certain facets of EGFR signaling [70]. One of the keys to understanding how recruitment of EGFR to CCPs controls receptor signaling outcome is the measurement of EGFR recruitment to CCPs under various conditions. We previously used quantitative, systematic and unbiased methods to measure the magnitude of EGFR recruitment into CCPs, based on stimulation of cells expressing fluorescently labeled clathrin with fluorescently conjugated EGF to label ligand-bound (and thus active) EGFR in cells [70]. Using this method, we found that EGFR recruitment into CCPs is impaired in certain conditions (e.g., in cells treated with the clathrin inhibitor pitstop2) [70]. This method allows dissection of signals required for the recruitment of EGFR to CCPs, and resolves this effect from other perturbations that alter EGFR endocytosis by impacting a stage subsequent to receptor recruitment to CCPs. Here, we describe a method using rhodamine-EGF labeling of RPE cells stably expressing eGFP-LCa [51] to resolve EGFR recruitment to CCPs upon ligand binding. EGFR recruitment into CCPs can be observed as early as 1 min of stimulation with EGF (Fig. 3). Automated detection of CCPs and quantification of

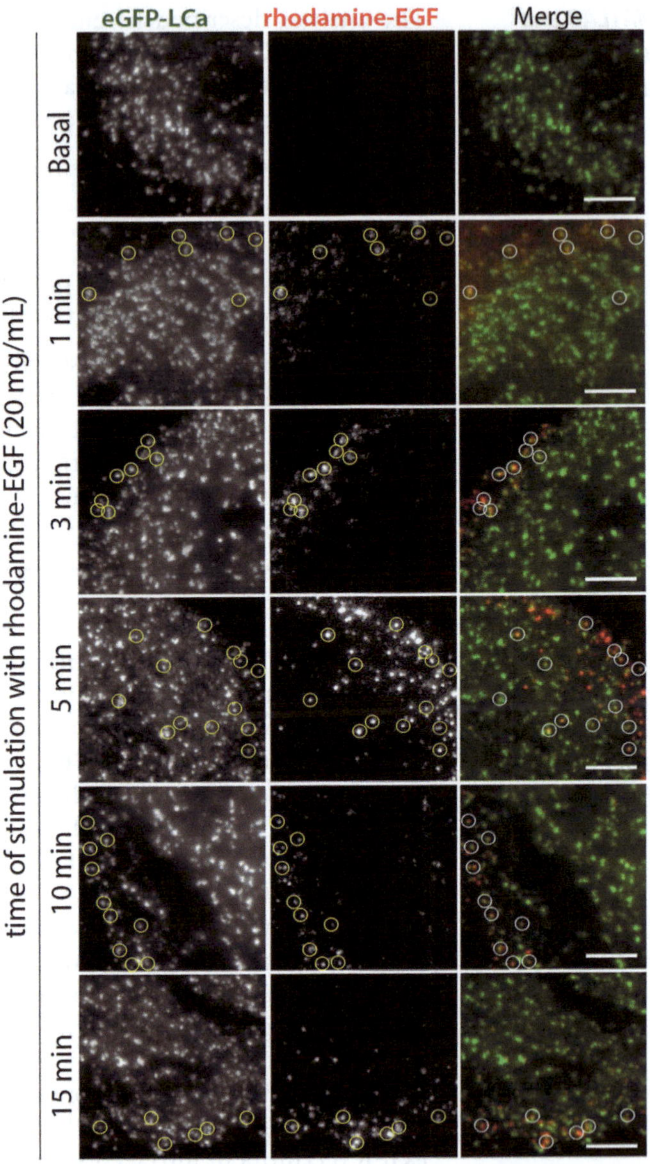

Fig. 3 Rhodamine-EGF is rapidly recruited to clathrin-coated pits (CCPs). RPE eGFP-LCa cells were treated with 20 ng/mL of rhodamine-EGF for the indicated times, then subjected to immediate fixation and sample preparation as in Subheading 3.1, followed by imaging with TIRF-M as in Subheading 3.3. Shown are representative fluorescence micrographs obtained by TIRF-M, showing eGFP-LCa (clathrin) and rhodamine-EGF (EGFR) channels separately and as a merged image. Circles depict CCPs that exhibit overlap with rhodamine-EGF, identified manually. Scale = 5 μm

receptor levels within CCPs revealed that EGFR enrichment with CCPs is maintained from 1–15 min following EGF stimulation (Fig. 5a), likely reflecting ongoing recruitment of EGFR to CCPs, as well as ongoing CCP disassembly or internalization.

3.1.1 Cell Culture, Cell Preparation and Experimental Stimulation

1. On days leading up to the experiment, place acid washed 22 mm-diameter round or square glass coverslips (*see* **Note 2**) in 6-well culture dishes (one coverslip per well) as needed.
2. Seed an appropriate amount of cells to achieve 50–60% confluence on the day of the experiment. For RPE GFP-LCa cells, this corresponds to seeding approximately 10^6 cells/well (of a 6-well plate) for an experiment to be conducted 24–48 h after seeding. To ensure even dispersal of cells on the bottom surface of the well, agitate cells by front-to-back and/or side-to-side movements, while avoiding circular motions, as the latter can propel cells to the periphery of the well.
3. On the day of the experiment, wash cells twice in PBS and replace existing media with exactly 2 mL of *serum starvation media*. Incubate RPE GFP-LCa cells in serum-free media at 37 °C with 5% CO_2 for 1 h. During this 1 h period, cells may also be treated with various inhibitors or other agents as required.
4. Following serum starvation, stimulate cells with rhodamine-EGF for defined periods of time. This is done by adding 40 μL of the *working stock fluorescent EGF solution* to each well (for a final concentration of 20 ng/mL EGF, this can be adjusted as needed). Allow EGF stimulation to proceed for the time period required (shown is the time course of 0–15 min EGF stimulation, Fig. 3 and 5). If multiple time points of EGF stimulation are required, these should be staggered in such a way that the stimulation for each condition ends at the same time. Once this is complete, proceed immediately to Subheading 3.1.2, **step 1** below.

3.1.2 Slide Preparation

1. Upon completion of EGF stimulation, place cells immediately on ice and wash 2× in ice-cold PBS to arrest EGFR membrane traffic. Then, again without delay, aspirate PBS solution from each well and quickly replace this with 2 mL *fixation solution*. Incubate the cells in this solution on ice for 10 min. During this and all other manipulations, protect samples from light, which can be achieved by placing samples in a drawer or covering with a box with opaque surfaces. After this 10 min fixation period, continue the fixation for an additional 30 min at room temperature, also protecting the cell samples from light at this stage.
2. Perform quenching of the fixative with a quick 2 mL wash in *quenching solution* and subsequently incubate the cells in 2 mL *quenching solution* at room temperature for 25 min.
3. Wash the cells 3× in 2 mL PBS.
4. As an optional step, the cells may be DAPI stained with the addition of 1 mL DAPI (final concentration of 0.27 μM) solution incubated at room temperature for 5 min.

5. Wash the cells twice in with 2 mL PBS and subsequently replace with 2 mL PBS to keep the cells from drying out.
6. At this stage, coverslips can be stored for 24–48 h. Store the 6-well plate in the dark (e.g., by wrapping the plate in tinfoil) at 4C until ready to proceed to imaging. Do not mount the slides in mounting medium, as the difference in the refractive indices between typical mounting medium (e.g., ~1.45–1.55) and the borosilicate glass of coverslips (~1.52) is not sufficient to support TIRF-M. As described in the section below, coverslips must be imaged in aqueous-based media (such as PBS, refractive index of ~1.3–1.35), allowing for a difference in refractive indices of the coverslip and mounting media compatible with TIRF-M. *See* **Note 3** for alternate wet mounting protocols.
7. When appropriate, proceed with imaging (Subheading 3.3) and analysis (Subheading 3.4)

3.2 Labeling to Detect Enrichment of Specific EGFR-Derived Signals Within CCPs

The recruitment of certain receptor-proximal signaling intermediates to EGFR occurs concomitantly with EGFR residence within CCPs and activation of some specific signals (phosphorylation of Gab1 and Akt) requires clathrin but not receptor endocytosis [70].

The detection of Gab1 and phosphorylated Gab1 at the plasma membrane requires microscopy techniques for the selective illumination of cell-surface clathrin structures and signals, for which total internal reflection fluorescence microscopy (TIRF-M) is ideal. This also requires the ability to systematically detect and analyze CCPs in order to quantitatively determine the mean intensity corresponding to Gab1 or phosphorylated Gab1 within CCPs. As we previously reported, using such a method reveals that EGF stimulation elicits a robust increase in phosphorylated Gab1 visible within CCPs (Fig. 4), and an increase in the mean phosphorylated Gab1 within CCPs determined by systematic image analysis (Fig. 5b). Here, we describe the method for the detection of endogenous, phosphorylated Gab1 within CCPs.

3.2.1 Cell Culture, Cell Preparation, and Experimental Stimulation

1. On days leading up to the experiment, place acid washed 22 mm-diameter round or square glass coverslips in 6-well culture dishes (one coverslip per well) as needed.
2. Seed appropriate amount of cells to achieve 50–60% confluence on the day of the experiment. For RPE RFP-LCa cells, this corresponds to seeding approximately 10^6 cells/well (of a 6-well plate) for an experiment to be conducted 24–48 h after seeding. To ensure even dispersal of cells on the bottom surface of the well, agitate cells by front-to-back and/or side-to-side movements, while avoiding circular motions, as the latter can propel cells to the periphery of the well.
3. On the day of the experiment, wash cells twice in PBS and replace cell culture media with *serum starvation media.*

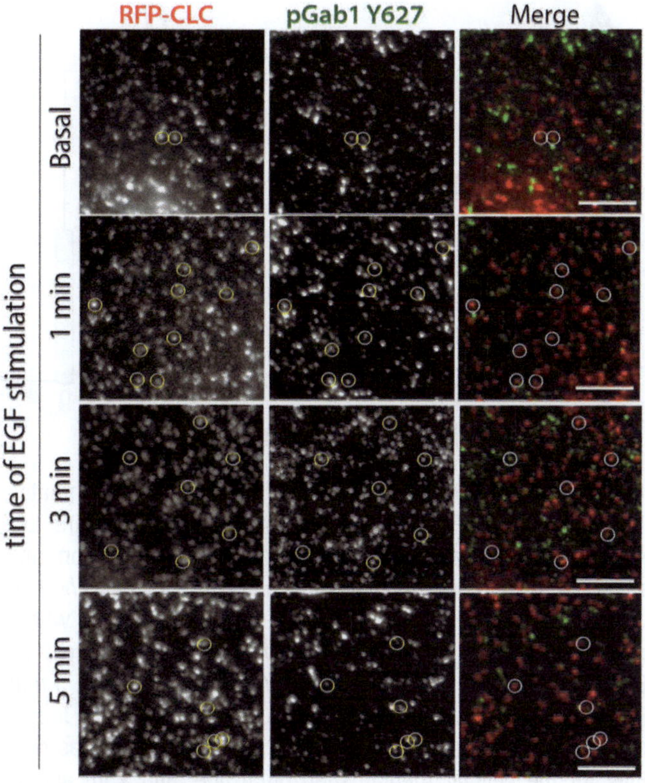

Fig. 4 Phosphorylated Gab1 is enriched within CCPs upon EGF stimulation. RPE RFP-LCa cells were stimulated with 20 ng/mL (unlabeled) EGF for the indicated times and subjected to immunostaining to detect phosphorylated Gab1 (pGab1) as in Subheading 3.2, followed by imaging by TIRF-M as in Subheading 3.3. Shown are representative fluorescence micrographs obtained by TIRF-M, showing RFP-LCa (clathrin) and A488-stained phosphorylated Gab1 channels separately and as a merged image. Circles depict CCPs that exhibit overlap with phosphorylated Gab1, identified manually. Scale = 5 μm

Incubate RPE RFP-LCa cells in this media at 37 °C with 5% CO_2 for 1 h. During this 1 h period, cells may be treated with various inhibitors as required.

4. Begin the experiment by stimulating cells with human recombinant EGF for required time in a humidified incubator at 37 °C with 5% CO_2. EGF stimulation conditions (time of stimulation, concentration of EGF) should be tailored to specific experimental needs. For example, pGab1 can be detected in CCPs upon stimulation with 5 ng/mL EGF for 3 min [70]. At the end of the duration of stimulation with EGF, proceed immediately to the Subheading 3.2.2, **step 1**, without delay.

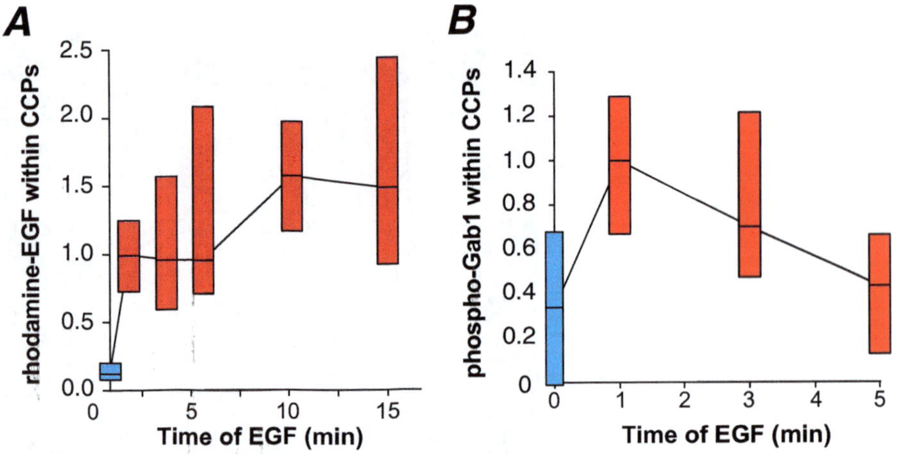

Fig. 5 Systematic detection and analysis of CCPs allows quantitative measurement of signal enrichment within CCPs. (**a**) RPE eGFP-LCa cells were treated with rhodamine-EGF, processed and imaged as in Fig. 3. The resulting images were subjected to automated detection and analysis of CCPs as described in Subheading 3.4. Shown are the values for the cellular median (*horizontal line*) and 75th and 25th percentile values (*boxes*) for the rhodamine-EGF fluorescence intensities within all detected CCPs within each EGF-stimulated condition. The number of CCPs (*n*) and cells (*k*) for each condition are basal: $n = 4104$, $k = 40$; 1 min EGF: $n = 5278$, $k = 45$; 3 min EGF: $n = 4208$, $k = 33$; 5 min EGF: $n = 4600$, $k = 42$, 5 min EGF. (**b**) RPE RFP-LCa cells were treated with EGF, processed and imaged as in Fig. 4. The resulting images were subjected to automated detection and analysis of CCPs as described in Subheading 3.4. Shown are the values for the cellular median (*horizontal line*) and 75th and 25th percentile values (*boxes*) for the pGab1 fluorescence intensities within all detected CCPs within each EGF-stimulated condition. The number of CCPs (*n*) and cells (*k*) for each condition are basal: $n = 49{,}838$, $k = 46$; 1 min EGF: $n = 59{,}185$, $k = 45$; 3 min EGF: $n = 69{,}413$, $k = 46$; 5 min EGF: $n = 57{,}003$, $k = 46$; 10 min EGF: $n = 54{,}341$, $k = 46$; 15 min EGF: $n = 49{,}245$, $k = 46$

3.2.2 Immuno-fluorescence Labeling and Slide Preparation

1. Immediately following EGF stimulation, place cells on ice and quickly wash the cells with 2 mL of ice-cold PBS.
2. Upon aspiration of PBS, fix cells by adding 1 mL of ice-cold *fixation solution* to each well. Incubate the cells on ice for 20 min. Ensure that this and every subsequent incubation is done in the dark (by placing the cell plate in a drawer or covering the plate on the laboratory bench with a box or tinfoil).
3. Aspirate the fixative from each well and quickly wash the cells with 2 mL of *quenching solution*. Following this, incubate cells in 2 mL of *permeabilization solution* for 20 min at room temperature in the dark.
4. Upon solution aspiration, wash the cells 3× with 2 mL of PBS. Block nonspecific antibody binding sites within with 2 mL Superblock Blocking Buffer (Thermo), or alternatively 3% BSA in PBS. Incubate the cells for 20 min at room temperature

in the dark. Following this, wash the cells again 3× in PBS, leaving the PBS in each well following the final wash.

5. For each coverslip sample to be labeled with anti-phospho-Gab1 antibody, pipette a 100 μL spot of *primary antibody solution* on top of a flat sheet of Parafilm. Using fine forceps, carefully flip each coverslip onto a spot of antibody solution (the surface of the coverslip with cells adhered is cells facing downward, toward the Parafilm). Incubate the coverslips for 1 h at room temperature in the dark.
6. At the end of this antibody incubation, carefully lift the coverslip off the Parafilm using forceps and place each back into its original well on the 6-well plate, with the surface of each coverslip with cells adhered facing up.
7. Upon aspiration of PBS from each well, wash each well 10× with 2 mL of PBS over the span of 10 min, with continuous gentle shaking to remove any antibodies that are bound nonspecifically.
8. To label the samples with fluorescent secondary antibodies, add 750 μL of *secondary antibody solution* to each well. Incubate for 1 h at room temperature in the dark.
9. Upon aspiration of secondary antibody solution from each well, wash cells with PBS as in **step** 7.
10. At this stage, coverslips can be stored for 24–48 h. To each well of a 6-well plate containing a processed coverslip, add 2 mL of PBS. Store the 6-well plate in the dark (e.g., by wrapping the plate in tinfoil) at 4 °C until ready to proceed to imaging. As described in Subheading 3.1.2, do not mount the slides in specialized mounting medium.
11. When appropriate, proceed with Imaging (Subheading 3.3) and Analysis (Subheading 3.4)

3.3 Imaging by Total Internal Reflection Fluorescence Microscopy (TIRF-M)

Total internal reflection fluorescence microscopy (TIRF-M) achieves selective illumination of fluorophores within a short distance from the coverslip (~100–200 nm), and is thus very useful for selective imaging of structures at the plasma membrane with high signal to noise [110, 111]. TIRF-M requires the illumination of a sample at an angle (above a critical angle, determined by many factors specific to illumination conditions), which results in total reflection, generating an evanescent wave within the sample that decays exponentially with distance from the coverslip [110, 111]. Imaging by TIRF-M can be done on any number of instruments, and requires training specific to each instrument that cannot be provided here. For detection of fluorescent EGF or phosphorylated Gab1 within CCPs, we previously described the use of an Olympus IX81 instrument equipped with a 150× (NA 1.45) TIRF objective and CellTIRF modules (Olympus Canada, Richmond Hill,

Canada) using 491- and 561-nm laser illumination and 520/30 and 624/50 emission filters; images were acquired using a C9100-13 EM-CCD camera [70]. All imaging performed in this chapter was done on an Quorum (Guelph, ON) Diskovery total internal reflection fluorescence microscope, comprised of a Leica DMi8 microscope equipped with a 63× (NA 1.49) TIRF objective with a 1.8× camera relay (total magnification 108×). Imaging was done using 488 nm, 561 nm and 637 nm laser illumination and 527/30, 630/75, and 700/75 emission filters and acquired using a Zyla 4.2Plus sCMOS camera (Hamamatsu Corp., Shizuoka Pref., Japan). Similar imaging experiments can be performed on any number of instruments from various manufacturers. Some important considerations for TIRF-M for the detection of EGFR signals within CCPs are as follows:

1. As stated above, imaging must be done with coverslip samples immersed in an aqueous solution (e.g., PBS), and not a self-drying oil or aqueous-based mounting medium.
2. Quantification of the position and amplitude of signals (corresponding to signal intensity) within detected CCPs (as well as additional channels) is done by Gaussian-based modeling (to model the point-spread function) of each signal within primary detections of CCP objects. For this to be effective, the signal of each channel within each object should be spread out over multiple pixels in the resulting image. As CCPs typically range in size from 50 to 200 nm, an image pixel resolution <120 nm, preferably ~60 nm is optimal to allow subsequent detection and analysis of fluorescence intensity of clathrin and other markers within CCPs.
3. Information about the TIRF objective (numerical aperture and magnification) as well as the camera pixel size must be noted as these parameters will be required to effectively detect and analyze CCPs using a Gaussian modeling approach [51].
4. The quality of the signal, in particular for the "First (primary) channel" fluorescence is critical, as this will form the basis of the detection of CCPs (Subheading 3.4). The requirements for the quality of signal for successful detection and tracking of CCPs in time-lapse image series was previously described [51], and many of these considerations apply for analysis of single images in fixed samples (*see* **Note 1**).
5. The usefulness of TIRF-M imaging is the selective illumination of fluorophores within a short distance (~100–200 nm) from the coverslip, and care must be taken to ensure conditions for optimal TIRF illumination, as described in **Note 4**.

3.4 Automated Detection of CCPs and Analysis of pGab1 Enrichment Within CCPs

Detection of CCPs and analysis of the enrichment of specific signals therein makes use of Matlab. We provide instructions for how to use the primary CCP detection and quantification software package developed by Danuser and colleagues (Subheading 3.4.1), as well as additional functions for the extraction and visualization of data resulting from this analysis pertinent to analysis of EGFR signals within CCPs that we have described previously [70, 71, 103], as follows: the mean EGFR signal intensity within CCPs (Subheading 3.4.2), and a 2D histogram (or heat map) of clathrin and EGFR signal intensity within each structure (Subheading 3.4.3), as follows:

3.4.1 Detection of CCPs

1. If in a different format, export or convert all image files to 8- or 16-bit grayscale .tiff files. Arrange files into a 4-level folder structure, as described within the instructions of the *cmeAnalysis* package. Briefly, each file must be placed in folders as follows: "condition/date/cell_number/channel/image.tiff". Do not use spaces in folder names. An example of such a folder structure would be: "pGab1_basal/10-31-16/cell1/TIRF_red/image.tiff". Thus, when performing 2-channel imaging (e.g., clathrin and an EGFR-derived signal, respectively), each "cell_number" folder will have two folders (one for each channel).
2. In Matlab, use the *loadConditionData.m* function to load each condition to be analyzed, using the following command: "[condition_name]=*loadConditionData();*". As an example, "pGab1_basal=*loadConditionData();*". To quantify signals within CCPs, the "First (primary) channel" designation must be given to the folder containing the clathrin channel image, and the "Channel #2" designation to that of the EGFR signal (e.g., fluorescent EGF or phosphorylated Gab1). Complete the prompts for the number of channels (e.g., 2), and the fluorophore for each channel (e.g., "GFP", "RFP").
3. Open the data structure loaded with the "*loadConditionData.m*" function and verify that the objective magnification (default: 100×), the objective numerical aperture (default: 1.49) and the camera pixel size (default 6.7×10^{-6} m) fields are accurate for the TIRF microscope used for imaging.
4. Run the object detection function for each data structure using the *runDetection.m* function. For example, run the command "*runDetection*(pGab1_basal);" to run the detection algorithm on the "pGab1_basal" image dataset previously loaded using the *loadConditionData.m* function in **step 2**. This may take a few minutes per data structure, depending on the number of cell images in each.
5. Once this is complete, it is advisable to save the Matlab workspace, otherwise, **steps 1–3** will have to be repeated each time new analyses are required.

3.4.2 Extraction of Mean Fluorescent EGF or Phosphorylated Gab1 Signal Intensity Values Within CCPs

The result of the *runDetection.m* function is the creation of new folders and files within each primary channel image folder to store a wide range of data related to the position, intensity and distribution of signal within each detected object [51]. This information is stored within a file named "detection_v2.mat" in a folder with the name "Detection" in the folder corresponding to the primary channel image in each cell. To obtain the mean fluorescence of the EGFR signal (secondary channel) as well as that of clathrin (primary channel) in each detected object, the *extractCCPdata_2chan.m* function may be used, as follows:

1. For each image, use the *extractCCPdata_2chan.m* function on each image data set, and store the resulting information in a new structure of appropriate name. To obtain the [result], the function call on a [dataset] has the format: "[result]=*extractCCPdata_2chan*([dataset],threshold)", where *threshold* is the minimum intensity of clathrin for an object to be used in this analysis. For example: "pGab1_basal_analysis=*extractCCPdata_2chan*(pGab1_basal,0)".
2. The resulting data structure ([result]) will have several fields, including the mean intensity of clathrin ([result]. Avg_clathrin_intensity) and mean intensity of the secondary channel ([result]. Avg_slave_channel_intensity).
3. As previously described, many CCPs detected by TIRF-M coupled to Gaussian modeling of the point-spread function of CCPs correspond to small and very short lived structures [51]. When analyzing time-lapse image series, these low intensity, transient structures (<10 s lifetimes) can be recognized by their intensity profile in the first few frames of their lifetime [51]. This allows these transient, very short lived structures to be discarded in order to study bona fide CCPs. Clearly, it is not possible to identify the highly transient, weakly fluorescent CCPs in single image frames in the same manner. If necessary for a specific analysis, it is possible to estimate that *bona fide* CCPs are enriched within a population of clathrin objects above the 50th percentile in clathrin intensity. As such, the *extractCCPdata_2chan.m* function can be run with a threshold value, to only obtain data from the objects detected with an intensity above a particular clathrin threshold. Furthermore, each run *of extractCCPdata_2chan.m* displays within the command line log the fluorescence value for the median clathrin intensity (50th percentile). To obtain a population of clathrin structures enriched in *bona fide* CCPs, the *extractCCPdata_2chan.m* function may be run again for all conditions using this intensity value. It is essential to note that the same threshold value must be used for all experimental conditions in the

analysis of a single experiment in order to allow comparison of the mean signal intensity values between conditions.

The mean rhodamine-EGF fluorescence intensity, corresponding to activated, ligand-bound EGFR, within CCPs at various times following the start of stimulation (corresponding to the analysis of the images obtained in Subheading 3.1) and the mean phosphorylated Gab1 (pGab1) within CCPs at various times following the start of EGF stimulation (corresponding to the analysis of the images obtained in Subheading 3.2) are shown in Fig. 5.

3.4.3 Two-Dimensional Histogram (Heat Map) for Clathrin and EGFR Signal Intensities Within Detected Objects

Obtaining an ensemble average of the intensity of specific signals within CCPs (as in Subheading 3.4.2 and Fig. 5) can be very informative, as it simplifies a large amount of information into a single parameter. However, given the heterogeneity of CCPs (as described in the introduction), it is often useful to examine the data in a manner that does not require this type of averaging. To do so, a heat map of the intensity of clathrin and EGFR signals (in this case, phosphorylated Gab1) can be generated to examine how the intensities of specific signals change within a population of CCPs upon EGF stimulation, using the *runCCPheatmap.m* function. This heat map is a two-dimensional (2D) histogram of the intensity of each channel.

A signal heat-map was generated from the analysis of the basal and 5 min EGF conditions of the TIRF image datasets labeled for clathrin and phosphorylated Gab1 (pGab1), as obtained from the experiment described in Subheading 3.2. This analysis shows a 2D histogram (heat map) depicting the frequency of each clathrin and pGab1 intensity cohort for the basal and 5 min EGF-stimulated conditions (Fig. 6a). More informative than the simple 2D histogram of the distribution of clathrin and pGab1 signals within an image dataset is how EGF stimulation *regulates* or changes the distribution of clathrin and pGab1 intensities within detected CCPs, as we describe in [70]. To do so, a heat map of the *difference* between these 2D histograms of EGF-stimulated and basal conditions is generated (called a *Δ-CCP-heatmap*), as shown in Fig. 6b. In the result of the *Δ-CCP-heatmap* analysis, any intensity cohort with a positive value represents a profile of CCP and pGab1 intensities increased in abundance upon EGF stimulation, while any cohort with a negative value is decreased in abundance upon EGF stimulation. Of note, the mean intensity of pGab1 within CCPs after 5 min of EGF stimulation is not different from that of the basal (Fig. 5c), yet the *Δ-CCP-heatmap* shown in Fig. 6b indicates remodeling of CCPs at 5 min EGF stimulation relative to the basal condition, as evinced by an increase in the frequency of pGab1-positive CCPs as well as pGab1-negative CCPs.

Hence, the use of 2D histograms (which does not rely on large ensemble averaging) to analyze the frequency distribution of CCPs

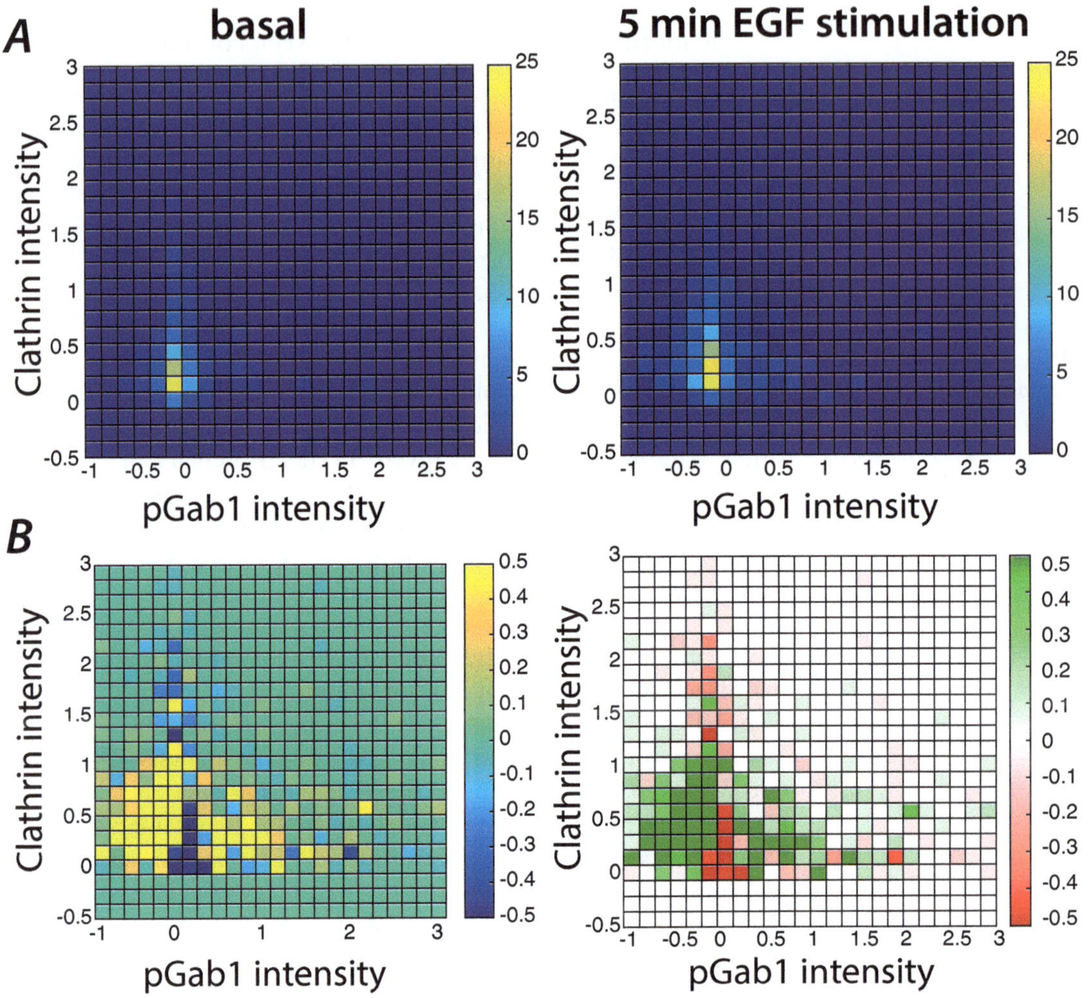

Fig. 6 Two-dimensional histograms reveal changes in phosphorylated Gab1 and clathrin fluorescence intensities within CCPs upon EGF stimulation. The image dataset in Figs. 4 and 5b (phosphorylated Gab1 enrichment within CCPs) was subjected to CCP heat map analysis as described in Subheading 3.4.3. The resulting two-dimensional histogram of clathrin (*y*-axis) and phosphorylated Gab1 (*x*-axis) intensity values within detected CCPs obtained using the *runCCPheatmap.m* is shown in the basal (*left panel*) and 5 min EGF conditions (*right panel*), using 25 cohorts. (**b**) Shown is the resulting *Δ-CCP-heatmap* analysis of the difference in various intensity cohorts between 5 min EGF and basal conditions. Positive values indicate ranges of clathrin and phosphorylated Gab1 intensity occurring at higher frequency in the 5 min EGF stimulated condition relative to basal, and negative values indicate ranges of clathrin and phosphorylated Gab1 intensity occurring at lower frequency in the 5 min EGF stimulated condition relative to basal. This *Δ-CCP-heatmap* is shown using the default color index (*left panel*), which can be adjusted to the red-green color index (*right panel*) that we described previously [70] using the Colormap Editor in Matlab

harboring specific EGFR signals (as in Fig. 6) can very useful, to complement analyses based on ensemble averages (as in Fig. 5). To generate these 2D histograms (heat maps), including the *Δ-CCP-heatmap*, with each dimension of the histogram having *c* number of cohorts can be done as follows:

1. To compare signal intensity properties of the entire population of CCPs between two conditions (e.g., basal vs EGF-stimulated), the two conditions to be compared are first loaded using the *loadConditionData.m* function as described in Subheading 3.4.1 (**steps 1–3**), and subjected to object detection as in Subheading 3.4.1 (**step 4**).
2. Run the function *runCCPheatmap.m* as follows: [result] = *runCCPheatmap*([dataset1], [dataset2], *c*).
3. The function will return a series of two-dimensional histograms, entitled Fig. 1–4. Figure 2 corresponds to the 2D histogram of fluorescence intensities of each channel in [dataset1], while Fig. 3 is a similar plot for [dataset2]. Figure 4 is the result of the *Δ-CCP-heatmap* representing the difference in frequency of each intensity cohort between dataset1 and dataset2, such that positive values represent intensity ranges that are more frequent in dataset2, while negative values represent intensity ranges that are more frequent in dataset1.
4. It may be helpful to rescale the color range using the "Open Colormap Editor" function for each figure in Matlab.
5. Files can be saved as .fig files to allow further editing in Matlab, or as .eps files to allow incorporation into a figure file for presentation or publication.

3.5 Resolving Unique Signaling Features of Specific CCP Subsets

CCPs are highly heterogeneous, with respect to their lifetime, size, and composition as defined by the recruitment of specific proteins from the cytosol [40, 42–46, 51–55]. Moreover, specific receptors are recruited to distinct subpopulations of CCPs, including EGFR, TfR, and Low Density Lipoprotein Receptor (LDLR) [56] and β1-integrin and TfR [112]. Importantly, the incorporation of specific receptors within particular CCPs imparts upon the latter certain properties and functions. For example, the incorporation of δ-opioid receptors (δ-OR) within CCPs results in the formation of longer-lived CCPs, a property of CCPs that is required for δ-OR signaling [55]. Furthermore, the incorporation of LDLR into CCPs increases the rate of assembly of clathrin and results in the formation of larger CCPs [43]. This phenomenon has indeed been revealed to also impact EGFR, for example as EGFR-containing CCPs are uniquely enriched in the lipid GM1 compared to CCPs harboring TfR [86].

As a result, understanding how specific subgroups of CCPs may uniquely function to control signaling is an important dimension to understanding how the spatiotemporal organization of signaling within the plasma membrane controls receptor signaling outcome. Indeed, while we observe robust quantitative recruitment of pGab1 within CCPs upon EGF stimulation (Figs. 4 and 5b), it is evident just by visual inspection that most CCPs do not exhibit enrichment of phospho-Gab1 within CCPs. We previously used automated detection of CCPs, followed by parsing CCPs into two cohorts: a population enriched in CCPs harboring EGFR (EGFR+) and one without detectable EGFR (EGFR−). This allowed us to resolve that the EGF-stimulated enrichment of pGab1 within CCPs was only detected within the EGFR+ CCP population, illustrating the selective behavior of CCPs harboring EGFR in the spatiotemporal organization of this signaling pathway [70].

Here, we present a method for three-color TIRF-M coupled to image analysis that allows parsing of CCPs into two cohorts following object detection. We describe the use of this method to study the relationship between CCPs (detected by eGFP-LCa), fluorescent EGF ligand to detect activated, ligand-bound EGFR (as per Subheading 3.1) and phosphorylated Gab1 (as per Subheading 3.2), followed by TIRF-M imaging and analysis (as per Subheadings 3.3 and 3.4, respectively). We also describe modifications related to sample preparation for three-color imaging as well as the unique analysis required for this strategy. Using this method, we readily observe CCPs harboring both rhodamine-EGF and phospho-Gab1 as early as 1–3 min after addition of the fluorescent ligand (Fig. 7a). Automated detection of CCPs and sorting into cohorts using an arbitrary threshold of EGF fluorescence corresponding to the ~85th percentile of EGF fluorescence within CCPs results in separation of CCPs into EGFR+ and EGFR− populations that exhibit high and negligible mean EGF fluorescence, respectively (Fig. 7b). The proportion of CCPs within the cohort of CCPs enriched for EGFR (EGFR+ CCPs) increases between 1 and 3 min of EGFR stimulation (Fig. 7c). Importantly, the enrichment of pGab1 upon EGF stimulation is restricted to the EGFR+ cohort of CCPs (Fig. 7d), as we reported previously [70], demonstrating the importance of tracking the recruitment of specific signals (e.g., pGab1) to subsets of CCPs defined by specific markers (e.g., EGFR).

3.5.1 Cell Preparation, Stimulation, Slide Preparation and TIRF-M

1. Prepare cells (culture, seeding), perform serum starvation and stimulate with fluorescently conjugated EGF as described in Subheading 3.1.1 (**steps 1–4**) and then proceed immediately to the next stage below.
2. Without delay, proceed with fixation and immunostaining to detect phosphorylated Gab1 as described in Subheading 3.2.2, with a slight modification for **step 8**, in which a fluorescent secondary conjugated to a compatible fluorophore is used (e.g., Alexa 647 conjugated anti-rabbit secondary).

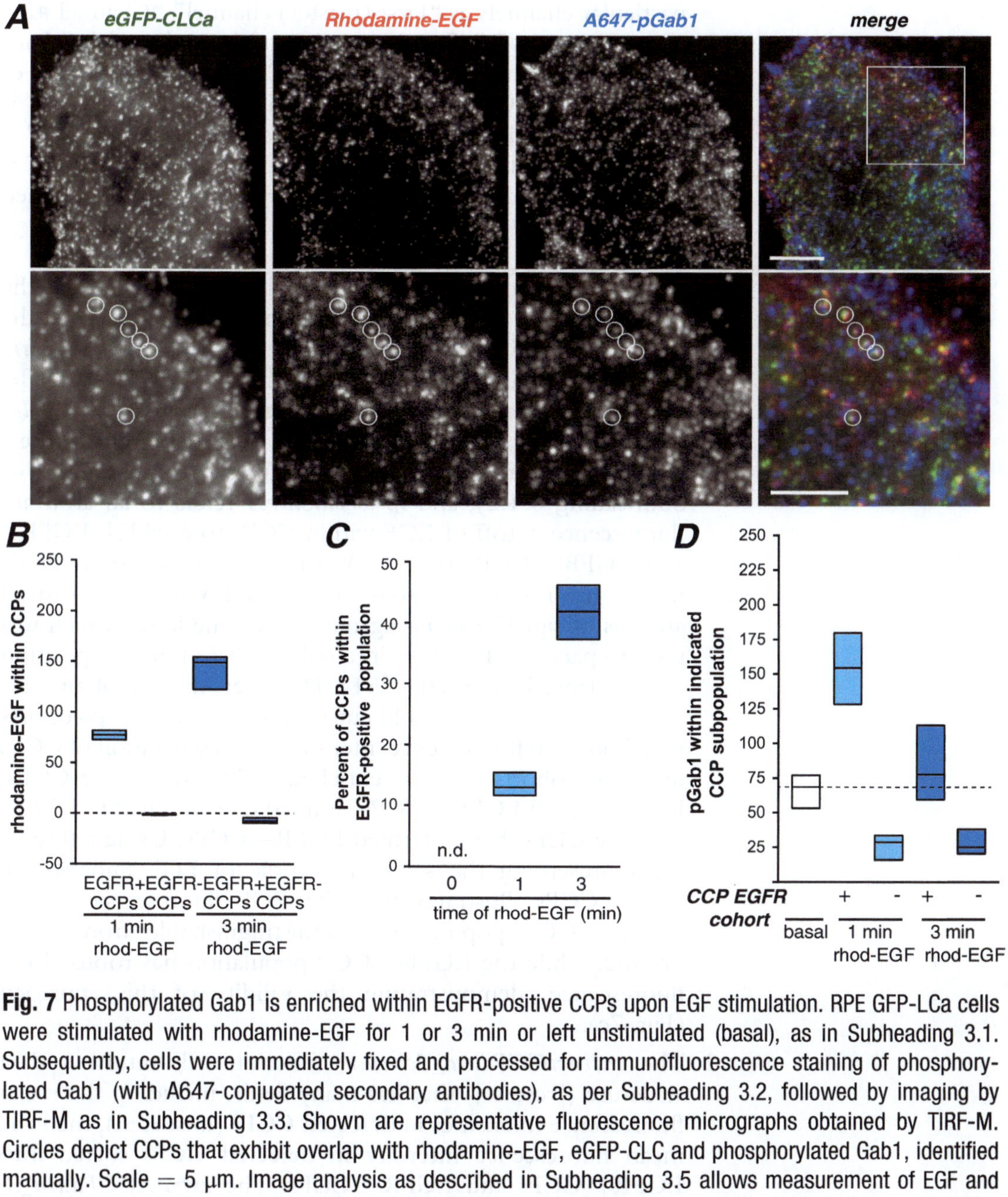

Fig. 7 Phosphorylated Gab1 is enriched within EGFR-positive CCPs upon EGF stimulation. RPE GFP-LCa cells were stimulated with rhodamine-EGF for 1 or 3 min or left unstimulated (basal), as in Subheading 3.1. Subsequently, cells were immediately fixed and processed for immunofluorescence staining of phosphorylated Gab1 (with A647-conjugated secondary antibodies), as per Subheading 3.2, followed by imaging by TIRF-M as in Subheading 3.3. Shown are representative fluorescence micrographs obtained by TIRF-M. Circles depict CCPs that exhibit overlap with rhodamine-EGF, eGFP-CLC and phosphorylated Gab1, identified manually. Scale = 5 μm. Image analysis as described in Subheading 3.5 allows measurement of EGF and phosphorylated Gab1 selectively within EGFR-positive and EGFR-negative CCPs. Shown are the median (*horizontal line*) and 75th and 25th percentile values (*boxes*) for rhodamine EGF fluorescence intensity per CCP cohort (**b**), the percent of CCPs that are classified as EGFR+ (**c**) and the phosphorylated Gab1 intensity per CCP cohort (**d**). The number of CCP (n) and cells (k) for each condition are basal: $n = 20{,}191$, $k = 13$; 1 min EGF: $n = 25{,}707$, $k = 14$; 3 min EGF: $n = 38{,}611$, $k = 22$

3. Perform TIRF microscopy (three channels), as described in Subheading 3.3.
4. Subject the images to CCP detection and analysis, as described in Subheading 3.4.1. The user will be prompted to designate

particular channels as "First (master) channel" "Channel #2," and "Channel #3." The designation of the order of the channels is consequential. To parse CCPs into populations of EGFR+ and EGFR– CCPs, the "First (master) channel" is that corresponding to clathrin fluorescence, "Channel #2" is that corresponding to EGF fluorescence and "Channel #3" is that corresponding to the particular signal being examined (i.e., pGab1).

3.5.2 Analysis of Three-Channel Images

1. Upon completion of the analysis for three-channel images, the mean intensity of various signals can be extracted from the stored analysis information using the *extractCCPdata_3chan.m* function. This function is called with the format: "[result]= *extractCCPdata_3chan*([dataset],*thresholdMaster*, *thresholdSlave1*)", where *thresholdMaster* refers to the minimum fluorescence of clathrin for a CCP to be counted in this analysis (*see* Subheading 3.4.2), and *thresholdSlave1* refers to an arbitrary fluorescence cutoff of EGF within CCPs to establish EGFR+ and EGFR– CCP cohorts. As this analysis occurs in single frames, it is not possible to use robust and systematic statistical analysis of signal over background in a time lapse, which was used to parse CCP objects into cohorts based on composition in time-lapse image series [51]. However, the use of an arbitrary fluorescence threshold cutoff to parse CCPs into cohorts based on EGF fluorescence intensity allows separation of CCPs into two cohorts: (a) one enriched in EGFR-positive CCPs, herein termed EGFR+ CCPs, and (b) one enriched in EGFR-negative CCPs, herein termed EGFR– CCPs. Using a threshold representing the ~85th percentile of EGF fluorescence within CCPs illustrates that the EGF fluorescence in the EGFR– CCP population is indistinguishable from background, while the EGFR+ CCP population has robust EGF fluorescence, demonstrating the validity of this approach (Fig. 7c).
2. The *extractCCPdata_3chan.m* function will return a data structure ([results]) that has many fields, including the mean fluorescence intensities in detected CCPs (unsorted), then the mean fluorescence intensity in detected CCPs sorted into EGFR+ CCPs and EGFR– CCPs. The *extractCCPdata_3chan.m* function is annotated with a description of each of these fields should further explanation be deemed useful. The principal results of interest may include the following:
   (a) *Fluorescence intensity of pGab1 within EGFR+ CCPs versus EGFR– CCPs.* To perform this comparison, extract the data within the results.Avg_tertiaryChannel_Intensity_EGFRpos (for pGab1 intensity in EGFR+ CCPs) and results.Avg_tertiaryChannel_Intensity_EGFRneg (for pGab1 intensity in EGFR– CCPs) fields in each of the

conditions examined (e.g., basal, EGF-stimulated). An example of this result for detection of pGab1 within these CCP populations is presented in Fig. 7d.

(b) *Fluorescence intensity of EGF within EGFR+ CCPs vs EGFR− CCPs.* To perform this comparison, extract the data within the results.Avg_EGF_Intensity_EGFRpos (for EGF intensity in EGFR+ CCPs) and results.Avg_EGF_Intensity_EGFRneg (for EGF intensity in EGFR− CCPs) in each of the conditions examined (e.g., basal, EGF-stimulated). An example of this result for detection of EGF (and thus ligand-bound EGFR) within these CCP populations is presented in Fig. 7c.

(c) *Fluorescence intensity of clathrin within EGFR+ CCPs vs EGFR− CCPs.* To perform this comparison, extract the data within the results.Avg_clathrin_Intensity_EGFRpos (for eGFP-CLC intensity in EGFR+ CCPs) and results.Avg_clathrin_Intensity_EGFRneg (for eGFP-CLC intensity in EGFR− CCPs) in each of the conditions examined (e.g., basal, EGF-stimulated).

4 Notes

1. *Importance of signal specificity for primary detection of CCPs or other structures.*

The protocol described here for the detection of CCPs can be adapted to study CCPs in cells that do not express fluorescent clathrin, or to study a number of different plasma membrane microdomains that also form diffraction-limited structures. Such strategies may involve primary detection of clathrin or other microdomains (e.g., specific tetraspanin proteins) by immunofluorescence labeling (e.g., of clathrin heavy chain to detect CCPs). When considering such approaches, it is important to validate the specificity of the antibodies and the effectiveness of the immunofluorescence staining protocol for the protein labeled to generate the "First (master) channel" signal that is used to detect objects (i.e., plasma membrane microdomains). For example, detection of CCPs by immunofluorescence staining with antibodies that specifically bind clathrin heavy chain yields reasonably good agreement of this signal with that of RFP-LCa expressed in these cells (Fig. 8). However, a number of puncta formed by the immunofluorescence labeling of clathrin heavy chain do not exhibit a corresponding RFP-LCa signal can be observed in these images. As RFP-LCa is incorporated into virtually all CCPs [51], this indicates that some, albeit a small minority, of the apparent structures observed by labeling with anti-clathrin heavy chain antibodies

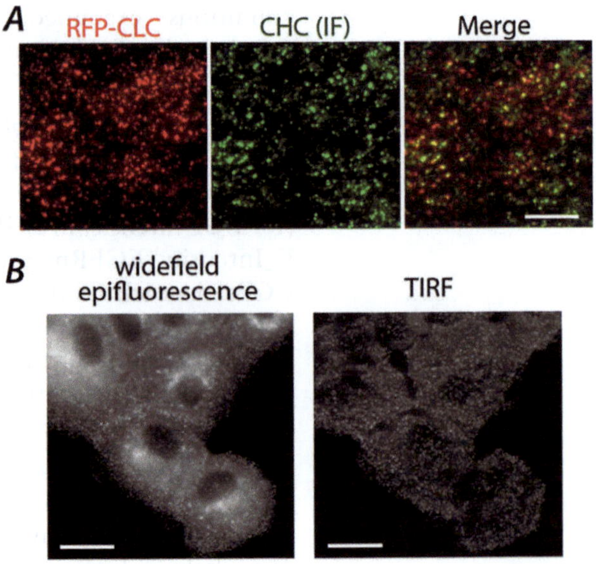

Fig. 8 Key considerations for TIRF-M imaging. (**b**) RPE RPE-LCa cells were fixed and stained with X22 anti-clathrin heavy chain antibodies [115], subsequently detected by A488-conjugated secondary antibodies, and imaged by TIRF-M, as described in Subheading 3.2. Shown are representative micrographs of RFP (clathrin) fluorescence and A488 (clathrin) fluorescence, scale = 10 μm (*top panels*) 10 μm (*lower panels*). Significant overlap of the two signals can be observed, but a number of structures positive for A488 (antibody-stained clathrin) fluorescence and negative for RFP-LCa fluorescence are also visible; the former likely correspond to nonspecific antibody labeling. (**b**) RPE GFP-LCa cells were fixed subjected to imaging by widefield epifluorescence and TIRF-M. Scale = 20 μm

may be nonspecific artifacts. When the resulting images are subjected to automated detection of diffraction-limited puncta, these nonspecific structures will be detected as CCPs and may skew the quantification of EGF or EGFR-derived signals (pGab1) within CCPs under various conditions. Indeed in our experience, many antibodies have some degree of nonspecific staining of the plasma membrane that manifests as punctate structures when observed by TIRF-M. Thus, refinement and validation of immunofluorescence staining for the detection of the "First (master) channel" protein is essential for the use of this approach.

It is also possible that the antibodies used to stain the EGFR-derived signals "Channel #2" or "Channel #3" may also exhibit nonspecific background labeling. Since the analysis only examines fluorescence in these secondary channels at or near the site of CCPs detected in the "First (master) channel", relative to the local background fluorescence intensity, the nonspecific staining in this situation is less of a concern. To determine if any of the detection of a secondary channel fluorescence within CCPs may be due to random overlap of nonspecific signals, it is possible to repeat the analysis described in Subheading 3.5 in a separate image set in which the secondary channel images have each been rotated 180° (to randomize the position of objects in that channel relative to CCPs in the "First (master) channel"), as we describe previously [70, 71]. This analysis allows one to distinguish signals specifically enriched within CCPs from those resulting from random, nonspecific fluorescence overlap in these structures.

2. *Acid washing coverslips to ensure quality of TIRF-M.*
The cleanliness of coverslips used in fluorescence microscopy is imperative in obtaining clean, resolvable and valuable images. Although coverslips appear visually clean when obtained directly from prepackaged commercial sources, each coverslip may in fact be coated with a thin film of grease due to manufacturing processes [113]. Any such deposits may interfere with cell adhesion or other aspects of imaging. This is particularly critical for TIRF-M, as uneven cell adhesion can result in a nonuniform apposition of the cell membrane to the coverslip and thus problems with maintaining uniform illumination of the cell surface. To achieve uniform imaging and the best quality images, coverslips should be routinely prepared by acid stripping, wherein the thin film of grease is effectively removed [113], as follows:
   (a) Several ounces of coverslips are removed from their packaging and placed in a 1 L glass beaker. For all stages, coverslips should be handled with gloves.
   (b) Coverslips are heated in a glass container to 50–60 °C for 4–16 h in 1 M HCl. Occasional agitation should be carefully administered to ensure even washing of the coverslips.
   (c) The coverslips are then washed 4× in distilled water.
   (d) The coverslips are then rinsed 3× with 100% ethanol.
   (e) The coverslips can either be kept dry in a clean container or placed in an ethanol filled jar for storage.

3. *Wet mounting of coverslips using VALAP seal.*
TIRF-M can be readily performed with slides mounted within imaging chambers. When these are not available, it is possible to perform a wet mount of the coverslips onto a glass slide, using a technique that involves construction of a small aqueous chamber between the coverslip and slide, and sealing of this chamber with an inert wax (VALAP), as follows:
   (a) Prepare slides by placing two thin strips (~2 mm wide, 1.5 cm long) of two-layered double-sided tape along each

edge of the glass slide, in the middle of the slide. This will serve as a platform for the coverslip to rest on.

(b) Using forceps, remove the coverslip from its well and lightly dry off the PBS from the non-cell side of the coverslip (e.g., using a Kimwipes). Place the coverslip gently on the slide, cell-side down, so that it rests as a bridge between the two strips of preplaced double-sided tape on the glass slide.

(c) Pipette about 100 μL of PBS in between the slide and the coverslip while minimizing air bubbles. Lightly dry the edges of the coverslip of any leaks, ensuring that PBS remains only under the coverslip.

(d) Seal the edges of the mounted coverslip with VALAP (a mixture of petroleum jelly, lanolin, and paraffin, 1:1:1 [w/w/w], prepared as in [114]) and heated to just above melting using a cotton swab applicator. Upon solidification of VALAP wax, check the slide for leaks by lightly tapping on the coverslip which forces PBS out of weak points of the VALAP seal. If leaks are present, simply dry the leak and reapply VALAP on the respective area.

4. *Determination of TIRF microscopy illumination.*
The usefulness of TIRF-M imaging is the selective illumination of fluorophores within a short distance (~100–200 nm) from the coverslip, and care must be taken to ensure conditions for optimal TIRF illumination, with limited direct illumination of fluorophores distant from the cell surface. Typically, this is achieved by increasing the angle of illumination during TIRF-M to achieve a low penetration depth of the illuminating light [110, 111], yet there is a limit to this as signal intensity also decreases with increased illumination angle. The actual penetration depth of the TIRF-M illumination depends on a few parameters, including excitation wavelength, the angle of incident light illumination, and the refractive index of the material being imaged. While some instruments allow TIRF illumination at specific penetration depths into the sample, some caution must be taken when using these settings, as there is variability in the refractive index of typical cellular material being imaged that can result in variation in the true penetration depth being imaged. TIRF-M penetration depth can be determined upon calibration through the use of synthetic fluorescent materials with specific sizes; however, this can be laborious. A useful guide to ensure imaging in TIRF is to examine the distribution of CCP punctate objects. A TIRF image will result in restricted illumination of CCPs at the plasma membrane, thus revealing a largely uniform distribution of CCPs throughout the ventral surface of the cell (Fig. 8b,

right panel). In contrast, illumination with significant levels of light directly entering the sample will result in visible fluorescent clathrin surrounding the perinuclear region (e.g. from clathrin localized to the trans-Golgi network) (Fig. 8a, left panel). This can be taken as a guide to determine if a particular illumination setting profile results in suitable TIRF illumination. Notably, when imaging coverslips that correspond to conditions that will be compared in a single experiment (e.g., basal, EGF-stimulated), identical TIRF settings should be used for imaging of each coverslip.

Acknowledgments

This was supported by an Operating Grant from the Canadian Institutes of Health Research (CIHR) to C.N.A (grant no. 125854).

References

1. Herbst RS (2004) Review of epidermal growth factor receptor biology. Int J Radiat Oncol 59:S21–S26
2. Hynes NE, MacDonald G (2009) ErbB receptors and signaling pathways in cancer. Curr Opin Cell Biol 21:177–184
3. Sanchez-Soria P, Camenisch TD (2010) ErbB signaling in cardiac development and disease. Semin Cell Dev Biol 21:929–935
4. Schneider MR, Werner S, Paus R, Wolf E (2008) Beyond wavy hairs: the epidermal growth factor receptor and its ligands in skin biology and pathology. Am J Pathol 173:14–24
5. Luetteke NC et al (1999) Targeted inactivation of the EGF and amphiregulin genes reveals distinct roles for EGF receptor ligands in mouse mammary gland development. Development 126:2739–2750
6. Jura N et al (2009) Mechanism for activation of the EGF receptor catalytic domain by the juxtamembrane segment. Cell 137:1293–1307
7. Zhang X, He X, X-Y F, Chang Z (2006) Varp is a Rab21 guanine nucleotide exchange factor and regulates endosome dynamics. J Cell Sci 119:1053–1062
8. Alvarado D, Klein DE, Lemmon MA (2010) Structural basis for negative cooperativity in growth factor binding to an EGF receptor. Cell 142:568–579
9. Oda K, Matsuoka Y, Funahashi A, Kitano H (2005) A comprehensive pathway map of epidermal growth factor receptor signaling. Mol Syst Biol 1:2005.0010
10. Mattoon DR, Lamothe B, Lax I, Schlessinger J (2004) The docking protein Gab1 is the primary mediator of EGF-stimulated activation of the PI-3K/Akt cell survival pathway. BMC Biol 2:24
11. Holgado-Madruga M, Emlet DR, Moscatello DK, Godwin AK, Wong AJ (1996) A Grb2-associated docking protein in EGF- and insulin-receptor signalling. Nature 8:560–564
12. Rodrigues GA, Falasca M, Zhang Z, Ong SH, Schlessinger J (2000) A novel positive feedback loop mediated by the docking protein Gab1 and phosphatidylinositol 3-kinase in epidermal growth factor receptor signaling. Mol Cell Biol 20:1448–1459
13. Matheny RW, Adamo ML (2009) Current perspectives on Akt Akt-ivation and Akt-ions. Exp Biol Med 234:1264–1270
14. Manning BD, Cantley LC (2007) AKT/PKB signaling: navigating downstream. Cell 129:1261–1274
15. Miettinen PJ et al (1995) Epithelial immaturity and multiorgan failure in mice lacking epidermal growth factor receptor. Nature 376:337–341
16. Sibilia M, Wagner EF (1995) Strain-dependent epithelial defects in mice lacking the EGF receptor. Science 269:234–238
17. Threadgill DW et al (1995) Targeted disruption of mouse EGF receptor: effect of genetic

background on mutant phenotype. Science 269:230–234

18. Sibilia M, Steinbach JP, Stingl L, Aguzzi A, Wagner EF (1998) A strain-independent postnatal neurodegeneration in mice lacking the EGF receptor. EMBO J 17:719–731
19. Gassmann M et al (1995) Aberrant neural and cardiac development in mice lacking the ErbB4 neuregulin receptor. Nature 378:390–394
20. Golub MS, Germann SL, Lloyd KCK (2004) Behavioral characteristics of a nervous system-specific erbB4 knock-out mouse. Behav Brain Res 153:159–170
21. Leu M et al (2003) Erbb2 regulates neuromuscular synapse formation and is essential for muscle spindle development. Development 130:2291–2301
22. Bhargava R et al (2005) EGFR gene amplification in breast cancer: correlation with epidermal growth factor receptor mRNA and protein expression and HER-2 status and absence of EGFR-activating mutations. Mod Pathol 18:1027–1033
23. Yewale C, Baradia D, Vhora I, Patil S, Misra A (2013) Epidermal growth factor receptor targeting in cancer: a review of trends and strategies. Biomaterials 34:8690–8707
24. Tebbutt N, Pedersen MW, Johns TG (2013) Targeting the ERBB family in cancer: couples therapy. Nat Rev Cancer 13:663–673
25. Ueno NT, Zhang D (2011) Targeting EGFR in triple negative breast cancer. J Cancer 2:324–328
26. Normanno N et al (2006) Epidermal growth factor receptor (EGFR) signaling in cancer. Gene 366:2–16
27. Sorkin A, Goh LK (2009) Endocytosis and intracellular trafficking of ErbBs. Exp Cell Res 315:683–696
28. Goh LK, Sorkin A (2013) Endocytosis of receptor tyrosine kinases. Cold Spring Harb Perspect Biol 5:a017459
29. Sigismund S et al (2008) Clathrin-mediated internalization is essential for sustained EGFR signaling but dispensable for degradation. Dev Cell 15:209–219
30. McMahon HT, Boucrot E (2011) Molecular mechanism and physiological functions of clathrin-mediated endocytosis. Nat Rev Mol Cell Biol 12:517–533
31. Kiyatkin A et al (2006) Scaffolding protein Grb2-associated binder 1 sustains epidermal growth factor-induced mitogenic and survival signaling by multiple positive feedback loops. J Biol Chem 281:19925–19938
32. Kholodenko BN, Hancock JF, Kolch W (2010) Signalling ballet in space and time. Nat Rev Mol Cell Biol 11:414–426
33. Conner SD, Schmid SL (2003) Regulated portals of entry into the cell. Nature 422:37–44
34. Schmid EM, McMahon HT (2007) Integrating molecular and network biology to decode endocytosis. Nature 448:883–888
35. Schmid EM et al (2006) Role of the AP2 beta-appendage hub in recruiting partners for clathrin-coated vesicle assembly. PLoS Biol 4:e262
36. Blondeau F et al (2004) Tandem MS analysis of brain clathrin-coated vesicles reveals their critical involvement in synaptic vesicle recycling. Proc Natl Acad Sci U S A 101:3833–3838
37. Borner GHH, Harbour M, Hester S, Lilley KS, Robinson MS (2006) Comparative proteomics of clathrin-coated vesicles. J Cell Biol 175:571–578
38. Posor Y et al (2013) Spatiotemporal control of endocytosis by phosphatidylinositol-3,4-bisphosphate. Nature 499:233–237
39. Perera RM, Zoncu R, Lucast L, De Camilli P, Toomre D (2006) Two synaptojanin 1 isoforms are recruited to clathrin-coated pits at different stages. Proc Natl Acad Sci U S A 103:19332–19337
40. Antonescu CN, Aguet F, Danuser G, Schmid SL (2011) Phosphatidylinositol-(4,5)-bisphosphate regulates clathrin-coated pit initiation, stabilization, and size. Mol Biol Cell 22:2588–2600
41. Hussain NK et al (1999) Splice variants of intersectin are components of the endocytic machinery in neurons and nonneuronal cells. J Biol Chem 274:15671–15677
42. Mettlen M et al (2009) Endocytic accessory proteins are functionally distinguished by their differential effects on the maturation of clathrin-coated pits. Mol Biol Cell 20:3251–3260
43. Mettlen M, Loerke D, Yarar D, Danuser G, Schmid SL (2010) Cargo- and adaptor-specific mechanisms regulate clathrin-mediated endocytosis. J Cell Biol 188:919–933
44. Loerke D, Mettlen M, Schmid SL, Danuser G (2011) Measuring the hierarchy of molecular

events during clathrin-mediated endocytosis. Traffic 12:815–825

45. Loerke D et al (2009) Cargo and dynamin regulate clathrin-coated pit maturation. PLoS Biol 7:e57
46. Nunez D et al (2011) Hotspots organize clathrin-mediated endocytosis by efficient recruitment and retention of nucleating resources. Traffic 12:1868–1878
47. Jaqaman K et al (2008) Robust single-particle tracking in live-cell time-lapse sequences. Nat Methods 5:695–702
48. Liu AP, Loerke D, Schmid SL, Danuser G (2009) Global and local regulation of clathrin-coated pit dynamics detected on patterned substrates. Biophys J 97:1038–1047
49. Reis CR et al (2015) Crosstalk between Akt/GSK3β signaling and dynamin-1 regulates clathrin-mediated endocytosis. EMBO J 13:2132–2146
50. Cocucci E, Aguet F, Boulant S, Kirchhausen T (2012) The first five seconds in the life of a clathrin-coated pit. Cell 150:495–507
51. Aguet F, Antonescu CN, Mettlen M, Schmid SL, Danuser G (2013) Advances in analysis of low signal-to-noise images link dynamin and AP2 to the functions of an endocytic checkpoint. Dev Cell 26:279–291
52. Ehrlich M et al (2004) Endocytosis by random initiation and stabilization of clathrin-coated pits. Cell 118:591–605
53. Antonescu CN, Danuser G, Schmid SL (2010) Phosphatidic acid plays a regulatory role in clathrin-mediated endocytosis. Mol Biol Cell 21:2944–2952
54. Taylor MJ, Perrais D, Merrifield CJ (2011) A high precision survey of the molecular dynamics of Mammalian clathrin-mediated endocytosis. PLoS Biol 9:e1000604
55. Puthenveedu MA, von Zastrow M (2006) Cargo regulates clathrin-coated pit dynamics. Cell 127:113–124
56. Tosoni D et al (2005) TTP specifically regulates the internalization of the transferrin receptor. Cell 123:875–888
57. Motley A, Bright NA, Seaman MNJ, Robinson MS (2003) Clathrin-mediated endocytosis in AP-2-depleted cells. J Cell Biol 162:909–918
58. Conner SD, Schroter T, Schmid SL (2003) AAK1-mediated micro2 phosphorylation is stimulated by assembled clathrin. Traffic 4:885–890
59. Wilde A et al (1999) EGF receptor signaling stimulates SRC kinase phosphorylation of clathrin, influencing clathrin redistribution and EGF uptake. Cell 96:677–687
60. Eisenberg E, Greene LE (2007) Multiple roles of auxilin and Hsc70 in clathrin-mediated endocytosis. Traffic 8:640–646
61. Vieira AV, Lamaze C, Schmid SL (1996) Control of EGF receptor signaling by clathrin-mediated endocytosis. Science 274: 2086–2089
62. Cohen S, Fava RA (1985) Internalization of functional epidermal growth factor: receptor/kinase complexes in A-431 cells. J Biol Chem 260:12351–12358
63. Wada I, Lai WH, Posner BI, Bergeron JJ (1992) Association of the tyrosine phosphorylated epidermal growth factor receptor with a 55-kD tyrosine phosphorylated protein at the cell surface and in endosomes. J Cell Biol 116:321–330
64. Di Guglielmo GM, Baass PC, WJ O, Posner BI, Bergeron JJ (1994) Compartmentalization of SHC, GRB2 and mSOS, and hyperphosphorylation of Raf-1 by EGF but not insulin in liver parenchyma. EMBO J 13:4269–4277
65. Wang Z, Tung PS, Moran MF (1996) Association of p120 ras GAP with endocytic components and colocalization with epidermal growth factor (EGF) receptor in response to EGF stimulation. Cell Growth Differ 7:123–133
66. Haugh JM, Huang AC, Wiley HS, Wells A, Lauffenburger DA (1999) Internalized epidermal growth factor receptors participate in the activation of p21(ras) in fibroblasts. J Biol Chem 274:34350–34360
67. Burke P, Schooler K, Wiley HS (2001) Regulation of epidermal growth factor receptor signaling by endocytosis and intracellular trafficking. Mol Biol Cell 12:1897–1910
68. Wang Y, Pennock S, Chen X, Wang Z (2002) Endosomal signaling of epidermal growth factor receptor stimulates signal transduction pathways leading to cell survival. Mol Cell Biol 22:7279–7290
69. Sousa LP et al (2012) Suppression of EGFR endocytosis by dynamin depletion reveals that EGFR signaling occurs primarily at the plasma membrane. Proc Natl Acad Sci U S A 109:4419–4424
70. Garay C et al (2015) Epidermal growth factor-stimulated Akt phosphorylation requires clathrin or ErbB2 but not receptor endocytosis. Mol Biol Cell 26(19):3504
71. Lucarelli S, Pandey R, Judge G, Antonescu CN (2016) Similar requirement for clathrin

in EGF- and HGF- stimulated Akt phosphorylation. Commun Integr Biol 9:e1175696

72. Delos Santos RC, Garay C, Antonescu CN (2015) Charming neighborhoods on the cell surface: plasma membrane microdomains regulate receptor tyrosine kinase signaling. Cell Signal 27:1963–1976
73. Grove J et al (2014) Flat clathrin lattices: stable features of the plasma membrane. Mol Biol Cell 25:3581–3594
74. Elenko E et al (2003) Spatial regulation of Gαi protein signaling in clathrin-coated membrane microdomains containing GAIP. Mol Pharmacol 64:11–20
75. Eichel K, Jullié D, von Zastrow M (2016) β-Arrestin drives MAP kinase signalling from clathrin-coated structures after GPCR dissociation. Nat Cell Biol 18:303–310
76. Ibach J et al (2015) Single particle tracking reveals that EGFR signaling activity is amplified in clathrin-coated pits. PLoS One 10: e0143162
77. Torreno-Pina JA et al (2014) Enhanced receptor-clathrin interactions induced by N-glycan-mediated membrane micropatterning. Proc Natl Acad Sci U S A 111:11037–11042
78. Simons K, Toomre D (2000) Lipid rafts and signal transduction. Nat Rev Mol Cell Biol 1:31–39
79. Anchisi L, Dessì S, Pani A, Mandas A (2013) Cholesterol homeostasis: a key to prevent or slow down neurodegeneration. Front Physiol 3:1–12
80. Pike LJ (2009) The challenge of lipid rafts. J Lipid Res 50(Suppl):S323–S328
81. Mukherjee S, Maxfield FR (2004) Membrane domains. Annu Rev Cell Dev Biol 20:839–866
82. Sonnino S, Prinetti A (2013) Membrane domains and the “lipid raft” concept. Curr Med Chem 20:4–21
83. Cambi A, Lidke DS (2014) Cell membrane and nanodomains: from biochemistry to nanoscopy, 1st edn. CRC Press, Boca Raton, FL
84. Pike LJ (2006) Rafts defined: a report on the keystone symposium on lipid rafts and cell function. J Lipid Res 47:1597–1598
85. Ringerike T, Blystad FD, Levy FO, Madshus IH, Stang E (2002) Cholesterol is important in control of EGF receptor kinase activity but EGF receptors are not concentrated in caveolae. J Cell Sci 115:1331–1340
86. Puri C et al (2005) Relationships between EGFR signaling-competent and endocytosis-competent membrane microdomains. Mol Biol Cell 16:2704–2718
87. Rappoport JZ, Simon SM (2009) Endocytic trafficking of activated EGFR is AP-2-dependent and occurs through preformed clathrin spots. J Cell Sci 122:1301–1305
88. Kazazic M et al (2006) EGF-induced activation of the EGF receptor does not trigger mobilization of caveolae. Traffic 7: 1518–1527
89. Roepstorff K, Thomsen P, Sandvig K, Van Deurs B (2002) Sequestration of epidermal growth factor receptors in non-caveolar lipid rafts inhibits ligand binding. J Biol Chem 277:18954–18960
90. Hofman EG et al (2008) EGF induces coalescence of different lipid rafts. J Cell Sci 121:2519–2528
91. Lin CY, Huang JY, Lo L-W (2015) Unraveling the impact of lipid domains on the dimerization processes of single-molecule EGFRs of live cells. Biochim Biophys Acta Biomembr 1848:886–893
92. Mineo C, Gill GN, Richard G, Anderson W, Anderson RGW (1999) Regulated migration of epidermal growth factor receptor from caveolae. J Biol Chem 274:30636–30643
93. Irwin ME, Mueller KL, Bohin N, Ge Y, Boerner JL (2011) Lipid raft localization of EGFR alters the response of cancer cells to the EGFR tyrosine kinase inhibitor gefitinib. J Cell Physiol 226:2316–2328
94. Hea YO et al (2007) Cholesterol level of lipid raft microdomains regulates apoptotic cell death in prostate cancer cells through EGFR-mediated Akt and ERK signal transduction. Prostate 67:1061–1069
95. Li YC, Park MJ, Ye S-K, Kim C-W, Kim Y-N (2006) Elevated levels of cholesterol-rich lipid rafts in cancer cells are correlated with apoptosis sensitivity induced by cholesterol-depleting agents. Am J Pathol 168: 1107–1118
96. Yáñez-Mó M, Barreiro O, Gordon-Alonso M, Sala-Valdés M, Sánchez-Madrid F (2009) Tetraspanin-enriched microdomains: a functional unit in cell plasma membranes. Trends Cell Biol 19:434–446
97. Danglot L et al (2010) Role of TI-VAMP and CD82 in EGFR cell-surface dynamics and signaling. J Cell Sci 123:723–735
98. Liu AP, Aguet F, Danuser G, Schmid SL (2010) Local clustering of transferrin receptors promotes clathrin-coated pit initiation. J Cell Biol 191:1381–1393
99. Ting AY, Kain KH, Klemke RL, Tsien RY (2001) Genetically encoded fluorescent

reporters of protein tyrosine kinase activities in living cells. Proc Natl Acad Sci U S A 98:15003–15008
100. Wang Y et al (2005) Visualizing the mechanical activation of Src. Nature 434:1040–1045
101. Tsou P, Zheng B, Hsu C-H, Sasaki AT, Cantley LC (2011) A fluorescent reporter of AMPK activity and cellular energy stress. Cell Metab 13:476–486
102. Fekri F, Delos Santos R, Karshafian R, Antonescu C (2016) Ultrasound microbubble treatment enhances clathrin-mediated endocytosis and fluid-phase uptake through distinct mechanisms. PLoS One 11:e0156754
103. Bone LN et al (2017) The acyltransferase LYCAT controls specific phosphoinositides and related membrane traffic. Mol Biol Cell 28(1):161–172
104. Crupi MJ et al (2015) Distinct temporal regulation of RET isoform internalization: roles of clathrin and AP2. Traffic 16:1155–1173
105. Pinilla-Macua I, Watkins SC, Sorkin A (2016) Endocytosis separates EGF receptors from endogenous fluorescently labeled HRas and diminishes receptor signaling to MAP kinases in endosomes. Proc Natl Acad Sci U S A 113:2122–2127
106. Nakatsu F et al (2010) The inositol 5-phosphatase SHIP2 regulates endocytic clathrin-coated pit dynamics. J Cell Biol 190:307–315
107. Goh LK, Huang F, Kim W, Gygi S, Sorkin A (2010) Multiple mechanisms collectively regulate clathrin-mediated endocytosis of the epidermal growth factor receptor. J Cell Biol 189:871–883
108. Wang Q et al (2005) Control of epidermal growth factor receptor endocytosis by receptor dimerization, rather than receptor kinase activation. EMBO Rep 6:942–948
109. Wang Q, Chen X, Wang Z (2015) Dimerization drives EGFR endocytosis through two sets of compatible endocytic codes. J Cell Sci 128:935–950
110. Mattheyses AL, Simon SM, Rappoport JZ (2010) Imaging with total internal reflection fluorescence microscopy for the cell biologist. J Cell Sci 123:3621–3628
111. Axelrod D (2001) Total internal reflection fluorescence microscopy in cell biology. Traffic 2:764–774
112. Teckchandani A, Mulkearns EE, Randolph TW, Toida N, Cooper JA (2012) The clathrin adaptor Dab2 recruits EH domain scaffold proteins to regulate integrin β1 endocytosis. Mol Biol Cell 23:2905–2916
113. Fischer AH, Jacobson KA, Rose J, Zeller R (2008) Preparation of slides and coverslips for microscopy. CSH Protoc 2008:pdb.prot4988
114. Cold Spring Harbor Protocols (2010) Valap for agar mounts. Cold Spring Harb Protoc 2010:pdb.rec12380–pdb.rec12380
115. Blank GS, Brodsky FM (1986) Site-specific disruption of clathrin assembly produces novel structures. EMBO J 5:2087–2095

Part IV

Methods for Studying EGFR Signaling in Drosophila

Chapter 16

Studying Nonproliferative Roles for Egfr Signaling in Tissue Morphogenesis Using Dorsal Closure of the Drosophila Embryo

Bruce Reed and Nicholas Harden

Abstract

For several decades, genetic analysis in Drosophila has made important contributions to the understanding of signaling by Egfr. Egfr has been well characterized with regard to its oncogenic potential but is also being studied for its roles in organismal development. We have recently developed dorsal closure of the Drosophila embryo as a system for characterizing Egfr regulation of events that do not involve proliferation, as no cell divisions occur during this process. Dorsal closure is essentially a developmental wound healing event with parallels to vertebrate developmental epithelial fusions such as neural tube closure and palate fusion. We describe here a set of materials and protocols for studying Egfr signaling during dorsal closure, including assessing effects of altering Egfr signaling on other pathways, gene expression and, using live imaging, morphogenesis and programmed cell death. Although this "tool kit" is designed for looking at Egfr, it can be readily adapted to look at the participation of any signaling molecule in dorsal closure.

Key words Drosophila, Egfr, Dorsal closure, Epithelial morphogenesis, In situ hybridization, Immunohistochemistry, Live imaging

1 Introduction

Although the Egfr-MAP kinase signaling cascade has been most thoroughly studied with regard to its promotion of cell proliferation in cancer, the pathway has numerous roles in both vertebrate and invertebrate development that in many cases do not involve cell division [1–3]. Understanding Egfr signaling in development is not only of interest to developmental biologists but also provides important insights into the participation of Egfr in disease. We recently established that Egfr functions in an epithelial fusion event during embryonic development in Drosophila [4]. Through several nonproliferative roles, Egfr regulates dorsal closure (DC), a developmental wound healing event in which a hole in the dorsal epidermis is sealed by migration of the epidermal flanks [4, 5]. There are no cell divisions occurring during DC, and it presents a

Zhixiang Wang (ed.), *ErbB Receptor Signaling: Methods and Protocols*, Methods in Molecular Biology, vol. 1652, DOI 10.1007/978-1-4939-7219-7_16, © Springer Science+Business Media LLC 2017

nice model for the genetic analysis of Egfr regulation of tissue morphogenesis that does not involve proliferation. Twenty-five years of genetic analysis of Egfr signaling in Drosophila has demonstrated that Egfr function in the fly has many parallels with its roles in vertebrates. Egfr participates in vertebrate wound healing and vertebrate developmental epithelial fusions and what is learned about Egfr in DC may aid in its positioning in the signaling regulating these processes [6–8].

DC involves the coordinated morphogenesis of two tissues, the epidermis and a transient epithelium occupying the dorsal hole, the amnioserosa. Genes participating in DC are turned on in either the dorsal most epidermal (DME) cells at the leading edge of the migrating epidermis, the amnioserosa or both [5]. The TGF-β family member Dpp is a key regulator of DC secreted by the DME cells that signals to the amnioserosa as part of a reciprocal communication network operating between the two tissues that ensures smooth closure [9–17]. Nonmuscle myosin heavy chain encoded by the *zipper* (*zip*) gene is expressed by both the DME cells and the amnioserosa and contributes to tissue morphogenesis as part of the actomyosin contractile apparatus [5, 18]. Another force contributing to DC is apoptosis of amnioserosa cells, which drives morphogenesis of the tissue through cell extrusion [19]. Using the techniques presented below, we demonstrated that Egfr signaling acts as a brake on DC, by repressing gene expression and amnioserosa apoptosis [4]. This tool kit will be useful for further characterizing Egfr function in tissue morphogenesis, including addressing interactions with other signaling pathways. In addition, the set of techniques presented here are readily applicable for the study of any other signaling molecule participating in DC, a process with striking parallels to both developmental and regenerative morphogenesis of epithelia across the metazoans.

2 Materials

2.1 Fly Stocks

Table 1 lists a collection of fly strains for manipulating Egfr function during DC and evaluating the consequences. At present there is a plethora of Drosophila stocks available at relatively little expense from various stock centers, most notably the Bloomington Drosophila Stock Center (BDSR), the Kyoto Stock Center, and the Vienna Drosophila Resource Center (VDRC), among several others (see Flybase link http://flybase.org/wiki/FlyBase:Stocks for links to most stock centers). The resources available from these stock centers have recently swelled to a point where many novice Drosophila researchers are overwhelmed by the task of navigating the myriad of lists, subheadings, and search queries. Even within the limited scope of searching for resources available for the analysis of the Egfr signaling pathway, in addition to

Table 1
Drosophila stocks useful in the analysis of Egfr signaling and live imaging dorsal closure

Stock or transgene description	General category	Utility	Source
cn $Egfr^{f2}$ *bw sp/CyO*	Egfr signaling	Strong loss-of-function allele of Egfr. Recessive lethal with defective germ band retraction.	Bloomington stock 2768
cn $Egfr^{1a15}$ *bw sp/CyO*	Egfr signaling	Loss-of-function allele of Egfr. Recessive lethal with defective germ band retraction.	B. Reed
cn $Egfr^{H25}$ *bw sp/CyO*	Egfr signaling	Loss-of-function allele of Egfr. Recessive lethal with defective germ band retraction.	B. Reed
cn $Egfr^{2C82}$ *bw sp/CyO*	Egfr signaling	Moderate loss-of-function allele of Egfr. Recessive lethal with defective germ band retraction.	T. Schüpbach
cn $Egfr^{1F26}$ *bw sp/CyO*	Egfr signaling	Conditional temperature sensitive loss-of-function allele of Egfr. Restrictive temperature 29 °C.	T. Schüpbach
UAS-Egfr- $RNAi^{VDRC43268}$	Egfr signaling	Reporter construct expresses dsRNA for RNAi mediated knockdown of Egfr expression under GAL4 control.	Vienna Drosophila Resource Center stock v43268
UAS-Egfr- $RNAi^{TRIP.JF01368}$	Egfr signaling	Reporter construct expresses dsRNA for RNAi mediated knockdown of Egfr expression under GAL4 control.	Bloomington stock 25781
UAS-λ-top4.2	Egfr signaling	Reporter construct expresses an activated form of Egfr under GAL4 control.	T. Schüpbach
UAS- $Egfr^{DN.B}$	Egfr signaling	Reporter construct expresses a dominant negative form of Egfr under GAL4 control.	Bloomington stock 5364 (carries two insertions, one on 2nd and one on 3rd)
UAS-Egfr-eGFP	Egfr signaling	Reporter construct expresses a Egfr-eGFP fusion protein that is biologically active.	J. Duffy

(continued)

Table 1
(continued)

Stock or transgene description	General category	Utility	Source
UAS-Ras85D.V12	Egfr signaling	Reporter construct expresses an activated form of Ras85D under GAL4 control. When expressed in the amnioserosa results in a persistent amnioserosa phenotype.	Bloomington stock 4847
UAS-Ras85D.N17	Egfr signaling	Reporter construct expresses a dominant negative version of Ras85D under GAL4 control.	Bloomington stock 4845
UAS-sSpi	Egfr signaling	Reporter construct expresses an activated form of the Egfr ligand "secreted Spitz" under GAL4 control.	B. Shilo
UAS-argos.M	Egfr signaling	Reporter construct expresses the Egfr antagonist encoded by *argos* under GAL4 control.	Bloomington stock 5363
Ubi-DEcadherin-GFP#5	Live imaging	Expression of GFP-DE-cadherin fusion protein under control of a ubiquitin promoter effectively marks the apical plasma membrane of all epidermal cells	H. Oda
UAS-GFP^{nls}	Live imaging	Reporter construct expressing GFP with a nuclear localization signal under GAL4 control.	Bloomington stock 4775 (2nd) stock 4776 (3rd)
UAS-mCherry.NLS	Live imaging	Reporter construct expressing mCherry with a nuclear localization signal under GAL4 control.	Bloomington stock 38425 (2nd) stock 38424 (3rd)
UAS-mCD8-GFP.L	Live imaging	Reporter construct expresses the mouse CD8 transmembrane protein as a GFP fusion protein under GAL4 control. A general GFP cell marker.	Bloomington stock 5130

UAS-actin5C-GFP#2.2	Live imaging	Reporter construct expressing actin-GFP fusion protein under GAL4 control. Useful for visualizing filopodia, but does sometimes show dominant negative phenotypes.	H. Oda
His2Av-mRFP	Live imaging	Ubiquitous expression of Histone H2A variant-mRFP fusion protein. Results in all nuclei being marked with mRFP.	Bloomington stock 23651 (2nd) stock 23650 (3rd)
His2Av-eGFP	Live imaging	Ubiquitous expression of Histone H2A variant-eGFP fusion protein. Results in all nuclei being marked with eGFP.	Bloomington stock 24163 (2nd) stock 32045 (X)
UAS-Apoliner	Live imaging/cell death	Caspase biosensor construct under GAL4 control. Expresses a GFP + RFP fusion protein that is cleaved by caspase activity. Uncleaved Apoliner remains colocalized to membranes. Cleaved Apoliner results in nuclear localized GFP. Works very well in cells of the amnioserosa.	Bloomington stock 32121 (X) stock 32122 (2nd) stock 32123 (3rd)
tub-Apoliner	Live imaging/cell death	Expression of the caspase biosensor Apoliner (described above) under the control of a ubiquitously expressed tubulin promoter.	P.L. Bardet
UASp-GFP-mCherry-Atg8a	Live imaging/autophagy	Expression of the autophagy marker "dual-tagged" Atg8a under GAL4 control. If expressed in a tissue that is autophagic, the subcellular distribution of the fusion protein is punctate. Live imaging and color shift of puncta permits observation and measurement of autophagic flux.	H. Stenmark
G289	Live imaging	Protein trap line reporting for the gene basigin. Useful for outlining all cells, but also useful for visualizing the yolk membrane beneath the amnioserosa during early stages of dorsal closure.	X. Morin

(continued)

Table 1 (continued)

Stock or transgene description	General category	Utility	Source
arm-GFP	Live imaging	Expresses GFP tagged β-catenin under the control of the arm (= β-catenin) promoter. Good for marking the apical membranes of all epithelial cells.	Bloomington stock 8556 (2nd) stock 8555 (3rd)
Yet1	Live imaging	A vital enhancer trap line that expresses YFP only in the outermost cells in the perimeter of the amnioserosa. Insert maps in the vicinity of the gene *slit*. Also expressed in mature enteroendocrine cells in the adult midgut.	A. Michelson (maintained by B. Reed)
Lac-GFP-644	Live imaging	Protein trap insertion reporting for expression of the gene *lachesin*. Marks septate junctions throughout the epidermis, but has no expression in the amnioserosa as these cells lack septate junctions.	*A. Jacinto*
UAS-src-eGFP	Live imaging	Expresses GFP containing 14 amino acid myristylation domain of v-src under GAL4 control. Good marker for outlining cell shapes.	Bloomington stock 5429 (X) stock 5432 (2nd)
UAS-eGFP-Clc	Live imaging	Expresses a clathrin light chain GFP fusion protein under GAL4 control. Very good for imaging endocytic vesicles.	Bloomington stock 7107 (2nd) stock 7109 (3rd)
UAS-p35	Cell death	Expression of baculovirus caspase inhibitor under GAL4 control. Can be used to suppress programmed cell death.	Bloomington stock 6298 (X) stock 5072 (2nd) stock 5073 (3rd)
UAS-hid.Z	Cell death	Expresses the proapoptotic gene *hid* under GAL4 control. Very effective for induced premature death of the amnioserosa. Effect can be suppressed by coexpression of *UAS-Ras85D.V12*.	Bloomington stock 65403

UAS-hid^{Ala5}	Cell death	Expresses a modified form of the proapoptotic gene *hid*, where phosphorylation sites have been mutated to alanine. Very effective for inducing premature death of the amnioserosa. Effect is not suppressed by coexpression of *UAS-Ras85D.V12*.	A. Bergmann
da-GAL4	Ubiquitous expression	GAL4 driver construct with ubiquitous expression throughout the early embryo. Inserts available on the second and third chromosomes.	Bloomington stock 55850 (2nd) stock 55851 (3rd)
Act5C-GAL4	Ubiquitous expression	GAL4 driver construct with ubiquitous expression driven by the actin5C promoter. Expresses throughout the early embryo.	Bloomington stock 4414
ftz-GAL4/CyO	Expression in stripes	GAL4 driver construct expressing GAL4 in the pattern of the pair-rule gene *ftz*. Stripes of ftz-GAL4 expression persist through the AS during dorsal closure.	J.P. Vincent
ptc$^{559.1}$*-GAL4*	Expression in stripes	Enhancer trap line expressing GAL4 in the stripes through the lateral epidermis in the pattern of *patched* gene expression.	Bloomington stock 2017
LE-GAL4	Expression in leading edge	Enhancer trap line expressing GAL4 in the dorsal most epidermal cells (the leading edge cells) during dorsal closure.	S. Noselli
69B-GAL4	Expression in lateral epidermis	Enhancer trap line expresses GAL4 throughout the lateral epidermis but not in the amnioserosa.	Bloomington stock 1774
dpp-GAL4.PS	Expression in lateral epidermis	Enhancer trap line expresses in pattern of *dpp* expression in lateral epidermis during dorsal closure.	Bloomington stock 7006
LP1-GAL4	Expression in amnioserosa	Enhancer trap line reporting for expression of the gene *calderon*. Good expression specific to the amnioserosa from extended germ band stage. Also known as *Orct2*$^{cald\text{-}GAL4}$.	Bloomington stock 58802

(continued)

Table 1 (continued)

Stock or transgene description	General category	Utility	Source
NP5328-GAL4	Expression in amnioserosa	Enhancer trap line reporting for expression of the gene *Atg1*, GAL4 expressed in the amnioserosa starting midway through dorsal closure.	Kyoto Stock Center stock 104954
NP3312-GAL4	Expression in amnioserosa	Enhancer trap line reporting for expression of the gene *hnt*, GAL4 expressed in the amnioserosa from extended germ band stage in addition to other *hnt*-expressing tissues.	Kyoto Stock Center stock 113216
c381-GAL4	Expression in amnioserosa	Enhancer trap line expressing GAL4 in the amnioserosa.	Bloomington stock 3734
VT056876-GAL4	Expression in amnioserosa	Reporter construct containing upstream regulatory sequences of the gene *hnt* driving expression of GAL4. Very early and highly specific amnioserosa expression.	Vienna Drosophila Resource Center stock v200380
$puc^{A251.1F3}$	JNK signaling	Enhancer trap line expresses β-galactosidase in the dorsal most epidermal cells (leading edge cells) during dorsal closure. Also a recessive lethal allele. Insert is also a loss-of-function *puc* allele.	Bloomington stock 11173
dpp^{10638}	TGF-β signaling	Enhancer trap line expresses β-galactosidase in pattern of *dpp* expression in lateral epidermis during dorsal closure. Expression in amnioserosa prior to dorsal closure. Can be used as a “readout” for JNK signaling pathway in dorsal most epidermal cells during dorsal closure. Insert is a weak loss-of-function allele of *dpp*.	Bloomington stock 12379
$UAS\text{-}dpp.GFP^{CH2}$	TGF-β signaling	Reporter construct expressing a Dpp-GFP fusion protein under GAL4 control.	S. Cohen

FM7c, B & Kr-GAL4 > UAS-GFP (short form name)	Live imaging: *Kr* based 2nd chromosome green balancer, balancer marked with *B* and is sn^+	*X* chromosome balancer associated with GFP expression in the pattern of the *Krüppel* gene expression (amnioserosa and thoracic segments) initiating at the onset of germ band retraction.	Numerous stocks available from stock centers. Bloomington stock 5193 is balanced over *Df(1)JA27*.
FM7c, B & twi-GAL4 > UAS-GFP (short form name)	Live imaging: *twi* based green balancer, balancer marked with *B* and is sn^+	Second chromosome balancer associated with GFP expression in the pattern of the *twist* gene expression (mesoderm) initiating at the onset of germ band retraction.	Numerous stocks available from stock centers. Bloomington stock 6873 is balanced over dominant marker *N*.
CyO, Kr-GAL4 > UAS-GFP (short form name)	Live imaging: *Kr* based 2nd chromosome green balancer, balancer marked with *Cy*	Second chromosome balancer associated with GFP expression in the pattern of the *Krüppel* gene expression (amnioserosa and thoracic segments) initiating at the onset of germ band retraction.	Numerous stocks available from stock centers. Bloomington stock 5194 is balanced over dominant markers L^2 and *Pin*.
CyO, twi-GAL4 > UAS-GFP (short form name)	Live imaging: *twi* based green balancer, balancer marked with *Cy*	Second chromosome balancer associated with GFP expression in the pattern of the *twist* gene expression (mesoderm) initiating at the onset of germ band retraction.	Numerous stocks available from stock centers. Bloomington stock 6662 is balanced over dominant marker associated with *In(2LR)Gla*.
TM3, Sb & Kr-GAL4 > UAS-GFP (short form name)	Live imaging: *Kr* based 3rd chromosome green balancer, balancer marked with *Sb*	Third chromosome balancer associated with GFP expression in the pattern of the *Krüppel* gene expression (amnioserosa and thoracic segments) initiating at the onset of germ band retraction.	Numerous stocks available from stock centers. Bloomington stock 5195 is balanced over the dominant marker *D*.
TM3, Sb Ser & twi-GAL4 > UAS-GFP (short form name)	Live imaging: *twi* based green balancer, balancer marked with *Sb* and *Ser*	Third chromosome balancer associated with GFP expression in the pattern of the *twist* gene expression (mesoderm) initiating at the onset of germ band retraction.	Numerous stocks available from stock centers. Bloomington stock 6663 is balanced over the dominant marker Dr^{Mio}.

live-imaging resources, or resources for analysis of DC—or programmed cell death, the task can seem truly daunting. We have, therefore, compiled a list of useful stocks regarding the aforementioned aims. The associated table (Table 1) does not list full stock descriptions in all cases, but, rather for convenience, lists the relative transgene or mutation associated with the stock (or stocks) of interest. Within the general category of Egfr signaling we have listed several loss-of-function *Egfr* alleles, a temperature sensitive allele of *Egfr*, in addition to *UAS-Egfr-RNAi* lines that have, in our experience, proven to be useful stocks that achieve the desired effect of reducing Egfr signaling. We also list various transgenes (*UAS-reporters*) that were designed to modulate Egfr signaling in a positive or negative manner and that have, in our experience, proven to be effective. For example, Egfr signaling can be upregulated through expression of a secreted version of the Egfr ligand Spitz (sSpi) [20], and expression of this transgene in *paired* (*prd*) stripes in the embryo was used to show that the activation of Egfr signaling led to downregulation of Dpp.

A number of stocks are also listed that are of general use for live imaging. While most of these stocks are also *UAS-reporter* based, a few are GAL4/UAS independent, such as *Ubi-DEcadherin-GFP*, the *His-2AvGFP/RFP* stocks, as well as additional protein trap based stocks or GFP-type enhancer trap lines. When recombined with *GAL4+ UAS-reporter* combinations, such *GAL4/UAS*-independent lines can be particularly useful for live-imaging embryonic mutant phenotypes—often at a cellular level.

When using the *GAL4/UAS* based method of inducible gene expression, finding the right *GAL4* driver for the experiment at hand can often be the most important factor. In many cases, when studying a particular process, one searches for a tissue-specific *GAL4* driver in order to disrupt or modulate a process of interest within a tissue of interest. One must, however, be alert to the possibility that a chosen enhancer trap or construct-based *GAL4* driver is also involved in the particular process under investigation, and that a positive or negative feedback loop could be induced between the *GAL4* driver and the *UAS-reporter*. For example, when using the *LP1-GAL4* driver to induce amnioserosa-specific expression, one must recognize that this driver reports for the expression of the gene *calderón*, which is itself a target of the insulin signaling pathway [21]. Therefore, when using the *LP1-GAL4* driver to express components of the insulin signaling pathway, one should be remember to test for the possibility of ectopic feedback loops, which in themselves could be informative, but could also present misleading phenotypes. Otherwise, it is important to use a variety of tissue-specific *GAL4* drivers for such experiments in order to recognize or rule out possible feedback loop effects that might be associated with a particular *GAL4/UAS* combination. For the analysis of the nonproliferative aspects of Egfr signaling during DC, we have, therefore, listed a variety of *GAL4* drivers specific to epidermal stripes, lateral epidermis, the dorsal most leading edge of the lateral epidermis, as well as a number of different amnioserosa *GAL4* drivers.

Also included within Table 1 are a number of stocks useful for analysis of cell death during DC. Egfr signaling is required for the maintenance of the extraembryonic tissue known as the amnioserosa prior to the completion of DC, and in the absence of Egfr this tissue undergoes premature cell death [4]. Moreover, in situations where Egfr signaling is overactive, the amnioserosa does not undergo its normal timely degradation, but will persist many hours beyond its normal time of degradation (Reed B., unpublished). Other than the premature disappearance or the persistence of the amnioserosa before, or after, its normal time of death, there are a number of different approaches for analyzing cell death in the amnioserosa, and these stocks are also listed in Table 1.

While the bulk of the amnioserosa undergoes programmed cell death at the time of, or shortly following, the completion of DC, approximately 10% of amnioserosa cells are actively extruded from this epithelium as it contracts [22]. These extruded cells are thought to contribute to the maintenance of tension within the contracting amnioserosa, and have been likened to an "apoptotic force" [19]. During early stages of DC, extruded cells are readily recognized in time-lapse live-imaging sequences of embryos carrying *Ubi-DE-cadherin-GFP*, which strongly labels the apical junctions of amnioserosa cells. Extrusion events can be recognized in compiled animations as cells that undergo a reduction in apical surface such that they disappear from the viewpoint of apical *Ubi-DE-cadherin-GFP* fluorescence. Direct observation of the apoptosis of such extruded cells can be achieved by combining the apical junction marker *Ubi-DE-cadherin-GFP* with a general cell marker combination such as *LP1-GAL4+UAS-GFP*^{nls}. Using this approach, as cells reduce their apical surface to the point where there is no longer an exposed apical surface area, the extruded cell can be observed to undergo canonical apoptotic blebbing and nuclear fragmentation [22].

Interestingly, the ***basigin*** protein trap line *G289*, is useful for observing the fate of extruded apoptotic cells. Apoptotic corpses of these extruded cells are rapidly engulfed by the underlying yolk membrane, and the points of engulfment show highly elevated GFP-basigin fluorescence [22]. Such yolk-engulfment events, however, only occur during early stages of DC, at which time the amnioserosa directly contacts the underlying yolk membrane, before the onset of midgut closure over the yolk sac.

There are presently several biosensor-based approaches for detecting cell death when using a live-imaging approach. We found the Apoliner system to be particularly effective for reporting caspase activation in the large, flat cells of the DC stage amnioserosa

[22, 23]. Briefly, Apoliner, which is available as either UAS- or tubulin promoter based transgenes, expresses an engineered protein containing an RFP domain including a transmembrane motif, a GFP with a nuclear localization signal sequence, and a linker region containing a consensus cleavage site for the caspase DIAP1. When expressed in the absence of caspase activity, the intact Apoliner protein remains membrane bound. When expressed in cells having caspase activity, however, the GFP^{nls} moiety is liberated and is imported into the nucleus.

Programmed cell death of the amnioserosa also involves an upregulation of autophagy [22, 24]. A particularly useful biosensor of autophagy, also listed in Table 1, is the *UAS-GFP-mCherry-Atg8a* reporter gene [25]. The Atg8a protein subcellular localization reflects the status of autophagy within a cell. If a cell has a low level of autophagy, Atg8a will appear homogeneous in its distribution with few localized punta. In highly autophagic cells, the subcellular localization of Atg8a is predominantly punctate. Using the GFP-mCherry "dual tag" fusion protein of Atg8a further permits analysis of autophagic flux rather than a mere blockage of the autophagy process. In the case of increased flux, Atg8a puncta can be observed using live imaging to initially reflect a colocalization of GFP and mCherry fluorescence, but as autophagy progresses, the progression of GFP-mCherry-Atg8a from the autophagosome to the more acidic environment of the lysosome extinguishes GFP fluorescence, leaving only mCherry fluorescence.

Last, we have included within our list of recommended stocks a number of stocks useful for identifying the desired genotypes of embryos collected from genetic crosses of balanced lines. In the same way that balancer chromosomes were designed to carry visible dominant markers, it is necessary when live imaging to have a dominant marker associated with the balancer chromosome, permitting accurate genotyping of mutant vs. nonmutant embryos or larvae. In the case of live imaging embryos during DC, we found that the *Kr-GAL+UAS-GFP* as well as the *twi-GAL4+UAS-GFP* based balancers for the *X*, *second*, and *third* chromosomes were most useful. In some cases, however, it was only through the progression of a particular live-imaging sequence that the "green balancer" expression would reveal itself—often midway through germ band retraction. Nevertheless, these so-called "green balancers" were found to be the most effective of the many that we tried for embryonic live-imaging applications.

2.2 Materials for Working with Embryos

Mini cages: These can be made from 100 mL plastic tricornered beakers together with polystyrene petri dishes as collection plates. For example, Fisher beakers (Fisher no. 02-593-50B) will fit

60 × 15 mm dishes (Falcon 351007). Poke holes in the beaker with a hot needle to provide ventilation.

Grape juice agar collection plates:

Agar—75 g.

Sucrose—83.5 g.

Water—2.5 L.

Grape Juice—830 mL.

Autoclave.

Cool to 60 °C (i.e., hand hot) before adding 5 g Tegosept.

Pour into plates and store inverted at 4 °C.

Baskets for embryo processing: Use a heated scapel to cut the end off a Falcon tube and to cut a hole in the screw cap of the tube. Cut a piece of Nitex Nylon mesh (57-102, Flystuff.com) big enough to fold over the end of the tube where the cap screws on and screw the cap back on. After processing, embryos can by collected off the piece of nylon mesh.

10× phosphate buffered saline (PBS): Dissolve 8 g NaCl, 0.2 g KCl, 1.44 g Na_2HPO_4, and 0.24 g KH_2PO_4 in 1 L distilled water and adjust to pH 7.4.

Dechorination solution: 50% household bleach in 0.01% Triton X.

Hoyer's medium: Dissolve 30 g of gum arabic in 50 mL of water. Add 200 g of chloral hydrate, followed by 16 mL of glycerol while stirring. Centrifuge the mixture at 10,000 g for 20 min to separate and remove the sediment. Store the medium in the dark at room temperature.

Washing solution for cuticle preparations: 0.01% Triton X in double distilled water.

20% Paraformaldehyde stock solution: Add 10 g paraformaldehyde (Anachemia, UN-2213) into 35 mL of double distilled water with 0.5 mL of 1 M NaOH in a 50 mL Falcon tube. Stir solution at 65 °C until the paraformaldehyde is dissolved. Add 5 mL of 10× PBS and double distilled water to bring final volume to 50 mL. The solution is stable at 4 °C for 1 month.

Fixing solution: 5 mL of heptane, 1 mL of 20% paraformaldehyde (freshly diluted from 20% paraformaldehyde), and 4 mL of 1× PBS.

Storage solution: 100% methanol or 80% ethanol.

PBTriton X: PBS + 0.1% Triton X.

PBB: PBTriton X + 1% BSA.

PBTween: PBS + 0.1% Tween 20.

Proteinase K solution: 3 μg/mL proteinase K solution freshly diluted from 3 mg/mL proteinase K stock (Sigma P6556).

Glycine solution: 20 mg glycine dissolved in 10 mL PBTween (can be stored at 4 °C for 1 month).

In situ hybridization buffer: 50% deionized formamide, 4× SSC [26], 1× Denhardts [26], 0.1% Tween 20, 5% dextran sulfate,

250 μg/mL salmon sperm DNA, and 50 μg/mL heparin. Stored at −20 °C.

Cyanine 3 tyramide substrate: 1:100 dilution of cyanine 3 tyramide reagent (PerkinElmer, SAT705A) in amplification diluent supplied by the manufacturer.

Mounting reagent: Vectashield mounting medium for fluorescence (Vector Laboratories, Inc., H-1000).

3 Methods

3.1 Cuticle Preparations (Modified from [27])

Preparations of the embryonic and first instar larval cuticle provide a fast, easy approach to quantifying morphogenetic defects in a population of embryos, including failure of DC. As can be seen in Fig. 1a, wild-type embryos show an intact, uniform dorsal cuticle indicating successful DC. In contrast, losses or gains of Egfr signaling in the two tissues undergoing morphogenesis during DC, the amnioserosa and the epidermis, results in a failure to close the dorsal epidermis, leading to a corresponding hole in the dorsal cuticle (Fig. 1b–g). In genetic interaction experiments, the effects of genetic modifiers on a DC phenotype can be evaluated initially by cuticle preparation. For example, as seen in Fig. 1c, making *Egfr* mutant embryos heterozygous for a *punt* allele leads to an increase in the size of the dorsal hole.

Experimental procedure:

1. Place adults in mini cages and collect embryos overnight or for a defined period on grape juice agar plates.
2. If scoring a particular genotype based on selection using chromosomes with fluorescent reporters, select the appropriate embryos using a fluorescence dissecting microscope.
3. Place embryos/larvae in baskets and dechorionate for 3 min in dechorionation solution, followed by three rinses in washing solution.
4. Mount embryos/larvae on slides in Hoyer's medium by transferring them from nylon mesh using a paint brush and bake at 65 °C until clear (typically several days) (*see* **Notes 1** and **2**).

3.2 Fixing Embryos for Immunohistochemistry and FISH

1. Place adults in mini cages and collect embryos overnight or for a defined period on grape juice agar plates.
2. Place embryos in baskets and rinse with washing solution.
3. Dechorionate embryos for 3.5 min in dechorionation solution.
4. Wash embryos three times, 3.5 min each time in washing solution.
5. Transfer embryos into 10 mL of fixing solution in a 20 mL scintillation vial.

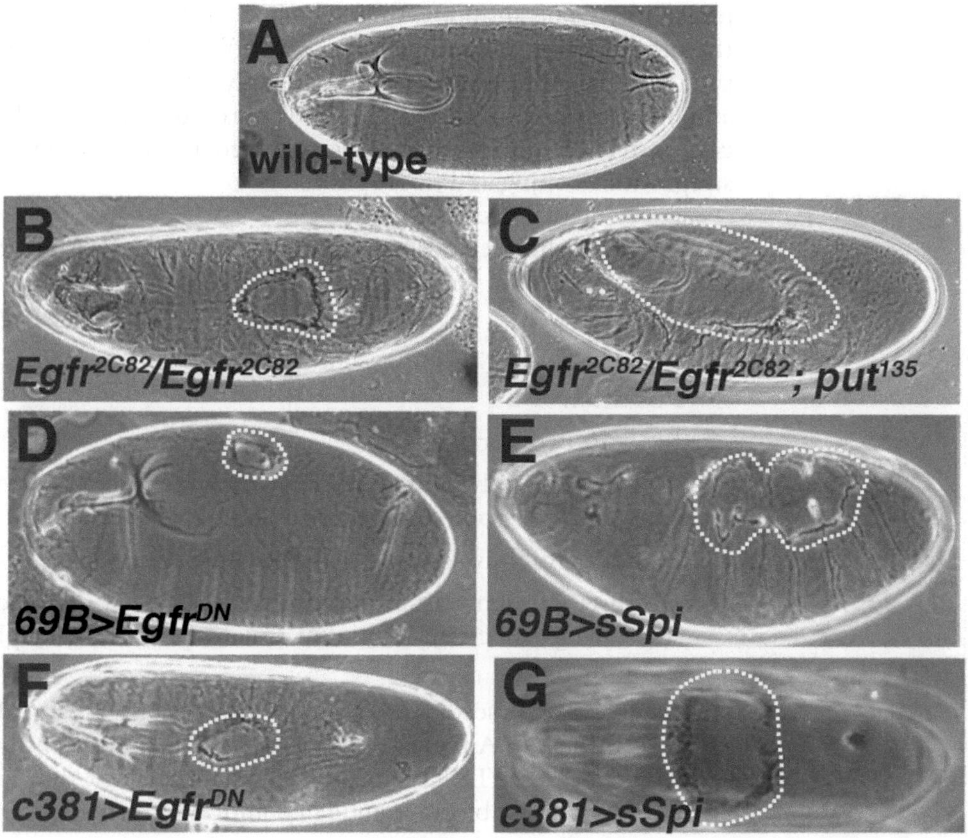

Fig. 1 (**a**) Wild-type embryo showing intact dorsal cuticle. (**b**) *Egfr*^{2c82} homozygous mutant embryo showing small dorsal hole (*dotted line*), indicating that disruption of Egfr signaling causes a defect in dorsal closure. (**c**). *Egfr*^{2c82} homozygous mutant embryo heterozygous for *put*^{135} showing large dorsal hole, incaing a genetic interaction between *put* and *Egfr* during dorsal closure. (**d**) Embryo expressing EgfrDN in the epidermis showing small dorsal hole, indicating that Egfr signaling is required in this tissue for correct dorsal closure. (**e**) Expression of the Egfr ligand sSpi in the epidermis causes a dorsal hole, indicating that excessive Egfr signaling causes failure of dorsal closure. (**f**, **g**) Loss (F) or gain (G) of Egfr signaling in the amnioserosa cause dorsal closure defects

6. Shake the vial vigorously for 25 min at room temperature and dispose of top layer (paraformaldehyde and PBS).
7. Add 5 mL methanol and shake vigorously for 1 min. As the vitelline membranes are ruptured the embryos will sink into the lower, methanol layer (*See* **Notes 3** and **4**).
8. Remove top layer (heptane) and rinse twice with methanol before storing embryos in methanol at −20 °C.

3.3 Immunostaining of Embryos

Much of the analysis of the participation of Egfr in DC is focused on the examination of fixed, antibody-stained embryos. Typically, we look at embryos immunostained with two antibodies, one revealing cell outlines and embryonic morphology (e.g., anti-phosphotyrosine antibody, Cell Signaling Technology #9411) and the other against an epitope or protein affected by Egfr signaling, such as p-Erk. When one antibody is clearly inferior to the other we amplify the signal from that antibody through the use of biotin-conjugated secondary antibodies followed by incubation with a streptavidin-conjugated fluorophore.

Experimental procedure:

Based on the previously described protocol [27]. All steps are performed at room temperature with rotation, unless otherwise stated.

1. Rehydrate fixed embryos using three 20 min washes in PBTriton.
2. Block in PBB for 1 h.
3. Add appropriately diluted primary antibodies in PBB and incubate for 2 h at room temperature or at 4 °C overnight.
4. Remove antibody solution and store at 4 °C for reuse.
5. Do three 20 min washes of embryos in PBB. For subsequent steps, wrap tubes in aluminum foil to protect fluorophores from the light.
6. Incubate with secondary antibodies (1:200, Vector Laboratories) for 2 h. A typical combination of secondary antibodies would be an FITC-conjugated secondary against one species together with a biotin-conjugated antibody against another.
7. If using a biotinylated secondary antibody, incubate for 30 min in appropriately tagged streptavidin, for example 1:1000 dilution of streptavidin-Texas Red (Vector Laboratories).
8. Wash embryos 3 × 10 min in PBS.
9. Resuspend embryos in Vectashield mounting medium (Vector Laboratories) and allow embryos to settle at least 1 h or preferably overnight at 4 °C before mounting on slides (*See* **Notes 5** and **6**).

3.4 Fluorescent In Situ Hybridization (FISH)

This sensitive, high-resolution technique, based on the technique of Lecuyer et al. [28], provides excellent visualization of transcript distributions during DC and can be combined with immunohistochemistry to simultaneously look at protein distribution. When evaluating the distribution of a transcript for the first time, it is useful to stain the embryos with a commercial anti-phosphotyrosine antibody (Cell Signaling Technology #9411) to reveal cell outlines and enable a determination of which cells a transcript is being expressed in and if there is any subcellular localization. For example, costaining with anti-phosphotyrosine confirms that transcripts for two important DC participants, *dpp* and *zip*, are elevated in the dorsal most epidermal cells during DC (Fig. 2a, e). FISH is an effective tool for evaluating the effects of Egfr signaling during

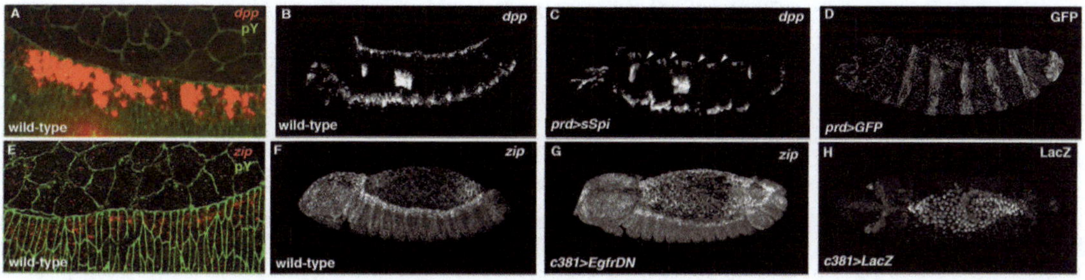

Fig. 2 Use of FISH to study the regulation of gene expression by Egfr signaling during dorsal closure. (**a**) *dpp* FISH costained with anti-phosphotyrosine showing *dpp* transcripts in DME cells. (**b**) *dpp* expression pattern in embryo at beginning of dorsal closure. Stripe of *dpp* expression at dorsal side of embryo is in the DME cells. (**c**) Expression of sSpi in *prd* stripes reduces *dpp* expression in DME cells in those stripes (*arrowheads*). (**d**) Expression pattern of *prd-Gal4* revealed by crossing to *UAS-GFP* reporter. (**e**) *zip* FISH costained with anti-phosphotyrosine showing *zip* transcripts in DME cells. (**f**) *zip* expression pattern in embryo at beginning of dorsal closure. Stripe of *zip* expression at dorsal side of embryo is in the DME cells. (**g**) Expression of EgfrDN in the amnioserosa cuases elevated *zip* expression. (**h**) Expression pattern of *c381-Gal4* revealed by crossing to a *UAS-LacZ* reporter

DC, and we used it to show that expression of both *dpp* and *zip* is inhibited by Egfr signaling [4] (Fig. 2). Activation of Egfr signaling by expression of sSpi in *prd* stripes leads to reduction of *dpp* transcripts in *prd* stripes (Fig. 2b–d), whereas inhibition of Egfr signaling in the amnioserosa through expression of dominant negative Egfr (EgfrDN) leads to elevation of *zip* transcript levels in the amnioserosa (Fig. 2f–h).

Experimental procedure:

To generate digoxigenin-labeled RNA probes use template cDNA subcloned into a vector with T3, T7 or Sp6 promoters. Linearize the cDNA-containing vector at the 5′ end with an appropriate restriction enzyme to prevent vector sequences from being transcribed into RNA probe.

1. Set up digestion reactions.

Template	15–25 μL (<100 ng/μL = 25 μL, 100–200 ng/μL = 20 μL, >200 ng/μL = 15 μL)
Enzyme	3 μL
10× buffer	10 μL
ddH_2O	62–72 μL
Total	100 μL

2. Mix thoroughly (i.e., finger vortex).
3. Spin down briefly.
4. Incubate at 37 °C for 3 h.

5. Purify the linearized templates.

Transcribe antisense RNA probe with polymerase according to the manufacturer's instructions.

Probe synthesis:

1. Set up transcription reactions.

Linearized template	26 μL
RNA polymerase	4 μL
Labeling mixture	4 μL
RNAse inhibitor	2 μL
10× buffer	4 μL
Total	**40 μL**

2. Incubate at 37 °C for 4 h.
3. Place reactions in freezer to stop reactions (or add 0.5 μL of 0.5 M EDTA).
4. Purify the probes using Illustra MicroSpin S-200 HR Columns according to manufacturer's instruction (GE Healthcare).
5. Measure probe concentration using a NanoDrop spectrophotometer. Store RNA probe at −80 °C.

FISH protocol:

Note: rinses/washes and incubations are done at room temperature with shaking unless otherwise stated.

1. Rinse embryos in methanol.
2. Rinse embryos in a 1:1 mixture of methanol and PBTween.
3. Rinse embryos twice in PBTween.
4. Post-fix embryos for 20 min in 4% formaldehyde (freshly made; diluted in PBT).
5. Wash embryos three times, 2 min/wash in PBTween.
6. Add 400 μL of 3 μg/mL proteinase K and incubate at room temperature for 2 min with shaking.
7. Transfer tube of embryos to ice and incubate for 1 h.
8. Remove proteinase K solution and stop digestion by washing twice, 2 min/wash with 2 mg/mL glycine in PBTween.
9. Rinse embryos two times in PBTween to remove the glycine.
10. Post-fix embryos again for 20 min in 4% formaldehyde in PBT.
11. Wash embryos five times, 2 min/wash in PBTween.
12. In the meantime, boil 400 μL of RNA hybridization solution at 100 °C for 5 min then cool on ice for 5 min.
13. Rinse embryos in a 1:1 mixture of PBTween and RNA hybridization solution.

14. Replace the mixture with 100% hybridization solution. Remove the hybridization solution, then add freshly boiled and cooled prehybridization solution and place embryos in a 56 °C water bath or hybridization oven and incubate for a minimum of 2 h.
15. Dilute 100–400 ng probe in hybridization buffer and heat to 80 °C or 3 min and cool on ice for 5 min.
16. Remove prehybridization solution and add at least 100 μL hybridization solution plus probe, making sure there is enough hybridization solution to cover embryos during incubation.
17. Hybridize embryos for 12–16 h without shaking at 56 °C.
18. Preheat ALL wash solutions to 56 °C.
19. Remove probe solution and store at −80 °C for reuse, and rinse the embryos once with 400 μL prewarmed hybridization buffer.
20. Replace and wash with another 400 μL prewarmed hybridization buffer for 20 min.
21. Wash for 15 min each with 400 μL of 3:1, 1:1, and 1:3 mixtures of hybridization buffer and PBTween.
22. Wash four times, 5 min/wash with 400 μL prewarmed PBTween, with last wash done at room temperature to cool embryos.
23. Incubate embryos in 3% BSA for 10 min.
24. Incubate embryos with 1:200 sheep anti-DIG-POD (Roche: 11207733910) in 3% BSA for 2 h.
25. Wash three times, 10 min/wash in PBTween.
26. Wash three times, 5 min/wash in PBS.
27. Incubate embryos with 1:50 tyramide in amplification buffer at 4 °C overnight (keep the sample in the dark from now on). If doing FISH with RNA probe alone, proceed to **step 33**.
28. Wash embryos twice, 5 min/wash in 3% BSA.
29. Incubate with any other desired primary antibody in 3% BSA for 2 h.
30. Wash three times, 10 min/wash in 3% BSA.
31. Add appropriate secondary antibodies in 3% BSA for 2 h.
32. Wash three times, 10 min/wash in PBS.
33. Resuspend embryos in Vectashield mounting medium (Vector Laboratories) and allow embryos to settle at least 1 h or preferably overnight at 4 °C before mounting on slides.

3.5 Live-Imaging Time-Lapse Microscopy

3.5.1 General Comments on Live Imaging and Reagents

The process of DC is readily studied using live imaging based time-lapse microscopy. Preparation of embryos and live imaging can be performed with minimal disruption to embryogenesis. Embryogenesis failure during live imaging is most often caused by hypoxia. Other concerns include heat stress, desiccation, and exposure to toxic chemicals. We have developed systems for live imaging Drosophila embryos using either inverted or upright microscopes.

For most live imaging applications, it is necessary to first remove the outer waxy chorion. This can be done by treatment with 50% commercial bleach (as described above) or by hand dechorionation. For the latter procedure, which is preferred for minimal disruption to embryogenesis, embryos are collected from the surface of egg collection plates with a very fine paintbrush or a pair of fine-tipped jeweller's forceps (Dumont, style No. 5). When using forceps, embryos tend to adhere to the forcep tips and can be gathered in clumps without clamping the forceps. Alternatively, the chorionic dorsal appendages can be clamped with forceps to pick up an embryo. While viewing through a stereomicroscope, embryos are lowered onto the surface of double-sided tape mounted onto a standard microscope slide. By gently nudging an embryo with the side of the forcep tips, or stroking the surface of the chorion with the forcep tips, the chorion layer can be cracked open. Once the chorion is open, the chorion can be further peeled away, and the dechorionated embryo can be teased from the chorion. The dechorionated embryo adheres to the forcep tips without any need to clamp the forceps. Care should be taken not to allow the dechorionated embryo to touch the surface of the tape, as once adhered, it is very difficult to recover without damaging the embryo. Depending on the microscopy system being used, the dechorionated embryo is either placed onto a thin strip of tape mounted onto a Teflon membrane (for inverted microscopes—details below) or placed in a small drop of halocarbon oil on a coverslip (for upright microscopes—details below). Before attempting to dechorionate another embryo, the forceps tips should be cleaned and be free of any halocarbon oil.

Key to both live imaging systems is the use of gas-permeable halocarbon oil. We use a 1:1 mixture of viscosity series 56 and series 700. These halocarbon oils are available in large volumes from Halocarbon Products Corp., New Jersey. Smaller 100 mL volumes are available from Sigma (H8898 Halocarbon oil series 700, H8773 Halocarbon oil series 27), although adapting the ratio in favor of increasing the amount of series 700 might be necessary when using the less viscous series 27 oil. The apparatus used for inverted microscope also uses a Teflon type of gas-permeable halocarbon membrane (PTFE). Small packets of this membrane are available as a YSI probe service kit for dissolved oxygen probes (YSI cat.# 1329883). The apparatus for use with inverted microscopes requires immobilization of embryos on double-sided tape, and some types of tape can be toxic to embryos. Scotch 3 M brand tape type 415 is suitable. It is generally recommended, however, that tape be tested for toxicity by mounting several dozen dechorionated embryos on the tape, covering with halocarbon oil, placing in a moist chamber and determining the survival rate.

Most live imaging is performed using embryos that express fluorescent proteins, and our methods are based on the premise of using fluorescence microscopy (where incident light passes through the objective). With some adaptations, however, it is also possible to use these systems for transmitted light imaging such as DIC or phase contrast. For laser scanning confocal microscopy, optimal embryo survival during DC requires fast acquisition times and limited exposure to laser excitation. In general, we use a fast laser scan setting, limit our average function to three passes, use Z-stacks of less than 12 slices with a 2 μ step size, and acquire Z-stacks at 4–5 min intervals. Based on these parameters, using a motorized X-Y stage permits simultaneous imaging of four embryos in one imaging session (typically 4–6 h). The optimal imaging parameters for any given confocal system, however, will need to be worked out in order to determine which settings result in a minimal loss of viability. In the case where settings are too severe or prolonged, an affected embryo shows a sudden cessation of morphogenetic movement, frequently preceded by a collapse in the number and activity of filopodia. In healthy embryos filopodia are prominent along the leading edge of the dorsal most epidermal cells as well as the apical surface of the amnioserosa. Imaging using a fluorescence stereomicroscope is readily achieved but requires that the system be equipped with a shutter that will shut off or block the incident light source between image acquisition timepoints. Continuous exposure of embryos to incident light on a fluorescence microscope is often lethal.

3.5.2 Stepwise Instructions for the Construction of Live-Imaging Chambers for Use with Inverted Microscopes

1. Mark the bottom surface of a 15 × 600 mm polystyrene petri dish to create a hole that is smaller than the size of the coverslip to be used. In the case of using 22 × 22 mm coverslips, the hole is approximately 17 × 17 mm (Fig. 3a).
2. Use a utility knife heated in a flame to cut out the marked region. Use a single edge razor blade or scalpel to smoothen the melted edges of the cuts (Fig. 3b).
3. Place double-sided tape around the perimeter of the hole and remove the backing (Fig. 3c).
4. Attach a Teflon membrane (described above) to the tape surrounding the hole cut in the bottom of the petri dish. If the Teflon membrane is not tight, it can be heat-shrunk using a hair drier, or by holding the dish above a flame.

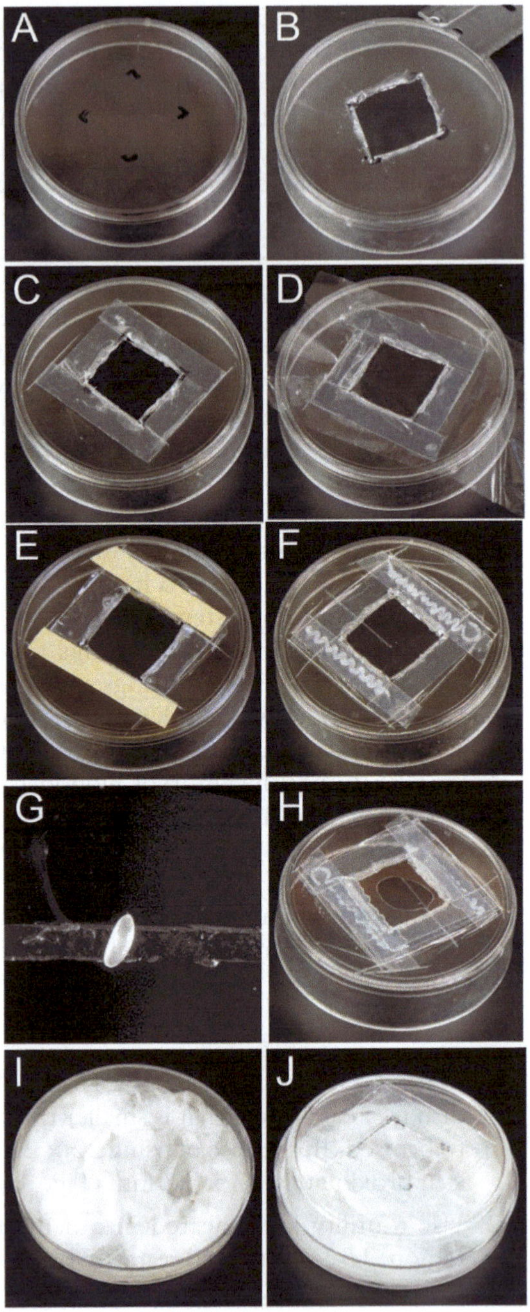

Fig. 3 Construction of live imaging chambers for use with inverted microscopes. (**a**) 60 × 15 mm polystyrene petri dish marked to create a hole that is approximately 5 mm smaller than the size of the coverslip to be used. (**b**) Petri dish with a hole created using a heated utility knife and edges trimmed flat with a single edge razor blade. (**c**) Double-sided tape mounted around perimeter of the hole. (**d**) Teflon membrane secured to tape. (**e**) Teflon membrane trimmed and additional strips of tape added to two sides. (**f**) Petroleum jelly on tape and a thin strip of double-sided tape secured over the centre of the Teflon membrane. (**g**) Hand dechorionated embryo mounted on the thin strip of tape. (**h**) Embryo in halocarbon oil with coverslip in place. (**i**) Petri dish lid with moistened tissue paper. (**j**) Completed live imaging chamber for use with an inverted microscope

5. Attach two additional strips of double-sided tape on opposite sides of the Teflon membrane. Remove the backing from the tape.
6. Cut a thin strip of double-sided tape using two single edge razor blades secured to one another by two layers of double-sided tape. Place the tape from which strips are to be cut on a smooth surface and make straight cuts along the length of the tape using the razor blade sandwich. Ideally, the thin strip should not be wider than the length of the embryo (*see* Fig. 3g) as a wider strip can limit gas exchange and result in failure due to hypoxia. Remove an appropriate length of the thin strip of tape and lay across the Teflon membrane using forceps. Remove the backing from the thin strip (Fig. 3f).
7. Place petroleum jelly along the two additional strips of tape that were previously placed on top of the Teflon membrane. This is most easily done using a hypodermic syringe loaded with petroleum jelly. The petroleum jelly serves to control the mounting of the coverslip and allows for removal of the coverslip, facilitating the reuse of the imaging chamber (Fig. 3f).
8. Position a dechorionated embryo on the thin strip of tape oriented dorsal side upward. Immediately cover the mounted embryo in a small drop of halocarbon oil. To avoid desiccation, keep the apparatus in a sandwich box lined with moistened tissues until ready for imaging.
9. When ready for imaging place a coverslip over the embryo, contacting the petroleum jelly on either side. While observing through a stereomicroscope gently tap the coverslip down, compressing the petroleum jelly until the coverslip contacts the drop of halocarbon oil. Care should be taken not to compress or flatten the embryo as this distortion can affect the rate of DC (Fig. 3h).
10. Place a piece of tissue paper in the polystyrene petri dish lid and soak the tissue with tap water. Squeeze the tissue to remove excess water and to flatten it into the lid (Fig. 3i).
11. Place the mounted embryo onto the petri dish lid. The dish can now be inverted onto a petri dish microscope stage adapter and imaged with an inverted microscope (Fig. 3j).

3.5.3 Stepwise Instructions for the Construction of Live-Imaging Chambers for Use with Upright Microscopes

The protocol for live-imaging embryos using upright microscopes or stereomicrosope, which we call the "hanging drop method," has been reported in detail elsewhere [29] and is here briefly summarized. The hanging drop method is simple and quick, has excellent embryo survival, and avoids any distortion of DC associated with compression of the embryo. In this technique embryos are simply placed in a small drop of halocarbon oil on a coverslip and the coverslip is inverted over a moist chamber. Since the embryos are buoyant in the halocarbon oil, they float upward and come to rest against the lower surface of the inverted coverslip. Overall this method improves gas exchange and avoids the need to immobilize embryos on potentially toxic tape.

1. Cut a 5 mm thick plexiglass sheet to the dimensions of a standard microscope slide (25 × 75 mm). Use a rotor to create a 3 mm deep channel in the plexiglass slide (Fig. 4a).
2. Trim tissue paper to fit the well. Two or three layers can be used, and these are soaked with tap water (Fig. 4b).
3. Using a micropipettor, place small drops of halocarbon oil on a coverslip (22 × 50 mm No.1) (Fig. 4c). Droplets should not exceed 2 mm in diameter.
4. Place dechorionated embryo in droplet of halocarbon oil (Fig. 4d). If hand dechorionating embryos, an embryo adhering to the tip of forceps is easily dislodged by touching the tip of the forceps into the halocarbon oil droplet. If dechorionating using bleach, thoroughly rinse the embryos in a stream of room-temperature tap water, and subsequently disassemble the modified Falcon tube filtration device to obtain embryos on the piece of Nitex nylon mesh (described above). Blot all water from the embryos by placing the Nitex nylon mesh onto absorptive paper or tissue paper. Working quickly to avoid desiccation of the embryos, use a very fine paintbrush previously dipped in halocarbon oil to scoop embryos from the Nitex nylon mesh. Embryos on the brush are then placed in a generous drop of halocarbon oil on a microscope slide and dislodged by swirling the brush in the halocarbon oil. Embryos can subsequently be examined using a stereomicroscope to identify embryos of the correct developmental stage for imaging. Embryos can also be examined and selected at this point based on fluorescent protein expression. Embryos to be imaged can be collected using the very fine paintbrush and transferred to the small halocarbon droplets on the coverslip. If there is to be any delay in imaging, the embryos in halocarbon oil on the microscope slide can be placed in a sandwich box lined with moistened tissues to avoid desiccation.
5. Once embryos have been placed in the halocarbon oil droplets, they are pushed to the bottom of the droplets and rolled into the desired orientation using jeweller's forceps. The coverslip is quickly inverted and placed over the well of the custom-made plexiglass slide. In general, if the embryos have been pushed against the coverslip, and if the halocarbon oil droplet is not too large and does not make contact with anything, the embryos are completely stable and will not drift or move during time-lapse imaging.

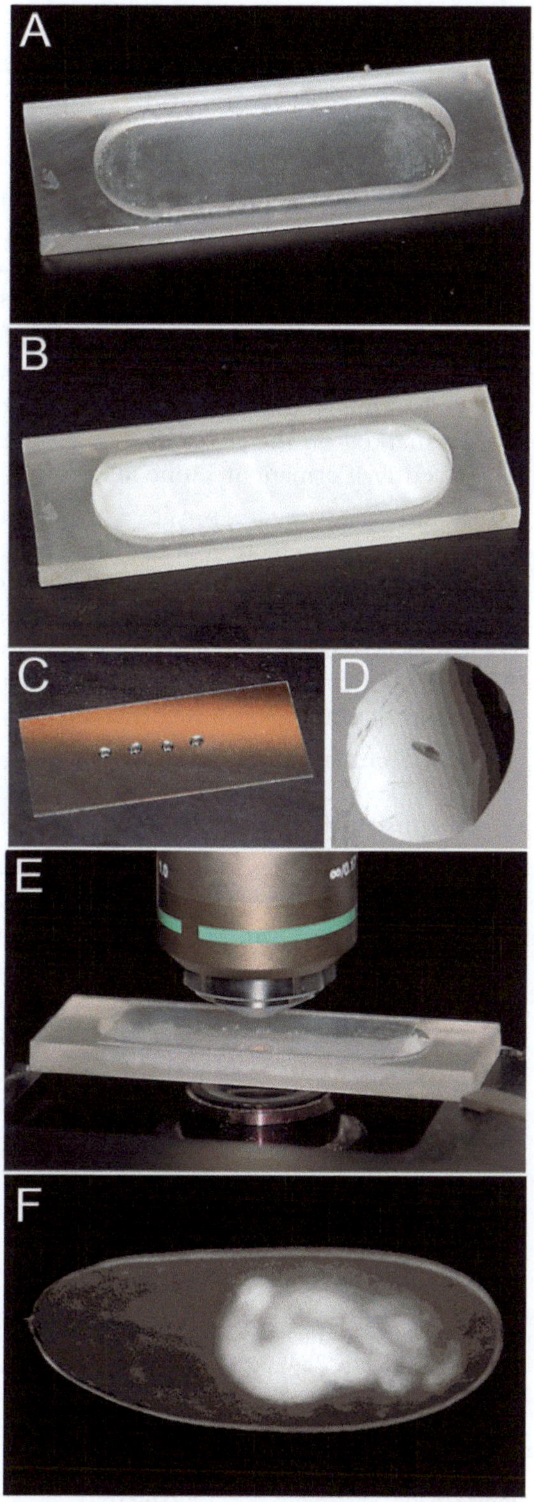

Fig. 4 The hanging drop live-imaging apparatus for use with upright microscopes. (**a**) Custom made 5 mm thick plexiglass (PMMA) base cut to the same size as a standard microscope slide (75 × 25 mm) and rotored to create a 3 mm deep well. (**b**) Moistened tissue paper lining the base of the well. (**c**) Droplets 1–2 mm in diameter of halocarbon oil on a 22 × 50 mm No.1 coverslip. (**d**) Hand dechorionated embryo placed in drop of halocarbon oil. (**e**) Coverslip inverted over live-imaging chamber mounted, and chamber mounted on an upright compound fluorescence microscope. (**f**) Stereomicroscopic view of embryo mounted using the hanging drop method showing autofluorescence of the late stage embryo (filter set optimized for detection of GFP or FITC)

4 Notes

1. When doing a large-scale screen for cuticle phenotypes, embryo processing for cuticle preparation can be accelerated by performing it directly on the grape-juice agar collecting plate. Various solutions are poured on and off the plate, taking care that not too many embryos are lost.
2. A convenient way to quantify cuticle phenotypes is to examine slides on an inverted, phase-contrast tissue culture microscope such as the Nikon TMS. As each embryo or larva is scored it can be marked on the coverslip using a fine-tipped marker such as Staedtler Permanent Lumocolor.
3. Some epitopes are sensitive to methanol and when fixing embryos for detection of such epitopes, an alternative to methanol must be used for removal of vitelline membranes. The membranes can be removed manually as described [30] or by using 80% ethanol in place of methanol. While 80% ethanol is not as effective as methanol at removing membranes, decent yields of embryos can be obtained.
4. Fixed embryos are very stable when stored in methanol at −20 °C. We have successfully used embryos for immunohistochemistry after 7 years storage.
5. In preparing slides for examining fixed embryos, use nail polish to glue square coverslips (22 mm × 22 mm) to either end of each slide to form two supports for the rectangular coverslip (22 mm × 40 mm) that will be laid over the sample. The supports provide some clearance of the rectangular coverslip over the embryos when examining them on the microscope, and will enable rolling of the embryos to an appropriate position by gently stroking the coverslip.
6. Primary antibody solutions for immunostaining of embryos can be re-sed several times, as can FISH probes.

References

1. Shilo BZ (2003) Signaling by the Drosophila epidermal growth factor receptor pathway during development. Exp Cell Res 284 (1):140–149
2. Sibilia M, Kroismayr R, Lichtenberger BM, Natarajan A, Hecking M, Holcmann M (2007) The epidermal growth factor receptor: from development to tumorigenesis. Differentiation 75(9):770–787. doi:10.1111/j.1432-0436.2007.00238.x
3. Schmid T, Hajnal A (2015) Signal transduction during *C. elegans* vulval development: a NeverEnding story. Curr Opin Genet Dev 32:1–9. doi:10.1016/j.gde.2015.01.006
4. Shen W, Chen X, Cormier O, Cheng DC, Reed B, Harden N (2013) Modulation of morphogenesis by Egfr during dorsal closure in Drosophila. PLoS One 8(4):e60180. doi:10.1371/journal.pone.0060180. PONE-D-12-29869 [pii]
5. Harden N (2002) Signaling pathways directing the movement and fusion of epithelial sheets: lessons from dorsal closure in *Drosophila*. Differentiation 70:181–203

6. Sanchis A, Bayo P, Sevilla LM, Perez P (2010) Glucocorticoid receptor antagonizes EGFR function to regulate eyelid development. Int J Dev Biol 54(10):1473–1480. doi:10.1387/ijdb.103071as. 103071as [pii]
7. Miettinen PJ, Chin JR, Shum L, Slavkin HC, Shuler CF, Derynck R, Werb Z (1999) Epidermal growth factor receptor function is necessary for normal craniofacial development and palate closure. Nat Genet 22(1):69–73. doi:10.1038/8773
8. Repertinger SK, Campagnaro E, Fuhrman J, El-Abaseri T, Yuspa SH, Hansen LA (2004) EGFR enhances early healing after cutaneous incisional wounding. J Invest Dermatol 123 (5):982–989. doi:10.1111/j.0022-202X.2004.23478.x. JID23478 [pii]
9. Glise B, Noselli S (1997) Coupling of Jun amino-terminal kinase and decapentaplegic signaling pathways in *Drosophila* morphogenesis. Genes Dev 11(13):1738–1747
10. Lamka ML, Lipshitz HD (1999) Role of the amnioserosa in germ band retraction of the *Drosophila melanogaster* embryo. Dev Biol 214(1):102–112
11. Reed BH, Wilk R, Lipshitz HD (2001) Downregulation of Jun kinase signaling in the amnioserosa is essential for dorsal closure of the *Drosophila* embryo. Curr Biol 11:1098–1108
12. Stronach BE, Perrimon N (2001) Investigation of leading edge formation at the interface of amnioserosa and dorsal ectoderm in the *Drosophila* embryo. Development 128 (15):2905–2913
13. Conder R, Yu H, Ricos M, Hing H, Chia W, Lim L, Harden N (2004) dPak is required for integrity of the leading edge cytoskeleton during Drosophila dorsal closure but does not signal through the JNK cascade. Dev Biol 276 (2):378–390
14. Scuderi A, Letsou A (2005) Amnioserosa is required for dorsal closure in Drosophila. Dev Dyn 232(3):791–800
15. Fernandez BG, Arias AM, Jacinto A (2007) Dpp signalling orchestrates dorsal closure by regulating cell shape changes both in the amnioserosa and in the epidermis. Mech Dev 124(11-12):884–897
16. Wada A, Kato K, Uwo MF, Yonemura S, Hayashi S (2007) Specialized extraembryonic cells connect embryonic and extraembryonic epidermis in response to Dpp during dorsal closure in Drosophila. Dev Biol 301(2):340–349
17. Zahedi B, Shen W, Xu X, Chen X, Mahey M, Harden N (2008) Leading edge-secreted Dpp cooperates with ACK-dependent signaling from the amnioserosa to regulate myosin levels during dorsal closure. Dev Dyn 237 (10):2936–2946. doi:10.1002/dvdy.21722
18. Franke JD, Montague RA, Kiehart DP (2005) Nonmuscle myosin II generates forces that transmit tension and drive contraction in multiple tissues during dorsal closure. Curr Biol 15 (24):2208–2221
19. Toyama Y, Peralta XG, Wells AR, Kiehart DP, Edwards GS (2008) Apoptotic force and tissue dynamics during Drosophila embryogenesis. Science 321(5896):1683–1686. doi:10.1126/science.1157052. 321/5896/1683 [pii]
20. Schweitzer R, Shaharabany M, Seger R, Shilo BZ (1995) Secreted Spitz triggers the DER signaling pathway and is a limiting component in embryonic ventral ectoderm determination. Genes Dev 9(12):1518–1529
21. Herranz H, Morata G, Milan M (2006) Calderon encodes an organic cation transporter of the major facilitator superfamily required for cell growth and proliferation of Drosophila tissues. Development 133(14):2617–2625. doi:10.1242/dev.02436
22. Cormier O, Mohseni N, Voytyuk I, Reed BH (2012) Autophagy can promote but is not required for epithelial cell extrusion in the amnioserosa of the Drosophila embryo. Autophagy 8(2):252–264
23. Bardet PL, Kolahgar G, Mynett A, Miguel-Aliaga I, Briscoe J, Meier P, Vincent JP (2008) A fluorescent reporter of caspase activity for live imaging. Proc Natl Acad Sci U S A 105(37):13901–13905. doi:10.1073/pnas.0806983105.. 0806983105 [pii]
24. Mohseni N, McMillan SC, Chaudhary R, Mok J, Reed BH (2009) Autophagy promotes caspase-dependent cell death during Drosophila development. Autophagy 5(3):329–338
25. Nezis IP, Shravage BV, Sagona AP, Lamark T, Bjorkoy G, Johansen T, Rusten TE, Brech A, Baehrecke EH, Stenmark H (2010) Autophagic degradation of dBruce controls DNA fragmentation in nurse cells during late *Drosophila melanogaster* oogenesis. J Cell Biol 190 (4):523–531. doi:10.1083/jcb.201002035
26. Sambrook J, Fritsch EF, Maniatis T (1989) Molecular cloning: a laboratory manual. Cold Spring Harbor Laboratory Press, Cold Spring Harbor, NY
27. Ashburner M (1989) Drosophila: a laboratory manual. Cold Spring Harbor Laboratory Press, Cold Spring Harbor, NY
28. Lecuyer E, Parthasarathy N, Krause HM (2007) Fluorescent *in situ* hybridization

protocols in *Drosophila* embryos and tissues. In: Dahmann C (ed) Drosophila: methods and protocols, Methods in Molecular Biology. Humana Press, Totowa, NJ

29. Reed BH, SC MM, Chaudhary R (2009) The preparation of Drosophila embryos for live-imaging using the hanging drop protocol. J Vis Exp 25:e1206. doi:10.3791/1206

30. Wieschaus E, Nusslein-Volhard C (1998) Looking at embryos. In: Roberts DB (ed) Drosophila: a practical approach. Oxford University Press, Oxford, pp 179–214

Part V

Methods Related to the Translational Research of ErbB Receptor Signaling

Chapter 17

Analysis of Epithelial–Mesenchymal Transition Induced by Overexpression of Twist

Jing-Wen Bai, Yong-Qu Zhang, Yao-Chen Li, and Guo-Jun Zhang

Abstract

Breast cancer, the most common malignancy among women worldwide, is a heterogeneous disease, and it therefore has remarkably different biological characteristics and clinical behavior. Breast cancer has been divided into several different molecular subtypes based on the status of estrogen receptor (ER), progesterone receptor (PR), human epidermal growth factor 2 (HER2, also named as ErbB2) status. Her2 is a member of EGFR family of transmembrane tyrosine kinase-type receptors, and is involved in the activation of its downstream signaling cascades, which could promote cell proliferation, metastasis, and angiogenesis in tumors. In addition, Twist, a transcriptional factor has been shown to associate with ErbB2 signaling to increase the proliferation and the number of cells, and to induce epithelial–mesenchymal transition. Deregulated cell proliferation can result in hyperplasia and even malignancies. Actually, the proliferative or survival ability of cells can be measured by a variety of methods. Clonogenic assay and CCK8 assay can serve as useful tools to test whether the clonogenic survival ability of tumor cells can be enhanced or reduced upon stimulation of appropriate mitogenic signals or a given cancer therapy respectively. A colony is defined as a cluster of at least 50 cells that can often only be determined microscopically. Moreover, migration and invasion assay, in some degree, represents the potential for EMT promotion. Here, we introduce colony formation assay; CCK8 proliferation assay; soft agar; and migration and invasion assay using overexpression of ErbB2 and EGFR receptors as an example.

Key words Epithelial–mesenchymal transition, Twist, Her2 signaling, Colony formation assay, Proliferation, Migration, Invasion assay

1 Introduction

A series of growth disorders can occur at the cellular level and these consequently underpin much of the subsequent course in cancer, in which a group of cells display uncontrolled growth and division beyond the normal limits, invasion (intrusion on and destruction of adjacent tissues), and sometimes metastasis (spread to other locations in the body via lymph or blood).

There are four members in the ErbB receptors family, including epidermal growth factor receptor (EGFR; ErbB1), HER-2 (ErbB2), ErbB3, and ErbB4. The members in this family receptors share a

Zhixiang Wang (ed.), *ErbB Receptor Signaling: Methods and Protocols*, Methods in Molecular Biology, vol. 1652, DOI 10.1007/978-1-4939-7219-7_17, © Springer Science+Business Media LLC 2017

common structural organization consisting of an extracellular ligand-binding domain, a transmembrane domain, and an intracellular tyrosine kinase domain [1–3]. They are typical receptor tyrosine kinases (RTKs). The EGF-family of peptides bind the ErbB to initiate signaling by causing specific homodimeric or heterodimeric receptor formation and activation of the cytoplasmic kinase domain that phosphorylates tyrosines in the tail region of each receptor. Phosphorylation triggers the association of specific signaling molecules whose binding initiates downstream signaling events.

Each of the pathways that constrain the proliferative response in normal cells is perturbed in most cancers. The activation of ErbB2 and EGFR receptors has been implicated in the development of many types of human cancer [4, 5]. In this section, we concentrate on ErbB2 signaling because the ErbB2 downstream signaling pathways have been identified to contribute to epithelial transformation, and ErbB2 receptor activation drives breast cancer development [6]. Many mechanisms can promote constitutive ErbB2 receptors activation. The most important mechanism is ERBB2 gene amplification and receptor overexpression. HER-2/neu is a proto-oncogene located on chromosome 17. ERBB2 gene amplification is the first consistent alteration found in breast cancer. This gene is amplified and therefore the protein (HER-2) overexpressed in around 15–30% of sporadic invasive breast cancers [7, 8].

The mechanism of c-erbB2-induced carcinogenesis is believed to involve overexpression-induced homodimerization of c-erbB2 subunits, leading to constitutive and ligand-independent signalling from the activated receptor. ErbB2 receptors control key intracellular pathways, such as proliferation, metabolism and survival. However, ErbB2 protein overexpression plays a pivotal role in oncogenic transformation and tumorigenesis. Studies using NIH 3T3 cells implicate HER2 overexpression in malignant transformation and tumorigenesis [9–11]. Transfection of the HER2 gene into human breast and ovarian tumor cell lines produced more aggressive growth characteristics, such as increased DNA synthesis; cell migration; cell growth; growth in soft agar in vitro; and tumorigenicity and metastatic potential in mice [12, 13].

EMT has also been shown to cause organ fibrosis and to promote carcinoma progression and chemoresistance through a variety of mechanisms. It is worth noting that, in a previous study, the capacity of prolonged ErbB2 homodimer signaling to cause epithelial–mesenchymal transition (EMT) was reported. It is well known that Twist is a highly conserved basic helix–loop–helix transcription factor and is one of the most important factors regulating EMT [14, 15]. Twist acts as an oncogene and is overexpressed in a variety of cancers, including breast cancer [16]. Elevated expression of Twist in breast tumor samples from patients correlates strongly with high-grade invasive carcinoma and with chromosome instability to promote an EMT-like transition that is pivotal for the transformation into an aggressive breast cancer phenotype [17], consistent with its role in driving mouse mammary carcinoma cell metastasis [18]. The downregulation of epithelial markers is a typical consequence of the activity of EMT-associated transcriptional repressors like Twist, Snail, Slug, ZEB1, and/or ZEB2 [19]. Twistl depletion by RNA interference blocks mesenchymal transformation, partially reverses multidrug resistance, and abolishes invasion induced by Adriamycin. Furthermore, Twistl RNA interference may show efficacy in Adriamycin-based chemotherapies for breast cancer [20]. In our study, Twist expression was observed in 54% (220/408) of breast cancer patients and was positively associated with tumor size, and Ki67, VEGF-C, and HER2 expression.

In addition to cell proliferation assay and colony formation assay, migration and invasion assays have also been used in the studies on EMT and ErbB signaling, and these assays were performed in our study that overexpressed Twist associates with markers of EMT and predicts poor prognosis in breast cancers via ERK and Akt activation. Herein, we describe the experimental procedure of these assays in detail.

2 Materials

2.1 Equipment

- Cell culture hood (i.e., laminar-flow hood or biosafety cabinet).
- Incubator (humidified CO2 incubator recommended).
- Water bath.
- Centrifuge.
- Refrigerator and freezer (−20 °C).
- Cell counter (e.g., Countess).
- Automated Cell Counter or hemacytometer.
- Inverted microscope.
- Liquid nitrogen (N2) freezer or cryostorage container.
- Sterilizer (i.e., autoclave).

2.2 Cell Lines, Cell Culture Vessels, and Consumables

The major requirement of a cell culture laboratory is the need to maintain an aseptic work area that is restricted to cell culture work.

In this experiment, MCF-7 breast cancer cell lines characterized as ER-positive/PgR-positive luminal mammary carcinoma and lower invasive and MDA-MB-231 characterized as triple negative breast cancer cells are mainly used.

12 × 25 cm^2 tissue culture flasks; 6-well cell culture plates; and 100 μL, 200 μL, and 1 mL pipettes.

2.3 Reagents

Dulbecco's modified Eagle's medium (DMEM); fetal bovine serum; L-glutamine; penicillin–streptomycin (P/S) solution; phosphate buffered saline (PBS); Lipofectamine 2000; trypsin–EDTA; G418 (for stable expression); 37% formaldehyde (1%); and Gentian Violet.

2.3.1 Proliferation Assay (CKK8)

Cell counting kit-8 (Dojindo Molecular Technologies, Kumamoto, Japan); plate reader (450 nm filter); 96-well plate; 10 μL, 100–200 μL, and multichannel pipettes; and CO_2 incubator.

2.3.2 Colony Formation Assay

Culture plates (6-well plates in this protocol), culture medium such as DMEM with 10% FBS containing 1% penicillin–streptomycin; cell counter; methanol and 0.1% Crystal Violet (Sigma, St. Louis, MO, USA) (or other suitable fixative and staining solution); microscope and camera; and CO_2 incubator.

2.3.3 The Soft Agar Colony Formation Assay

Culture plates (6-well plates in this protocol); culture medium such as DMEM; fetal bovine serum; penicillin–streptomycin; Difco noble agar (BD Science); sodium bicarbonate; trypsin–EDTA; sterile bottle-top filters; 37 °C/5% CO_2 incubator; ice bucket; microwave; cell counter; PBS; and nitroblue tetrazolium chloride.

2.3.4 Migration Assay in Transwell

BD Falcon™ Cell Culture Inserts (8 μm pore size; Corning, NY, USA), 10 cm tissue culture dish, culture plates (24-well plates in this protocol); culture medium such as DMEM with 10% FBS containing 1% penicillin–streptomycin; bovine serum albumin; trypsin inhibitor of choice; cell counter; cotton swabs; and methanol and Wright-Giemsa solution (Sigma-Aldrich) (or other suitable fixative and staining solution).

2.3.5 Invasion Assay in Transwell with Matrigel

BioCoat Matrigel invasion chambers with culture plates (BD Biosciences; we use 24-well plates in this protocol); culture medium DMEM; chemoattractants such as 20% FBS in the culture medium; bovine serum albumin; trypsin inhibitor of choice; cell counter; cotton swabs; and methanol and 0.1% Crystal Violet (Sigma, St. Louis, MO, USA) (or other suitable fixative and staining solution).

2.3.6 In Vitro Wound-Healing (Scratch) Assay

Culture medium with supplements (serum, antibiotics); PBS; culture dishes (60 mm, 35 mm, or 6-well dishes work well); sharpie marker, and p200 or p1000 pipette tips (or other appropriate scratching device); phase-contrast microscope with camera; and image analysis software.

3 Methods

In vitro analysis of EMT is done by characterizing specific changes to the epithelial cell phenotype. More specifically, it principally includes the analysis of loss of epithelial features and acquisition of mesenchymal, fibroblastic-like traits. This includes (1) the dissolution of the cell–cell junctions and the subsequent loss and/or delocalization of molecular components of cell junctions (E-cadherin and β-catenin), which can be monitored by immunodetection followed by chemical or fluorescent revelation performed on fixed cells and tissue sections, (2) the decrease in total expression levels of specific epithelial proteins (E-cadherin, claudins, and occludins) and the increase in expression levels of mesenchymal proteins (N-cadherin and Vimentin), which can be analyzed at the mRNA (RT-qPCR) and protein (Western blotting) levels, (3) the gain of cell motility characterized by cell migration assay such as wound healing experiments, or by invasion assay (Boyden chamber or transwell Matrigel invasion assays) and cell behavior in 3D collagen gels.

This chapter assumes that the readers are familiar with several common laboratory techniques including cellular proliferation assay; colony formation assay; soft agar assay; and migration and invasion assay using Twist overexpressed breast cancer cells as an example. We found Twist overexpression involved in EMT by activating Akt and ERK signaling. Once Akt and ERK signaling pathways are activated, the cell-cycle progression will be accelerated, and the colony number will be increased.

3.1 Cell Culture and Experimental Setup

1. MCF-7 breast cancer cells are maintained as monolayers in 12×25 cm^2 tissue culture flasks containing 5 mL of DMEM supplemented with 10% fetal bovine serum, L-glutamine (2 mM), and 20 mg/mL P/S. The cells are grown in a humidified 5% CO_2 environment at 37 °C.
2. When the cells are 90% confluent, single cell suspensions are prepared by trypsinization. The cells are washed with PBS and incubated with a 0.05% trypsin–EDTA solution for 5–10 min. When the cells start to become rounded and ~30% are detached, 3 volumes of DMEM containing 10% fetal bovine serum is added to neutralize the trypsin. The cells are detached by pipetting up and down (20 times).
3. The cells are counted using a hemacytometer.
4. MCF-7 cells are inoculated at 5×10^5 cells/mL in a 6-well plate and allowed to attach for 24 h (according to the doubling time of the cell line, approximately 50–85 h for human MCF-7). The aim is to achieve ~90% confluency on the day of the experiment.
5. The cells are then cotransfected with 0.1 μg of pFlag-Twist (experimental group) and 1 μg of pFlag-CMV (negative control), using Lipofectamine 2000.
6. For stable transfection, at 48 h post-transfection, the concentration of G418 (Sangon Biotech, Shanghai, China) is

determined by drawing a killing curve, and 1 mg/mL of G418 is used to kill nontransfected cells. Afterward, the cells are maintained in medium contained 1 mg/mL of G418.

3.2 Cell Proliferation Assay (CKK8)

The Cell Counting Kit-8 (CKK-8) is a colorimetric assay kit used to measure cell proliferation and cytotoxicity. It is a ready-to-use solution that does not require radioisotopes and correlates with the [^{3}H]-thymidine incorporation assay. CCK-8 is based on WST-8, which is an MTT-like compound that can be reduced by some dehydrogenases in the mitochondria to produce yellow formazan in the presence of electron-coupling reagents. It can be added directly to the cell media for fast, high-throughput screening without a solubilization process obtaining highly reproducible and accurate results. For the same cells, the color darkness correlates with the number of cells. The detection sensitivity of CCK-8 is higher than other tetrazolium salts such as MTT, XTT, MTS, or WST-1.

1. Prepare MCF-7 cells following stable transfection of pFlag-Twist, and pFlag-CMV empty vector used as the negative control. Detach the cells by pipetting up and down (20 times), and single cell suspensions are obtained.
2. Trypsinize and count cells in each sample carefully using a hemacytometer, and make the cells at 1×10^5 cells/mL in culture media. This number needed per well will vary depending on cell type.
3. Plate 100 μL media of cell suspension into each well of a 96-well plate at a density of 1×10^4 per well and incubate in a 5% CO2 incubator at 37 °C until the cell proliferation assay (*see* **Note 1**).
4. Thaw the frozen CCK-8 on the bench top or in a water bath at 37 °C.
5. Add 10 μL of the CCK-8 solution to each well of the plate. Be careful not to introduce bubbles to the wells since the bubbles interfere with the O.D. reading.
6. Incubate the plate for 1–4 h in the incubator. The length of incubating time depends on the type of cells and the density of the cells, etc. (*see* **Note 2**).
7. Measure the absorbance at 450 nm using an ELX800 microplate reader at Day 1 to 6 after plating.
8. To measure the absorbance later, add 10 μL of 1% w/v SDS to each well, cover the plate and store it with protection from light at room temperature. No absorbance change should be observed for 48 h.
9. Experiments are performed three times in triplicate. A representative result is shown in Fig. 1.

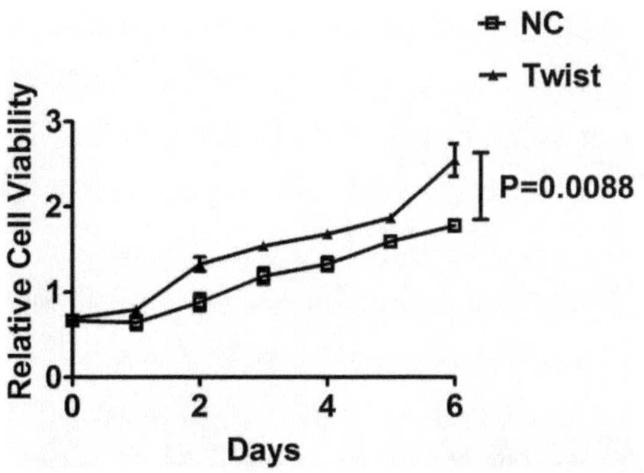

Fig. 1 Overexpression of Twist enhanced the proliferation of MCF-7 cells. CCK8 assays were used to estimate the cell proliferation at different time points. Data shown are means ± SD of at least three independent experiments. *P*-values were obtained using the Student's *t*-test analysis

3.3 Colony Formation Assay

Clonogenic assay or colony formation assay is an in vitro cell survival assay based on the ability of a single cell to grow into a colony.

1. Grow MCF-7 cells stably expressing Twist until more than 90% confluence in 10 cm culture dish.
2. Detach and resuspend cells by trypsinization.
3. Count the number of cells carefully using a hemacytometer, and diluted such that appropriate cell numbers are seeded into petri dishes or multi-well plates (three replicates of each in 6-well plate). In this case, dilute MCF-7 cells at the concentration of /mL (*see* **Note 3**).
4. Plate a total of 500 cells into each well of a 6-well plate.
5. Incubate cells in a 5% CO_2 environment at 37 °C. The incubation time for colony formation varies from 1 to 3 weeks for different cell lines. In this example, the control dishes for MCF-7 cells (control cells) require 8 days to form sufficiently large clones consisting of 50 or more cells. The experiment is repeated three times independently (*see* **Note 4**).

Complete the following steps in a fume hood.

6. Prepare 1 L Fixing/Staining solution consisting of 0.5 g Crystal Violet (0.05% w/v), 27 mL 37% formaldehyde (1%), 100 mL 10× PBS (1×), and 10 mL methanol (1%), and add 863 mL dH_2O to 1 L.
7. Remove media (do not wash cells), and add fixing/staining solution to cover dish, and stain for 20 min at room temperature (*see* **Note 5**).

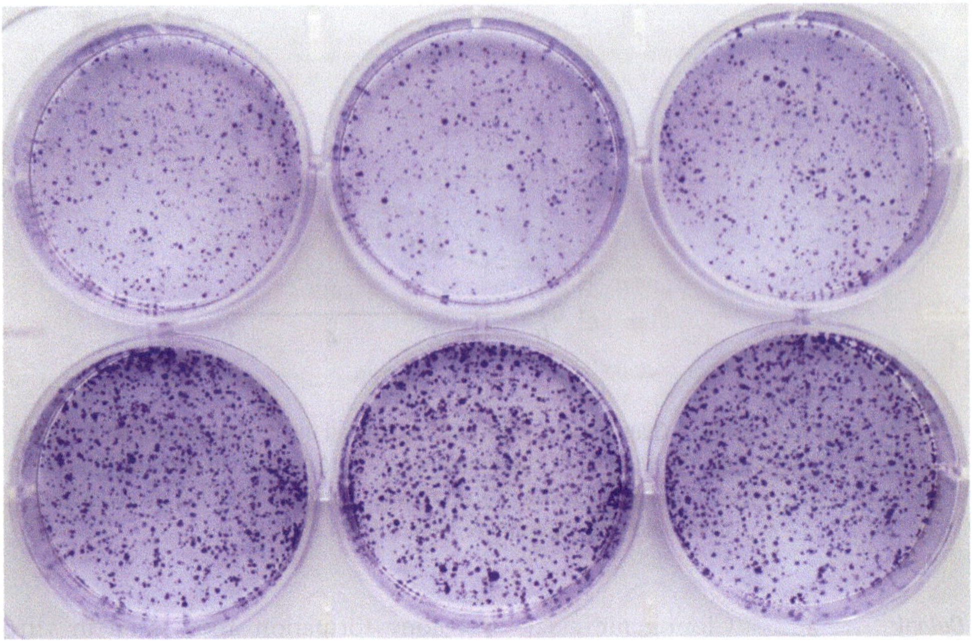

Fig. 2 Representative colony formation assay with monolayer culture to assess the tumorigenesis function of Twist. Twist-negative MCF-7 breast cancer cell were then transfected with pFlag-Twist. The cell were transfected with pFlag-CMV as negative control. Twist greatly increases the colony formation of tumor cells. Quantitative analyses of colony numbers and the colony formation rate were from three independent experiments

8. Remove fixing/staining solution.
9. Wash dishes one at a time by dipping into bucket of water in the sink with the water continuing to run.
10. Air-dry dishes. Count colonies with >50 cells by eye under a Stereomicroscope or with automated colony counter. Colonies containing more than 50 individual cells are counted using a stereomicroscope (*see* **Note 6**). A representative result is shown in Fig. 2.

Alternatively, digital imaging and counting using imaging software

1. Digital images of the colonies are obtained using a camera or scanning device.
2. Colonies are counted using imaging analysis software packages as described below.
3. Count colonies using ImageJ (Fiji Version 1.44a). Open the image file in Fiji, go to File → Open.
4. If required convert the image to 8-bit format, go to Image → Adjust... → Threshold.

5. Adjust threshold to reduce levels of nonspecific background so that only the colonies are detected.
6. Count colonies using the following: go to Process → Binary → Find maxima (*see* **Note 7**).

3.4 The Soft Agar Colony Formation Assay

In the soft agar colony formation assay, cells are grown in a layer of soft agar mixed with cell culture medium that rests on another layer of soft agar, also mixed with cell culture medium, but containing a higher concentration of agar. This prevents cells from adhering to the culture plate, yet allows transformed cells to form visible colonies. Normal cells depend on cell–to–extracellular matrix contact to be able to grow and divide. Conversely, transformed cells have the ability to grow and divide irrespective of their surrounding environment. Therefore, cells able to form colonies in an anchorage-independent manner are considered to be transformed and carcinogenic.

1. Prepare 2× and 1× cell culture medium and pass these medium through a 0.2 μm filter to sterilize. Add additional components needed for normal culture of the cell line. For example, grow MDA-MB23-1 cell line in DMEM supplemented with 10% FBS and 1% penicillin–streptomycin solution. Warm medium to 37 °C in hot water bath prior to use.
2. Prepare 1% and 0.6% noble agar and autoclave the noble agar mixtures to sterilize in microwave for about 1–2 min until agar is completely dissolved and the solution is clear.
3. Prepare nitroblue tetrazolium chloride solution by making a 1 mg/mL stock solution in 1× PBS.
4. Place melted agar solution and prewarmed 2× culture medium in an ice bucket filled with hot tap water (42 °C).
5. Mix 6 mL 2× culture medium with 6 mL 1% noble agar solution. Depositing 1.5 mL of this mixture into each well of 6-well plates without any air bubbles into the plate wells (*see* **Note 8**).
6. Cover the plates and allow agar mixture to solidify at room temperature, in cell culture hood, for 30 min. Once the lower layer of agar has solidified, begin preparation of the upper layer.
7. Trypsinize and count the cells. This number needed per well will vary depending on cell type. Use 5000 cells/well as a starting point and adjust as needed (*see* **Note 9**).
8. Melt 0.6% agar solution in a microwave as above and place into ice bucket containing hot water along to keep around 42 °C to avoid premature hardening and to maximize cell survival (*see* **Note 10**).
9. Mix 0.6% agar and cell suspension in a 1:1 ratio.

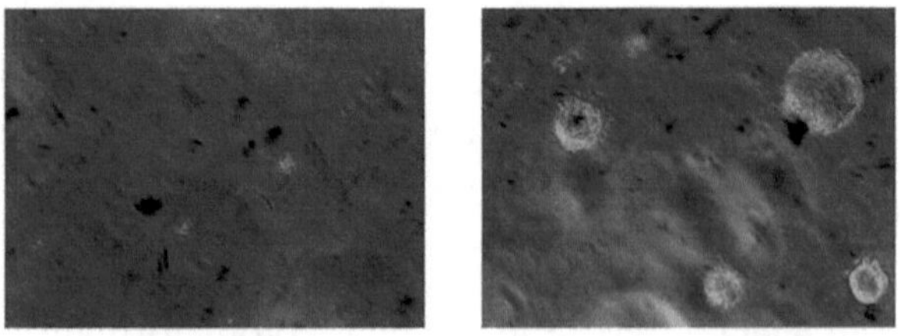

Fig. 3 Representative soft agar colony formation assay from article "twist is a potential oncogene that inhibits apoptosis" by Roberta Maestro. Twist promotes colony formation in soft agar. C8 MEFs (mouse embryo fibroblasts) were infected with retroviruses that direct the expression of LacZ or Twist. Infected cells were plated in soft agar, and colony formation was assessed after 2 weeks (as indicated). The LacZ-expressing cells form only a few, small colonies. Cells infected with Twist form large colonies. Expression of Twist enhances colony formation by approximately five- to sevenfold

10. Deposit 1.5 mL of this mixture into each well of 6-well plate. Use caution to avoid deposition of any air bubbles into the plate wells (*see* **Note 11**).
11. Allow cell–agar mixture to solidify at room temperature, in cell culture hood, for 30 min before placing into a 37 °C humidified cell culture incubator.
12. Add a layer of normal growth medium over the upper layer of agar to prevent desiccation. 100 μL of medium added twice weekly is sufficient for this purpose. The time required for adequate colony formation varies for each cell line, typically around 14–21 days (*see* **Note 12**).
13. Stain cells by adding 200 μL of nitroblue tetrazolium chloride solution per well and incubating plates overnight at 37 °C.
14. Once colonies are stained, take photographs of each well using an imager and count colonies using image analysis software as describe in "colony formation assay" part. A representative result is shown in Fig. 3.

3.5 Migration Assay in Transwell

The transwell migration assay is a commonly used test to study the migratory response of cancer cells. This assay is also known as the Boyden or modified Boyden chamber assay. In this experiment, cells are added to the upper chamber of an insert and a stimulus is added to the lower chamber. Due to these attractive forces, cells in the top compartment migrate through the pores into the bottom reservoir. Adherent cells will stick to the underside of the membrane. The membrane can be fixed, stained, and cells can be counted under the microscope.

1. Trypsinize the cells and suspend in 5 mL (per 10 cm dish) of medium containing 0.1% BSA and trypsin inhibitor.

2. Spin down cells and resuspend in 5 mL of medium with 0.1% BSA.
3. Count the cells. Dilute the cells in 0.1% BSA to a concentration of 1×10^5 cells/mL.
4. Add 500 μL per well of chemoattractant medium (such as 20% FBS) to 24-well culture dishes.
5. Place the BD Falcon™ Cell Culture Inserts on top of the culture dishes from **step 5**, ensuring no air bubbles are created (*see* **Note 13**).
6. Add 200 μL per well of the cell suspension to the top of the Falcon™ Cell Culture Inserts. The number of cells will depend on cell types and the experimental considerations (*see* **Note 14**).
7. Put covers on the plates and incubate at 37 °C for 24–48 h. The incubation time for the migration will depend on the experimental considerations (*see* **Note 15**).
8. Remove the cells from the upper side of the chamber by using a cotton swab and rinse with PBS.
9. Immerse the swabbed inserts into methanol for 10 min at room temperature to fix the cells that have invaded onto the lower side of the transwell filter.
10. Stain the cells with 0.1% crystal violet for 15 min at room temperature.
11. Wash away excess stain with water and invert and air-dry the stained inserts.
12. Cells on the underside of the membrane represent the number of cells that have migrated. Use a microscope to observe the migrated cells and five high-powered fields are counted for each well (*see* **Note 16**). A representative result is shown in Fig. 4.

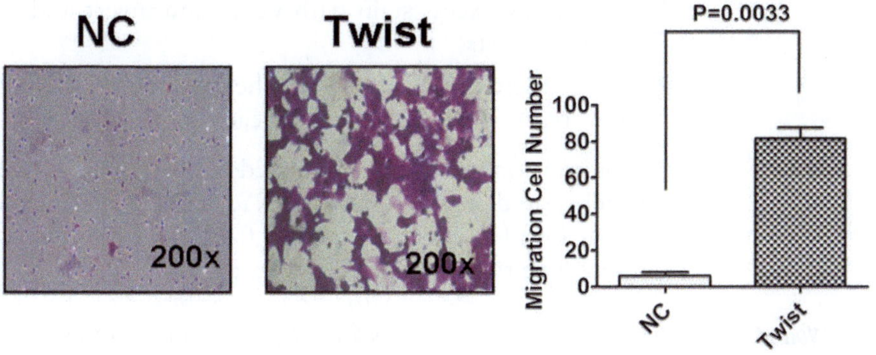

Fig. 4 Representative migration assay in Transwell. Overexpression of Twist enhanced cell migratory capability. Empty vector and Twist-overexpressing MCF-7 cell were subjected to transwell migration assay. Data shown are means ± SD of at least three independent experiments. *P*-values were obtained using the Student's *t*-test analysis

3.6 Invasion Assay in Matrigel

The invasion assay in Matrigel measures the capacity of cells to invade through a barrier of ECM, in vitro. In this experiment, cells are added to the upper chamber of a Matrigel coated insert and a stimulus is added to the lower chamber. Invasive cells will pass through the Matrigel barrier and the porous membrane to the underside of the insert. The cells that invade through the Matrigel can be quantified by staining and counting the cells under the microscope.

1. Trypsinize the cells and suspend in 5 mL (per 10 cm dish) of medium containing 0.1% BSA and trypsin inhibitor.
2. Spin down cells and resuspend in 5 mL of medium with 0.1% BSA.
3. Count the cells. Dilute the cells in 0.1% BSA to a concentration of 1×10^5 cells/mL (*see* **Note 17**).
4. Add 500 μL per well of chemoattractant medium (such as 20% FBS) to 24-well culture dishes.
5. Place the BioCoat Matrigel invasion chambers on top of the culture dishes from **step 5**, ensuring no air bubbles are created (*see* **Note 18**).
6. Add 200 μL (for MDA-MB-231) per well of the cell suspension to the top of the Matrigel chambers. The number of cells will depend on cell types and the experimental considerations.
7. Put covers on the plates and incubate at 37 °C for 24–72 h.
8. Remove the cells from the upper side of the chamber by using a cotton swab and rinse with PBS.
9. Immerse the swabbed inserts into methanol for 15 min at room temperature to fix the cells that have invaded onto the lower side of the transwell filter.
10. Stain the cells with 0.1% crystal violet for 15 min at room temperature.
11. Wash away excess stain with water and invert and air-dry the stained inserts.
12. Use a microscope to observe the invaded cells and five high-powered fields are counted for each well.

For invasion assay, cells are seeded in the upper compartment of Matrigel-coated inserts (Corning, NY, USA). After 72 h for MDA-MB-231, invaded cells are counted from five fields in each well. A representative result is shown in Fig. 5.

3.7 In Vitro Wound Healing Assay

The wound healing assay is an easy, economical, and well-developed method to study cell migration in vitro. In this experiment, the basic steps involve creating a "scratch" in a confluent cell monolayer, capturing the images at the beginning and at regular intervals

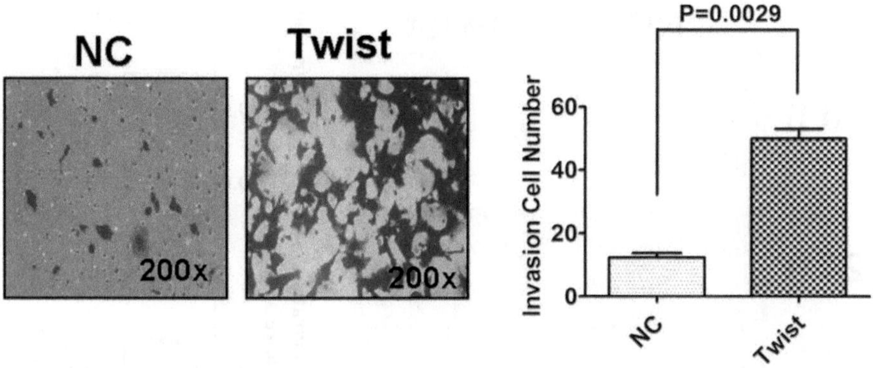

Fig. 5 Representative invasion assay in Matrigel. Overexpression of Twist enhanced cell invasive capability. Empty vector and Twist-overexpressing MCF-7 cell were subjected to Matrigel invasion assay. Data shown are means ± SD of at least three independent experiments. *P*-values were obtained using the Student's *t*-test analysis

during cell migration to close the scratch, and comparing the images to quantify the migration rate of the cells.

1. Trypsinize subconfluent growing cells and then mixing cells with medium containing serum.
2. Count the cells and plate cells onto the 6-well plates with 1×10^6 per well to create a confluent monolayer in 2 mL medium with 5% FBS (For MDA-MB-231). The number of cells will depend both on cell types and the size of dishes. It is recommended to use a lower percentage of serum to minimize cell proliferation, but just sufficient to prevent apoptosis and/or cell detachment (*see* **Note 19**).
3. Incubate the dishes properly for approximately 6–24 h at 37 °C, allowing cells to adhere and spread on the substrate completely (*see* **Note 20**).
4. Scrape the cell monolayer in a straight line to create a "scratch" with a p200 pipet tip. Remove the debris and smooth the edge of the scratch by washing the cells once with PBS and then replace with 2 mL DMEM with 5% FBS (*see* **Note 21**).
5. To obtain the same field during the image acquisition, create markings with an ultrafine tip marker. Place the dish under a phase-contrast microscope and acquire the first image of the scratch.
6. Every 6–12 h intervals, place the dish under a phase-contrast microscope, align the photographed region acquired in **step 4** and acquire a second image. Take pictures at regular intervals until scratch is closed.
7. The images acquired can be further analyzed quantitatively by using computing software of choice. By comparing the images

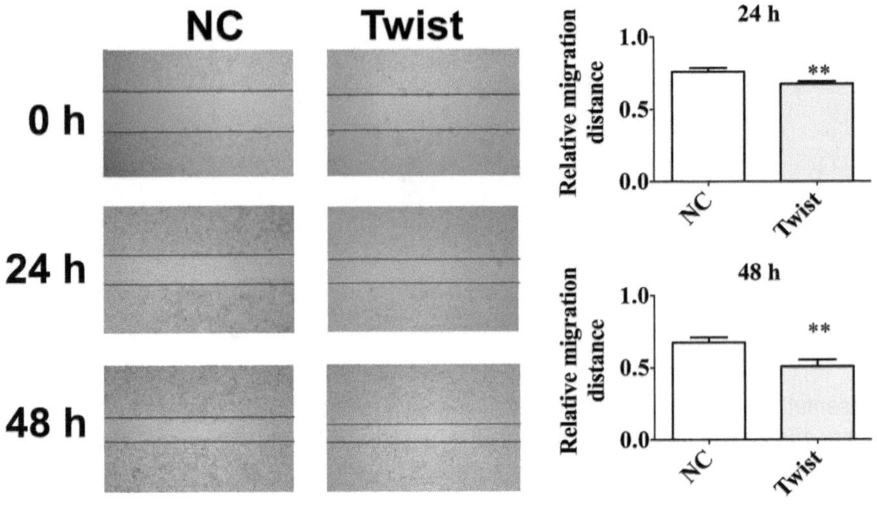

Fig. 6 Wound healing assay in MCF-7 cells. (**a**) MCF-7 breast cancer cell were transfected with pFlag-Twist, as well as its negative control pFlag-CMV. Twist greatly increases the motility of MCF-7 cells. (**b**, **c**) Quantitative analyses of relative migration distance at 24 and 48 h, respectively. Data shown are means ± SD of at least three independent experiments. ** $P < .01$, (Student's *t*-test) as compared to control cells

from time 0 (**step 4**) to the last time point, Obtain the distance of each scratch closure on the basis of the distances that are measured by software. Measure at least five readings of distance for each sample and repeat each experiment at least three times. A representative result is shown in Fig. 6.

4 Notes

1. The plating efficiency and/or surviving fraction should be anticipated when deciding the number of cells to seed per plate. The aim is to achieve a range of between 20–150 colonies. It is recommended to simultaneously perform a viability assay on day 0 to verify that the initial number of cells used is the same between the different samples.
2. The initial experiment can be used in 1, 2 and 4 h respectively with the microplate reader detection, and then select the appropriate time point for follow—up experiments.
3. The number of cells will need to be optimized as it will vary depending on the cell line used.
4. The incubation time for colony formation varies from 1 to 3 weeks for different cell lines. It is accepted that the time must be equivalent to at least six cell divisions.
5. It is advised not to wash cells after removing media because the colony might be washed out.

6. One should observe the colony under phase-contrast microscope by setting up appropriate the number of cells. In this case, the number of cells is determined as more than 50 cells.
7. It is recommended to apply computer software for digital imaging and counting as an alternative approach.
8. Avoid air bubbles. These can sometimes be removed by using a pipette tip and it is easier to do when the collagen has solidified (30 min after the plate is placed in the incubator).
9. The number of cells will need to be optimized as it will vary depending on the cell line used. In this case, we use 5000 cells/well as a starting point.
10. It is advised to use hot water around 42 °C to avoid hardening of agar and to maximize cell survival.
11. It is recommended to fill the mixture of agar and cells slowly and continuously by leaning the culture plate, in order to avoid air bubble.
12. A layer of normal growth medium should be maintained over the upper layer of agar to prevent desiccation. 100 μL of medium added twice weekly is sufficient for this purpose.
13. Avoid the formation of air bubbles.
14. The number of cells should be optimized because the ability of colony formation varies depending on cell types.
15. The incubation time is advised to be optimized depending on each type of cells.
16. Alternatively, the computer software could be used to count the migrated cells.
17. The optimal concentration of cells as well as length of time for migration will vary depending on the cell type and experimental conditions used. This will need to be optimized.
18. BioCoat Matrigel invasion chambers are applied for invasion assay but instead of Transwell chamber in migration assay. This is the only difference between two experimental procedures.
19. It is generally advised to use a lower concentration of serum than full growth medium (or use serum-free medium) to minimize the effect caused by cell proliferation.
20. It is important to create scratches of equal width. Up to three different scratches can be easily performed in the same well of a six-well culture dish. This is helpful for statistical comparisons when the migration rates are quantified.
21. The time frame for migration will need to be optimized depending on the cell type and serum concentrations. This can be monitored by checking the cells periodically under a microscope.

References

1. Yarden Y, Sliwkowski MX (2001) Untangling the ErbB signalling network. Nat Rev Mol Cell Biol 2(2):127–137
2. Mendelsohn J, Baselga J (2003) Status of epidermal growth factor receptor antagonists in the biology and treatment of cancer. J Clin Oncol 21(14):2787–2799
3. Olayioye MA et al (2000) The ErbB signaling network: receptor heterodimerization in development and cancer. EMBO J 19 (13):3159–3167
4. Epis MR et al (2009) miR-331-3p regulates ERBB-2 expression and androgen receptor signaling in prostate cancer. J Biol Chem 284 (37):24696–24704
5. Holbro T, Hynes NE (2004) ErbB receptors: directing key signaling networks throughout life. Annu Rev Pharmacol Toxicol 44:195–217
6. Hynes NE, Lane HA (2005) ERBB receptors and cancer: the complexity of targeted inhibitors. Nat Rev Cancer 5(5):341–354
7. Slamon DJ et al (1987) Human breast cancer: correlation of relapse and survival with amplification of the HER-2/neu oncogene. Science 235(4785):177–182
8. Berger MS et al (1988) Correlation of c-erbB-2 gene amplification and protein expression in human breast carcinoma with nodal status and nuclear grading. Cancer Res 48(5):1238–1243
9. Di Fiore PP et al (1987) Overexpression of the human EGF receptor confers an EGF-dependent transformed phenotype to NIH 3T3 cells. Cell 51(6):1063–1070
10. Di Fiore PP et al (1987) erbB-2 is a potent oncogene when overexpressed in NIH/3T3 cells. Science 237(4811):178–182
11. Hudziak RM, Schlessinger J, Ullrich A (1987) Increased expression of the putative growth factor receptor p185HER2 causes transformation and tumorigenesis of NIH 3T3 cells. Proc Natl Acad Sci U S A 84(20):7159–7163
12. Benz CC et al (1992) Estrogen-dependent, tamoxifen-resistant tumorigenic growth of MCF-7 cells transfected with HER2/neu. Breast Cancer Res Treat 24(2):85–95
13. Chazin VR et al (1992) Transformation mediated by the human HER-2 gene independent of the epidermal growth factor receptor. Oncogene 7(9):1859–1866
14. Teng Y, Li X (2014) The roles of HLH transcription factors in epithelial mesenchymal transition and multiple molecular mechanisms. Clin Exp Metastasis 31(3):367–377
15. Soini Y et al (2011) Transcription factors zeb1, twist and snai1 in breast carcinoma. BMC Cancer 11:73
16. Martin TA et al (2005) Expression of the transcription factors snail, slug, and twist and their clinical significance in human breast cancer. Ann Surg Oncol 12(6):488–496
17. Mironchik Y et al (2005) Twist overexpression induces in vivo angiogenesis and correlates with chromosomal instability in breast cancer. Cancer Res 65(23):10801–10809
18. Yang J, Weinberg RA (2008) Epithelial-mesenchymal transition: at the crossroads of development and tumor metastasis. Dev Cell 14(6):818–829
19. Thiery JP et al (2009) Epithelial-mesenchymal transitions in development and disease. Cell 139(5):871–890
20. Li QQ et al (2009) Twist1-mediated adriamycin-induced epithelial-mesenchymal transition relates to multidrug resistance and invasive potential in breast cancer cells. Clin Cancer Res 15(8):2657–2665

Chapter 18

Assessment of Specificity of an Adenovirus Targeted to HER3/4

Sheena H. MacLeod, Kyle G. Potts, Shyambabu Chaurasiya, and Mary M. Hitt

Abstract

Gene therapy with viral vectors, such as adenovirus (Ad), targeted to the human epidermal growth factor receptors 3 and 4 (HER3/4) are potentially useful for cancer therapy. Testing the expression of a reporter gene from these viruses in target cells is essential to determine functionality of the targeted virus. A competition assay with a relevant ligand (heregulin, HRG) can provide convincing evidence that blocking binding to the HER3/4 receptor results in decreased reporter gene expression. Labeling individual viruses with a fluorescent molecule allows examination of the targeted virus in specific steps in the infection. Virus internalization into cell lines can be determined using antibody-labeled receptors, and the virus colocalization with receptors can also be visualized. Characterization of a targeted virus in this fashion is important to demonstrate that the targeting of the virus functions in an expected manner, and provides support for larger-scale testing of the virus. Information acquired in these experiments may also be useful to inform and improve on the design of future targeted viruses.

Key words HER3/4, Heregulin, Cancer, Adenovirus, Vector, Gene therapy, Targeting, Luciferase, Fluorescence, Internalization

1 Introduction

The human epidermal growth factor (EGF) receptor 3 (HER3, ErbB3) is potentially an excellent therapeutic cancer target because it is overexpressed in many types of tumors and is frequently associated with poor prognosis (reviewed in [1, 2]). Overexpression of another member of the EGF receptor family, HER4 (ErbB4), has been observed in some types of tumors, but association of HER4 expression with clinical outcome remains controversial [3–6]. We have developed an adenovirus (Ad) vector that targets these two receptors [7] by taking advantage of the fact that HER3 and HER4 receptors share common ligands (neuregulins or heregulins (HRG) [1, 4]). This protocol describes specific methods successfully used to characterize this targeted Ad vector [7]. These methods may also

Zhixiang Wang (ed.), *ErbB Receptor Signaling: Methods and Protocols*, Methods in Molecular Biology, vol. 1652, DOI 10.1007/978-1-4939-7219-7_18, © Springer Science+Business Media LLC 2017

be applied for use with other Ad vectors targeted to similar receptors, or for use with other viruses targeted to HER3/4.

Ad is the most common type of gene therapy vector used in clinical trials [8], for several reasons including its high efficiency of gene transfer, stability of the Ad virus particle, ability of the viral genome to accommodate large inserts, ease of manufacturing high titers of virus, and very low risk of insertional mutagenesis upon infection. For some of the same reasons, Ad has been investigated for its potential as an oncolytic virus, a class of cancer therapeutics that replicate selectively in cancer cells, resulting in tumor lysis. Ad has a double-stranded DNA genome of approximately 36 kb, with a nonenveloped icosahedral capsid [9]. Several methods of construction have been employed to generate Ad vectors for gene therapy (recently reviewed in [10]). In the following protocol, we have used a first generation vector rendered nonreplicative by deletion of genes encoded by the viral early region 1 (E1) [7, 11, 12].

Targeted Ad virotherapy can be achieved through multiple strategies that allow expression of the transgene in specific cells in the body while minimizing expression in nontarget cells that could potentially result in detrimental effects [10, 13–15]. One strategy is to use a target-cell-specific gene promoter to limit transgene expression, or replication of oncolytic viruses, to specific cells [16–19]. Alternatively, the virus capsid can be modified or coated to target specific cellular receptors. This is especially important when targeting tumor cells, since Ad receptors are not always highly expressed on tumor cells [20–23]. Some targeting methods have employed polymers, liposomes, or other molecules physically or covalently attached to the Ad virus [24–26] (reviewed in [14, 15]). Others have focused on genetic modification of Ad capsid proteins, including fiber [27–32], pIX [28, 33, 34], and hexon [35, 36] (reviewed in [13, 15, 37]). We have genetically modified an Ad vector by insertion of the EGF-like domain of HRG into the HI loop of the fiber knob [7]. This modification expands the host range of the vector to include not only cells displaying the wild-type Ad receptor, but also cells displaying HER3/4 receptors.

An important strategy to validate targeted viruses is to test for expression of a virally encoded reporter (e.g., green fluorescent protein (GFP) [38] or luciferase) in infected cells, tissues, or animals. One of the most commonly used reporter genes is luciferase, which catalyzes the oxidation of the substrate luciferin in the presence of ATP, oxygen and Mg^{++}, resulting in the production of light [39]. Luciferase activity can be detected both in vitro [39] and in vivo [40]. In Subheading 3.1, we describe the assay of luciferase in extracts of tissue culture cell lines. We have used this assay to quantify targeted Ad vector-encoded luciferase reporter gene expression in a series of cell lines with varying levels of cellular receptors (Fig. 1). We have described the detection of luciferase in extracts of animal tissues [7] or in vivo [41] elsewhere.

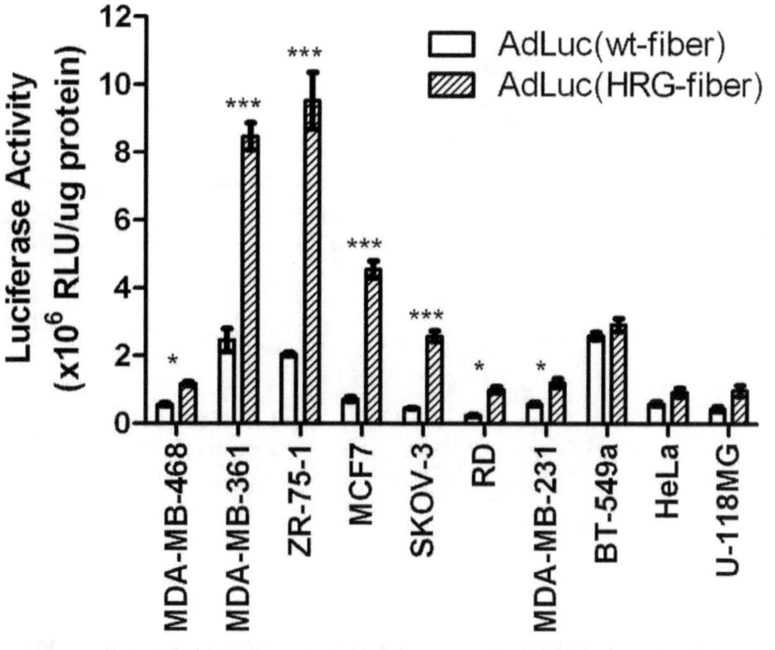

HER3/4	high	high	high	high	high	high	low	low	low	low
CAR	+++	+/-	+/-	-	+/-	+/-	+	++	+	+/-

Fig. 1 Luciferase expression in breast cancer cell lines 24 h after infection with AdLuc(HRG-fiber) or with AdLuc(wt-fiber). Comparison of viruses at a single multiplicity of infection (MOI) (1000 vp per cell). Luciferase activity is expressed as relative light units (RLU) per μg of total protein. The table below indicates receptor levels as previously determined by western blot analysis. Cells were infected in triplicate. Error bars are equal to 1 standard deviation. Significance is determined by analysis of variance (ANOVA) $^{*}P < 0.05$ and $^{***}P < 0.0001$. Ad, adenovirus; HRG, heregulin-α; vp, viral particles; wt, wild type. Adapted by permission from Macmillan Publishers Ltd.: Cancer Gene Therapy [7], copyright (2012)

An advantage of using a reporter gene assay system is that it models expression of a therapeutic gene in the target cell type. In other words, reporter gene analysis reflects the sum total of all the steps required in viral vector infection, from receptor binding to transgene expression (Fig. 2). In Subheading 3.2, we describe a competition assay between Ad vectors and soluble HRG-α for binding of the receptors HER3 and HER4 using luciferase expression as a readout. Using the assay described in Subheading 3.2 we demonstrated specificity by showing that competition for virus binding to HER3/4 by soluble HRG-α could reduce virus infection resulting in reduction of reporter gene expression [7]. We have also used soluble fiber knob to compete for viral binding to the Coxsackie virus-adenovirus receptor, CAR (described elsewhere [7]).

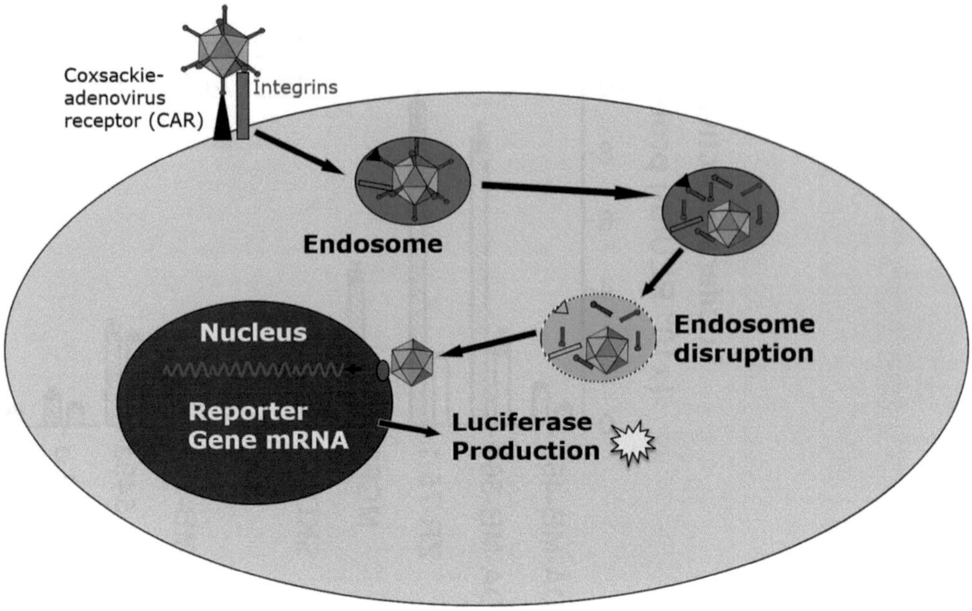

Fig. 2 Infection by untargeted adenovirus vectors. Adenovirus first binds to its primary cell surface receptor, CAR, through the knob domain of the fiber protein. The second interaction is binding of the viral penton base to integrins (primarily $\alpha_v\beta_3$ and $\alpha_v\beta_5$) on the cell surface, triggering clathrin-mediated endocytosis of the virus. The capsid is partially disrupted and escapes to the cytosol by lysis of the endosome, where it is transported rapidly to the nuclear pore complex. The capsid undergoes further disassembly and releases the genome into the nucleus. Inside the nucleus, the episomal viral genome is transcribed, and transcripts are processed, by cellular machinery. Adapted by permission from Macmillan Publishers Ltd.: Tumor Targeting [52], copyright (1998)

While the reporter gene assay provides a readout of overall transgene expression, which is influenced by multiple stages of viral infection, it is also valuable to examine early steps of viral infection in isolation. Ad binding and uptake has been studied by multiple methods, including radioactive labeling of virions [42] and genetic modification of capsid proteins, such as pIX, to fluorescently tag the virus particle [43, 44]. As an alternative to genetic modification to fluorescently tag Ad, Leopold et al. described a method of labeling an existing Ad with a commercial antibody labeling kit [45]. An advantage of this method is that it does not require further manipulations of the viral genome, and can be quickly adapted to any previously constructed virus. We have labeled our HER3/4 targeted virus in this manner, and we describe the method in Subheading 3.3. Since the label is covalently attached to amines on the surface of the viral capsid, it is important to ensure that the label will not interfere with the virus's ability to bind to receptors and infect cells. We confirmed by plaque assay that there was no significant difference in the ability to infect CAR-expressing HEK-293 cells when the labeled virus, generated using two different commercial labeling kits, was compared to a nonlabeled virus, indicating that labeling did not interfere with uptake via the wild-type receptor. Finally, fluorescence microscopy can be used to visualize and to quantify internalization of fluorescently labeled virus into cells, as well as to assess colocalization of the labeled vectors with their receptors (CAR and HER3) (*see* Subheading 3.4, and Figs. 3 and 4) [46].

It is important to note that, as is the case for many other viruses, a proportion of Ad viral particles in each preparation are not infectious [42]. Therefore, for experiments using fluorescently labeled virus (Subheadings 3.3 and 3.4) it is important to label

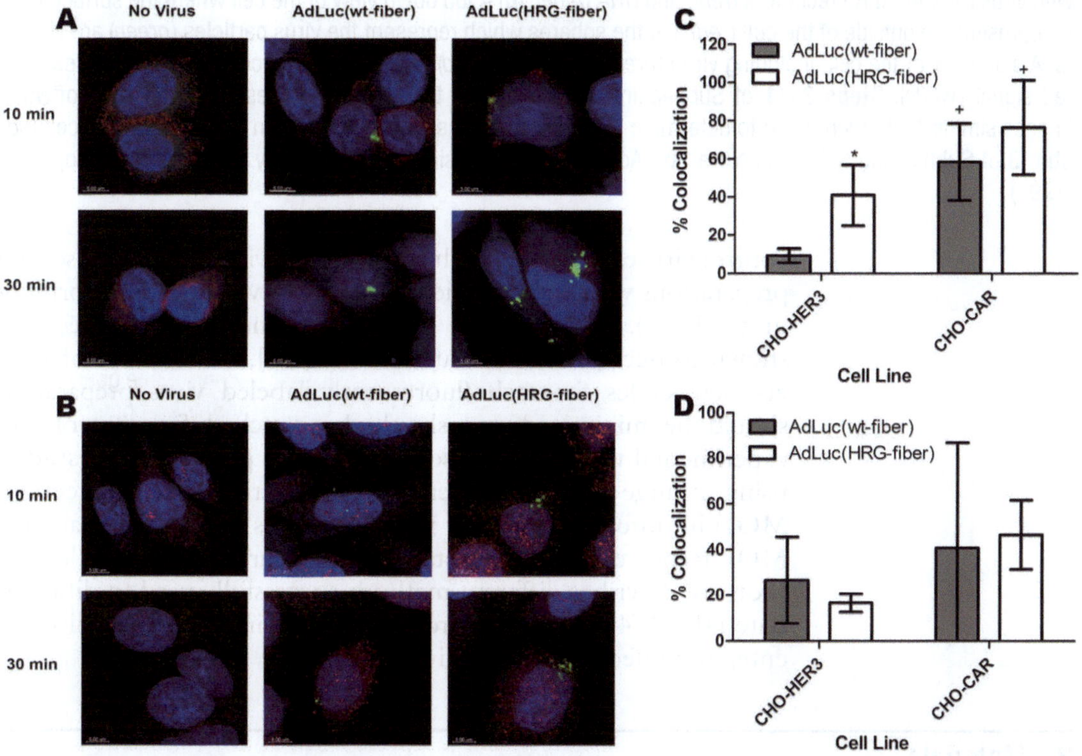

Fig. 3 Fluorescent virus uptake into CHO-CAR and CHO-HER3 cells and colocalization with receptors. Fluorescently labeled viruses were incubated with either (**a**) CHO-CAR cells or (**b**) CHO-al2/HER3 on coverslips for 30 min on ice, before incubation at 37 °C for either 10 min or 30 min. The cells were then washed three times with PBS before fixation and immunofluorescence staining for receptor expression [anti-CAR antibody (**a**) or anti-HER3 antibody (**b**)]. Images of three to six representative virus infected cells per coverslip were taken. Virus location relative to the surface defined by receptor expression was quantified (*see* **step 8** of Subheading 3.4.3 and **Note 20**). Representative 2D images of z-stack series are shown. *Green* is fluorescent virus, *blue* is DNA, and *red* is anti-CAR antibody (**a**) or anti-HER3 antibody (**b**). The percent of virus colocalized with either HER3 in CHO-al 2/HER3 cells or CAR in CHO-CAR cells after incubation at 37 °C for either (**c**) 10 min or (**d**) 30 min (*see* **steps 9–11** of Subheading 3.4.3). Significance is determined by analysis of variance (ANOVA) $P < 0.05$ *significantly different from AdLuc(wt-fiber) in the same cell line at the same time point; + significantly different from the same virus in the other cell line at the same time point. Adapted by permission from University of Alberta: [46] copyright (1998)

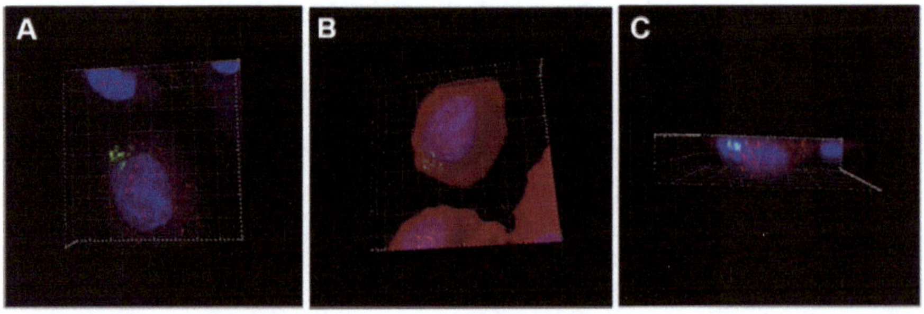

Fig. 4 3D confocal image of adenovirus entry. Images were generated from data acquired in the experiment shown in Fig. 3b, AdLuc(HRG-fiber) infection of CHO-al2/HER3 cells at 30 min. (**a**) Top down view of the cell with virus (*green*), HER3 receptors (*red*), and DNA (*blue*). (**b**) A top down view of the cell where the surface used to represent the outside of the cell (*red*) and the spheres which represent the virus particles (*green*) are visible. (**c**) A side view of the cell, including virus (*green*), receptors (*red*), DNA (*blue*), and colocalization of green and red signal (*white*). **Steps 3–11** of Subheading 3.4.3 describe the analysis process using Imaris Software. Images similar to (**b**) were used to determine if the virus was inside, outside or "on the edge" of the cell (*see* **step 8** of Subheading 3.4.3 and **Note 20**). Adapted by permission from University of Alberta: [46] copyright (1998)

control viruses in parallel with experimental viruses, and to use viral preparations with similar ratios of particles (vp) to plaque forming units (PFU). Furthermore, since freeze–thaw cycles have been shown to decrease adenovirus stability [47], the number of freeze–thaw cycles for each fluorescently labeled virus preparation should be minimized and should be matched for control and experimental viruses. It is also important to perform these studies using a range of virus concentrations (multiplicities of infection, MOI) in different cell lines, with duplicates or triplicates at each MOI, to prevent misinterpretations due to artifacts. The following protocol describes different methods successfully used to differentiate HER3/4-targeted Ad from wild type virus in their abilities to enter and infect HER3-positive cell lines.

2 Materials

2.1 Cell Culture

1. Chinese Hamster Ovary (CHO) cell line derivatives.
   - (a) CHO-NT (a nontransfected CHO cell line; a gift from Dr. Z. Wang, University of Alberta). Alternatively, a commonly used strain of CHO cells (CHO-K1) is available from American Type Culture Collection (ATCC) (CCL-61)].
   - (b) CHO-CAR (a CHO cell line overexpressing CAR; a gift from Dr. J. Bergelson, University of Pennsylvania).
   - (c) CHO-al2 (CHO overexpressing the human α2 integrin receptor; a gift from Dr. J. Bergelson, University of Pennsylvania).

(d) CHO-al2/HER3 (CHO overexpressing al2 and HER3) generated in-house.

(e) CHO-CAR/HER3 (CHO overexpressing CAR and HER3) generated in-house.

2. Dulbecco's Modified Eagle's Medium (DMEM, Gibco) supplemented with 10% fetal bovine serum (Gibco), PSF (Antibiotic-Antimycotic, Gibco), and 200 μM L-glutamine (Gibco).

2.2 Viruses

1. *Precautions on handling viral vectors:* Adenovirus and Ad vectors are biohazard level 2 agents, and must be handled using biosafety procedures in compliance with national laboratory biosafety guidelines [48] and as mandated by the institution in which the experiments are performed. All samples and materials containing, or potentially containing, infectious virus must be handled using appropriate personal protective equipment and clothing (laboratory coats, gloves, etc.). Culture dishes, centrifuge tubes, dialysis devices and other vessels containing live virus can only be opened in a certified biosafety cabinet. Biohazardous waste disposal procedures must be followed in compliance with the researcher's institution.
2. AdLuc(HRG-fiber) [7]: an adenovirus targeted to HER3/4, with luciferase reporter gene under control of human cytomegalovirus immediate early promoter inserted in place of the viral E1 region; typically at a concentration of 1×10^9 PFU/ml or 1×10^{12} vp/ml. Store at −70 °C in small aliquots.
3. AdLuc(wt-fiber) [49]: a control for AdLuc(HRG-fiber), matched for PFU/ml, vp/ml, and number of freeze–thaw cycles. Store at −70 °C in small aliquots.

2.3 General Solutions

1. Phosphate buffered saline (PBS): 137 mM NaCl, 2.7 mM KCl, 10 mM Na_2HPO_4, 2 mM KH_2PO_4.
2. PBS++: 137 mM NaCl, 2.7 mM KCl, 10 mM Na_2HPO_4, 2 mM KH_2PO_4, 0.1 g/l $CaCl_2$ $2H_2O$, 0.1 g/l $MgCl_2$ $6H_2O$.

2.4 Reporter Gene Assay and Competition Assay

1. Reporter lysis buffer (RLB, Promega). Store at −20 °C or room temperature away from direct sunlight.
2. Luciferase enzyme (QuantiLum® Recombinant Luciferase, Promega, cat #E170A) stock solution ($\geq 2.0 \times 10^{10}$ light units/mg protein). Store at −70 °C in small aliquots.
3. Luciferase Assay Reagent (Promega, cat #E1483), purchased ready to use. Can be aliquoted and stored at −70 °C; however, multiple freeze–thaw cycles decrease substrate activity.

4. HRG: recombinant human HRG1-α EGF like domain (R&D Systems, 296-HR) reconstituted at 100 μg/ml in PBS and stored at −20 °C and/or 4 °C.
5. White 96-well flat bottom plates (Corning™ 3912): a white plate is recommended for luciferase assays because it allows for more efficient detection of light, and prevents spillover from neighboring wells.
6. Luminometer: FLUOstar Optima (BMG Laboratories), or equivalent. We use a series of five gains to increase the opportunity to detect light within the linear range of the instrument without needing to repeat the addition of luciferase substrate.

2.5 Preparation of Fluorescently Labeled Virus

1. 10% glycerol in PBS++.
2. A488 dye, or similar: Alexa Fluor® 488 (A488) monoclonal antibody labeling kit (A20181, Molecular Probes). A488 has an absorption maximum at 494 nm and an emission maximum at 519 nm, and the kit chosen is optimized for small-scale labeling (100 μg of protein). Store at 2–8 °C.
3. 0.2 M sodium bicarbonate.
4. Dialysis buffer (~1500 ml per experiment, 4 °C): 10% glycerol, 50 mM Tris–HCl pH 7.5, 10 mM $MgCl_2$, 150 mM NaCl.
5. Glycerol (sterile).
6. Slide-A-Lyzer MINI dialysis device (10,000 molecular weight cutoff, 0.1 ml maximum volume; Cat# 69576, Thermo Scientific).

2.6 Internalization and Colocalization Determination by Immunofluorescence

1. 4% paraformaldehyde (Sigma-Aldrich) in PBS.
2. 20 mM glycine in PBS.
3. 0.4% Triton X-100 (BDH).
4. 4% bovine serum albumin (BSA, Sigma-Aldrich) in PBS. Prepare fresh.
5. Primary antibodies: HER3 antibody C-17 (ErbB3, SC-285, Santa Cruz) at 1:200 dilution and undiluted CAR antibody (obtained from the serum-free cell culture supernatant of a mouse hybridoma cell line stored in aliquots at −70 °C (IgG1, RmcB, CRL-2379, ATCC) [50]).
6. Secondary antibodies: Cy3-conjugated donkey anti-rabbit antibody (711-165-152, Jackson ImmunoResearch Laboratories, Inc.) and Cy5-conjugated donkey anti-mouse antibody (715-175-150, Jackson ImmunoResearch Laboratories, Inc.), both at 1:200 dilution.
7. Mounting medium (prepared in-house): 90% glycerol-PBS based medium containing anti-fade [1 mg/ml of paraphenylenediamine (Sigma-Aldrich)] and 0.5 mg/ml of DAPI

(diamidino-2-phenylindole dye, Thermo Fisher Scientific), used as a counterstain for nuclei.

8. Parafilm M (Sigma-Aldrich).
9. Microscope: Zeiss 710 LSM equipped with a Plan-Apochromat 40×/1.3 Oil DIC M27 objective, or equivalent.
10. Analysis software: Imaris software (7.6.0, Bitplane Scientific Software), or equivalent.

3 Methods

3.1 Determination of Transgene Expression by Reporter Gene Assay

3.1.1 Infection of CHO Cell Lines in Monolayer Cultures

1. Seed cells from an exponentially growing monolayer culture into a 24-well plate at sufficient density to achieve 80–90% confluency at the time of infection. We plated our cell lines between 1×10^5 and 2.5×10^5 cells per well, usually 24 h before infection. Culture at 37 °C in standard growth medium.
2. On the day of the infection, count the number of cells in 2–3 wells to determine an average number of cells per well.
3. Calculate the number of infectious virus required per well, based on the desired multiplicity of infection (MOI, in either PFU/cell or in vp/cell) and the average number of cells per well.
4. Dilute virus in PBS++ to a concentration that will give the number of virus required per well in a volume of 100 μl. (See biosafety precautions for handling viruses, Subheading 2.2, **item 1**) For ease of handling, prepare virus dilutions for replicate wells in a single tube, and include extra volume for loss in mixing and transfer (*see* **Note 1**).
5. Remove medium from cells in the 24-well plate and add 100 μl virus dilution to each well. Rock plates to ensure that the virus is evenly distributed over the cells, and the monolayer will not become dry (*see* **Note 2**).
6. Incubate plates at 37 °C for 10–30 min to allow internalization of virus (*see* **Note 3**).
7. Wash wells twice with room-temperature PBS (~500 μl per wash, *see* **Note 4**).
8. Add 500 μl standard growth medium to each well.
9. Incubate at 37 °C for the desired length of time (24–48 h, depending on the cell line).
10. Perform the luciferase assay as described in Subheading 3.1.2.

3.1.2 Luciferase Assay

1. Remove the medium from cells in the 24-well plate after the desired incubation period (Subheading 3.1.1, **step 9**). Do not allow the cells to dry out. We suggest working with a single plate at a time.

2. Wash wells twice with room-temperature PBS (*see* **Note 4**).
3. To lyse the cells, add 200 μl room-temperature reporter lysis buffer (RLB) to each well after the final PBS wash, then incubate plates at room temperature for 20 min (*see* **Note 5**).
4. Freeze plates at −70 °C (*see* **Note 6**).
5. Thaw plates at room temperature. Pipette the RLB against the bottom of the well 5–6 times to release all adherent cells and ensure complete cell lysis. Transfer the entire contents of the well to a labeled microfuge tube (*see* **Note 7**).
6. Repeat **step 5** for all wells on the plate.
7. Centrifuge the lysate at 12,000 × *g* for 2 min at 4 °C.
8. Transfer the supernatants to fresh microfuge tubes without disturbing the pellets. If desired, at this point the supernatant may be stored at −70 °C for future analysis.
9. Repeat **steps 5–8** for subsequent plates.
10. Add 20 μl of lysate supernatant in duplicate to wells of a white 96-well plate (*see* **Note 8**).
11. Prepare a luciferase enzyme standard curve from the stock solution by making tenfold serial dilutions in RLB. Add 20 μl of each standard to duplicate wells in the plate (*see* **Note 9**).
12. If necessary, prepare the luciferase assay reagent by adding the buffer to the substrate, shortly before use.
13. Set up the luminometer according to instrument instructions to inject 100 μl of luciferase substrate into each well, immediately before measuring light emission at five different gains from that specific well.
14. Plot the standard curve at each of the five gains to determine the linear range of relative light units (RLUs) at that gain.
15. The RLUs detected in duplicate sample wells can be averaged and plotted against the standard curve at the appropriate gain to determine the concentration of luciferase in μg/ml (*see* **Note 10**).

3.2 Competition Assay with HRG

1. Culture cells in 24-well plates and prepare for infection as described in **steps 1–3** of Subheading 3.1.1. It may be sufficient to carry out the competition assay using a single MOI (e.g., 100 vp per cell, or 1 PFU per cell); however, a range of MOIs would also be useful to examine.
2. Dilute HRG in PBS to 5 μg/ml (100 μl is required for each well to be treated with HRG).
3. Remove medium from wells and add either 100 μl PBS or 100 μl of HRG diluted in PBS.

4. Rock plates to evenly distribute the medium, then incubate for 30 min on ice.
5. During the incubation, dilute virus to the desired concentration in PBS++ (a total volume of 10 μl per well is required) (*see* **Note 11**). Dilute the virus for replicate wells in a single tube, including extra volume for loss in mixing and transfer.
6. Add 10 μl virus directly to the wells containing PBS or HRG.
7. Incubate the plates for an additional 30 min on ice.
8. Incubate for 10 min at room temperature to allow for virus internalization.
9. Wash wells twice with PBS (*see* also Subheading 3.1.1, **step 7** and **Note 4**).
10. Add 500 μl standard growth medium to each well.
11. Incubate at 37 °C for the desired length of time (24–48 h).
12. Prepare cell lysates and perform a luciferase assay as described in Subheading 3.1.2.

3.3 Preparation and Validation of Fluorescently Labeled Virus

3.3.1 Virus Labeling Reaction

1. Dilute virus to 1×10^{12} vp/ml with 10% glycerol in PBS++ (*see* **Note 12**).
2. Resuspend the A488 dye in 100 μl of 0.2 M sodium bicarbonate (*see* **Note 13**).
3. Mix 50 μl diluted virus with 50 μl dye solution. We generally label the virus in small batches to control conditions and to prevent later freeze–thaw steps for the virus.
4. As a control, also prepare a "mock-labeled" virus by substituting 0.2 M sodium bicarbonate for the dye, then continue with procedures described in Subheadings 3.3.1 and 3.3.2 in parallel with preparation of the labeled virus.
5. Incubate virus–dye mixture at room temperature for 1 h, mixing three times. Wrap the tube in foil to prevent exposure to light.

3.3.2 Dialysis to Remove Unbound Label

1. During the incubation, prepare for dialysis by transferring ~500 ml cold, sterile dialysis buffer into a 1 l beaker with a magnetic stir bar. Place the beaker on a stir plate at 4 °C and cover with foil to prevent light from damaging the fluorescent dye during dialysis.
2. Transfer each virus–dye mixture to a labeled Slide-A-Lyzer MINI dialysis device (*see* **Note 14**).
3. Insert the dialysis devices in the float provided then place in the dialysis beaker.
4. Turn on the stir plate and adjust to a level that allows gentle stirring of the buffer without creating a vortex. Check the stir

rate frequently the first few times this procedure is performed to ensure the dialysis device stays submerged in the buffer while stirring, and the float does not tip in the beaker.

5. Over a period of 24 h, change the dialysis buffer twice by removing the float, decanting the used buffer into a waste beaker, and adding ~500 ml fresh buffer to the dialysis beaker. Allow a minimum of 4 h between buffer changes. To decontaminate the waste buffer, add bleach to a final concentration of 10%.
6. After dialysis is complete, transfer the samples in the dialysis beaker to a biosafety cabinet, remove labeled virus carefully from the dialysis device with a pipette, and note the volume of virus solution recovered.
7. Add glycerol to a final concentration of 30% (v/v) to stabilize the virus for freezing.
8. Store labeled virus at −70 °C (wrapped in foil or in an opaque box to be kept in dark).

3.3.3 Determination of Labeling Efficiency and Infectivity

1. Measure the absorbance at 494 nm (A488) and 430 nm (background), diluting a sample of the virus in PBS if required for accurate measurement (*see* **Note 15**).
2. Subtract background (absorbance at 430 nm) from dye (absorbance at 494 nm) to correct the absorbance reading.
3. Calculate the molar concentration of the dye (A488) by multiplying the corrected absorbance from **step 2** by the dilution factor and dividing by the extinction coefficient of A488 provided by the manufacturer (71,000 cm^{-1} M^{-1}).
4. Convert the molar concentration of dye (**step 3**) to the number of molecules of dye in the dialyzed sample by multiplying the dialysis volume in liters by the molar concentration and by Avogadro's number (6.022×10^{23} molecules per mole).
5. Calculate the total number of virus particles used in the labeling reaction (Subheading 3.3.1) based on the dilution of the virus stock (e.g., 50 μl at 1×10^{12} vp/ml is 5×10^{10} vp total). Multiply the number of virus particles by 252 to obtain the total number of viral capsomeres (viral proteins in the outer capsid) in the labeling reaction.
6. Divide the total number of molecules of dye in the dialysate (from **step 4**) by the total number of viral capsomeres (from **step 5**) to calculate the minimum dye-to-protein (capsomere) ratio. The actual ratio would be higher if recovery of virus were less than 100%.
7. Compare the labeled and mock-labeled virus stock for infectivity using a limiting dilution assay (endpoint method [51]) or other test to measure virus infectivity, if desired (*see* **Note 16**).

3.4 Internalization and Colocalization Determination by Immunofluorescence

3.4.1 Infection of Cells in a Monolayer with Fluorescently Labeled Virus

1. The day before the infection, place a sterile coverslip into each well of a 24-well plate. Plate cells in the wells with coverslips such that the monolayer is 70% confluent on the day of the infection (*see* **Note 17**). Use the number of cells plated to calculate the amount of virus required to infect at the desired MOI. We use an MOI of 100 or 1000 vp/cell.
2. On the day of the infection, dilute fluorescently labeled virus to the desired MOI in a volume of PBS++ sufficient for 100 μl per well (*see* **Note 18**).
3. Remove the medium from the cells and add 100 μl of virus or PBS++ to each well (*see* **Notes 1** and **2**).
4. Incubate cells with virus for 30 min in the dark, on ice, to allow time for virus binding.
5. Incubate the virus-bound cells for 10–30 min at 37 °C to allow for viral internalization (*see* **Note 3**).
6. Wash the coverslips three times with PBS (*see* **Note 4**).

3.4.2 Immunofluorescence Staining

1. Fix cells on the coverslips with 4% paraformaldehyde in PBS at room temperature for 10 min.
2. Wash the coverslips with ~500 μl of 20 mM glycine in PBS for 5–10 min (*see* **Note 19**).
3. Permeabilize cells with 200 μl of 0.4% Triton X-100 for 15 min to allow antibody staining of intracellular receptors.
4. Wash the coverslips three times in ~500 μl PBS.
5. Block nonspecific antibody binding by treating with 200 μl of 4% BSA in PBS for at least 10 min.
6. Wash the coverslips three times in ~500 μl PBS.
7. Invert the coverslips onto a 20–30 μl drop of primary antibody in PBS + 4% BSA on Parafilm M.
8. Incubate the coverslips for 1 h at room temperature in the dark. Alternatively, incubate with the primary antibody for 30 min at 37 °C or overnight at 4 °C.
9. Return the coverslips to the 24-well plate and wash three times with ~500 μl PBS for 5 min.
10. Invert the coverslips onto a 20–30 μl drop of secondary antibody coupled with a fluorophore in PBS + 4% BSA on Parafilm M.
11. Incubate the coverslips in the dark for 1 h at room temperature or 30 min at 37 °C.
12. Wash the coverslips three times in ~500 μl PBS, then once with ~500 μl distilled water.

13. Mount the coverslips on slides with a 10–20 μl drop of mounting medium. Allow the mounting medium to harden in the dark overnight or for at least 2 h.
14. Store samples at 4 °C in the dark until imaging.

3.4.3 Imaging and Analysis of Infected Cells

1. Using a confocal microscope, scan the coverslip to find infected cells, where the virus (green) signal is present.
2. Image three to six infected cells per coverslip, ensuring that images are obtained at multiple focal planes, in order to generate a 3D image of the cells during analysis. This number of cells allows analysis of as many cells as possible within the time constraints for analysis of all the z-stack images. 3D imaging is required to determine the location of the virus inside or outside the cell.
3. Process images using appropriate analysis software, such as Imaris software. Create a surface based on the red channel signal (receptor) to represent the edge of the cell.
4. Smooth the surface to area level detail of 1.00 μm, adjusting the absolute threshold to a level appropriate to the size of the cell to be analyzed, ensuring all cells present in the image are included.
5. Set the surface to 70% transparency to allow clear visualization of viruses inside the cell, while maintaining the ability to determine the location of the cell surface.
6. Using the analysis software, create spheres based on the green channel to represent virus particles in later analysis steps. We used an estimated diameter of 0.5 μm for these spheres (*see* **Note 20**).
7. Repeat **steps 3–6** for all images in the experiment.
8. Determine the location of the virus spheres (particles) relative to the cell surface by viewing the cells in 3D from multiple angles. Assign each virus sphere to one of the following categories: inside, outside, or "on the edge" of a cell. Quantify the total number of virus particles in each category (*see* **Note 20**).
9. Determine the colocalization of virus (green) and receptor (red) using the colocalization module of the analysis software.
10. Automatically scale the two channels and choose a threshold value above which colocalization is determined. For our experiments, after a visual inspection of the data, we chose a threshold value of 10% of maximum intensity of each channel in the stack. This threshold was more stringent than the standard threshold of mean plus two standard deviations.
11. Count the number of virus spheres colocalized with receptor, and determine the percentage of colocalization. In this case,

colocalization represents virus and receptor which are near each other, but may not actually be bound. This analysis allows a comparison of different viruses or receptors under the same conditions.

4 Notes

1. 100 μl is the minimum volume to cover cells in a 24-well plate, preventing the cells from drying out. Minimizing the volume increases binding of virus to cells. Scale accordingly for different plate sizes. Prepare all virus dilutions prior to starting any cell infections, to allow for a smooth transition between plates at later steps.
2. Due to short incubation times, it is advisable to carry out **steps 5–8** of Subheading 3.1.1 (or **steps 3–6** of Subheading 3.4.1) one plate at a time.
3. We use a standard CO_2 incubator at 37 °C for cell culture. Incubation at room temperature may be performed instead, but internalization may not be as efficient. In addition, time required for internalization may vary depending on the types of receptors or cell lines studied [46].
4. Aspirate virus-containing medium from wells starting at the lowest concentration of virus (usually mock infection). Add ~500 μl of PBS to each well with a 5 or 10 ml pipette, in the same order used for medium removal. Add the PBS gently to the side of the well, in order to decrease the potential for disturbing the monolayer. Repeat the aspiration and addition of PBS for the second wash step, and finally aspirate the remaining PBS. Do not allow cells to dry out during washing steps.
5. RLB is a detergent and renders the virus noninfectious, and therefore, level 2 biosafety precautions are not required when handling these lysates.
6. Plates can be stored at −70 °C. For storage, wrap plates tightly with Parafilm M or a similar product to prevent sublimation while in storage.
7. Use of a 200 μl pipettor set to 150 μl for mixing and transfer allows repeated pipetting of the cells without producing bubbles that hinder sample recovery.
8. Infections with luciferase-encoding viruses can result in high levels of luciferase activity in some cell lines, thus dilution of lysates by 100- or 1000-fold in RLB may be required for detection within the linear range of the luminometer.

9. We have found a range of 10^{-3}–10^{-10} μg/μl of luciferase enzyme at 20 μl per well works well to cover the linear range of the FLUOstar plate reader at the five gains we use.
10. If desired, a bicinchoninic acid (BCA) assay could be performed on the supernatant to normalize luciferase activity to total protein in the sample.
11. Note that this volume is lower than Subheading 3.1.1, **step 4**, in order to minimize the volume in the well during the binding and internalization steps.
12. This step uses viral particles (vp) and not plaque forming units (PFU) or other measure of infectious virus, because the fluorescent label will be added to all viral particles, regardless of infectious capacity.
13. This step results in a sodium bicarbonate concentration of 0.1 M in the final virus–fluorophore solution (as instructed by the kit), in a sufficient volume for labeling two viruses (e.g., test and control virus). This step may vary if other kits are used.
14. This dialysis device has been chosen for its ease of use, small sample volume, and low binding. Alternative dialysis devices could be substituted, but should be tested to ensure that the virus does not bind excessively to the membrane.
15. We measure absorbance using a NanoDrop (Thermo Scientific) after 1:4 dilution of virus with PBS.
16. The limiting dilution assay (endpoint method [51]) determines the tissue culture infectious dose at 50% of maximum ($TCID_{50}$). We prefer this assay because it uses a minimal amount of labeled virus; however, other methods of measuring infectious units of virus (e.g., plaque assay) could also be used. We found the addition of dye made little difference in adenoviral infectivity in HEK-293 cells using the limiting dilution assay.
17. Cells are plated at a lower confluency than in Subheading 3.1.1 to allow better visualization for immunofluorescence and to conform with the standard protocols.
18. Keep fluorescently labeled virus on ice in the dark as much as possible to prevent degradation of the virus or the fluorophore. Keep track of the number of freeze–thaw cycles for each virus tube and use viruses with a similar number of thaws, to help keep the activity of the viruses consistent. Keep the coverslips in the dark as much as possible for the rest of the procedure, to help prevent fluorophore degradation.
19. Glycine is added to decrease background by binding free aldehyde groups that would otherwise bind the antibodies. At this point, cells can be stored in PBS at 4 °C in the dark until the

staining procedure is carried out (usually overnight, or up to 1 week).

20. The size of the "estimated" virus particle (500 nm) used for software analysis is bigger than an actual virus particle (90 nm), but is appropriate to use because the resolution for a conventional light microscope for green and red fluorophores is approximately 250–300 nm and 300–350 nm, respectively. Using a larger sphere allows us to be more accurate in judging a virus signal as inside or outside the cell. We included a category called "on the edge" to capture signals not clearly inside or outside the cell, since the signal for the virus cannot be determined with the level of precision required to make the distinction between inside or outside the cell with this type of microscope. Most viruses in our experiments have been found inside the cells.

References

1. Mujoo K, Choi B-K, Huang Z, Zhang N, An Z (2014) Regulation of ERBB3/HER3 signaling in cancer. Oncotarget 5(21):10222–10236
2. Sithanandam G, Anderson LM (2008) The ERBB3 receptor in cancer and cancer gene therapy. Cancer Gene Ther 15(7):413–448
3. Sergina NV, Moasser MM (2007) The HER family and cancer: emerging molecular mechanisms and therapeutic targets. Trends Mol Med 13(12):527–534
4. Koutras AK, Fountzilas G, Kalogeras KT, Starakis I, Iconomou G, Kalofonos HP (2010) The upgraded role of HER3 and HER4 receptors in breast cancer. Crit Rev Oncol Hematol 74 (2):73–78
5. Sassen A, Diermeier-Daucher S, Sieben M, Ortmann O, Hofstaedter F, Schwarz S et al (2009) Presence of HER4 associates with increased sensitivity to Herceptin™ in patients with metastatic breast cancer. Breast Cancer Res 11(4):R50
6. Okazaki S, Nakatani F, Masuko K, Tsuchihashi K, Ueda S, Masuko T et al (2016) Development of an ErbB4 monoclonal antibody that blocks neuregulin-1-induced ErbB4 activation in cancer cells. Biochem Biophys Res Commun 470(1):239–244
7. MacLeod SH, Elgadi MM, Bossi G, Sankar U, Pisio A, Agopsowicz K et al (2012) HER3 targeting of adenovirus by fiber modification increases infection of breast cancer cells in vitro, but not following intratumoral injection in mice. Cancer Gene Ther 19 (12):888–898
8. Wiley J and Sons (2016) Vectors used in gene therapy clinical trials. http://www.wiley.com/legacy/wileychi/genmed/clinical/. Accessed 22 Jun 2016
9. Berk AJ (2013) Adenoviridae. In: Knipe DM, Howley P (eds) Fields virology, 6th edn. Lipincott, Williams & Wilkins, Philadelphia, p 2664
10. Wold WSM, Toth K (2013) Adenovirus vectors for gene therapy, vaccination and cancer gene therapy. Curr Gene Ther 13(6):421–433
11. Bett AJ, Haddara W, Prevec L, Graham FL (1994) An efficient and flexible system for construction of adenovirus vectors with insertions or deletions in early regions 1 and 3. Proc Natl Acad Sci U S A 91(19):8802–8806
12. Danthinne X, Imperiale MJ (2000) Production of first generation adenovirus vectors: a review. Gene Ther 7(20):1707–1714
13. Yoon A-R, Hong J, Kim SW, Yun C-O (2016) Redirecting adenovirus tropism by genetic, chemical, and mechanical modification of the adenovirus surface for cancer gene therapy. Expert Opin Drug Deliv 13(6):843–858
14. Kasala D, Choi J-W, Kim SW, Yun C-O (2014) Utilizing adenovirus vectors for gene delivery in cancer. Expert Opin Drug Deliv 11 (3):379–392
15. Capasso C, Garofalo M, Hirvinen M, Cerullo V (2014) The evolution of adenoviral vectors through genetic and chemical surface modifications. Viruses 6(2):832–855
16. Chaurasiya S, Hew P, Crosley P, Sharon D, Potts K, Agopsowicz K et al (2016) Breast cancer gene therapy using an adenovirus

encoding human IL-2 under control of mammaglobin promoter/enhancer sequences. Cancer Gene Ther 23(6):178–187

17. Zhang J-F, Wei F, Wang H-P, Li H-M, Qiu W, Ren P-K et al (2010) Potent anti-tumor activity of telomerase-dependent and HSV-TK armed oncolytic adenovirus for non-small cell lung cancer in vitro and in vivo. J Exp Clin Cancer Res 29:52
18. Irving J, Wang Z, Powell S, O'Sullivan C, Mok M, Murphy B et al (2004) Conditionally replicative adenovirus driven by the human telomerase promoter provides broad-spectrum antitumor activity without liver toxicity. Cancer Gene Ther 11(3):174–185
19. Sadeghi H, Hitt MM (2005) Transcriptionally targeted adenovirus vectors. Curr Gene Ther 5 (4):411–427
20. Matsumoto K, Shariat SF, Ayala GE, Rauen KA, Lerner SP (2005) Loss of coxsackie and adenovirus receptor expression is associated with features of aggressive bladder cancer. Urology 66(2):441–446
21. Miller CR, Buchsbaum DJ, Reynolds PN, Douglas JT, Gillespie GY, Mayo MS et al (1998) Differential susceptibility of primary and established human glioma cells to adenovirus infection: targeting via the epidermal growth factor receptor achieves fiber receptor-independent gene transfer. Cancer Res 58 (24):5738–5748
22. Dmitriev I, Krasnykh V, Miller CR, Wang M, Kashentseva E, Mikheeva G et al (1998) An adenovirus vector with genetically modified fibers demonstrates expanded tropism via utilization of a coxsackievirus and adenovirus receptor-independent cell entry mechanism. J Virol 72(12):9706–9713
23. Douglas JT, Kim M, Sumerel LA, Carey DE, Curiel DT (2001) Efficient oncolysis by a replicating adenovirus (ad) in vivo is critically dependent on tumor expression of primary ad receptors. Cancer Res 61(3):813–817
24. van Beusechem VW, Grill J, Mastenbroek DCJ, Wickham TJ, Roelvink PW, Haisma HJ et al (2002) Efficient and selective gene transfer into primary human brain tumors by using single-chain antibody-targeted adenoviral vectors with native tropism abolished. J Virol 76 (6):2753–2762
25. Kashentseva EA, Seki T, Curiel DT, Dmitriev IP (2002) Adenovirus targeting to c-erbB-2 oncoprotein by single-chain antibody fused to trimeric form of adenovirus receptor ectodomain. Cancer Res 62(2):609–616
26. Dmitriev I, Kashentseva E, Rogers BE, Krasnykh V, Curiel DT (2000) Ectodomain of coxsackievirus and adenovirus receptor genetically fused to epidermal growth factor mediates adenovirus targeting to epidermal growth factor receptor-positive cells. J Virol 74 (15):6875–6884
27. Lord R, Parsons M, Kirby I, Beavil A, Hunt J, Sutton B et al (2006) Analysis of the interaction between RGD-expressing adenovirus type 5 fiber knob domains and alphavbeta3 integrin reveals distinct binding profiles and intracellular trafficking. J Gen Virol 87(Pt 9):2497–2505
28. Kurachi S, Koizumi N, Sakurai F, Kawabata K, Sakurai H, Nakagawa S et al (2007) Characterization of capsid-modified adenovirus vectors containing heterologous peptides in the fiber knob, protein IX, or hexon. Gene Ther 14 (3):266–274
29. Krasnykh V, Dmitriev I, Mikheeva G, Miller CR, Belousova N, Curiel DT (1998) Characterization of an adenovirus vector containing a heterologous peptide epitope in the HI loop of the fiber knob. J Virol 72(3):1844–1852
30. Piao Y, Jiang H, Alemany R, Krasnykh V, Marini FC, Xu J et al (2009) Oncolytic adenovirus retargeted to Delta-EGFR induces selective antiglioma activity. Cancer Gene Ther 16 (3):256–265
31. Koizumi N, Mizuguchi H, Utoguchi N, Watanabe Y, Hayakawa T (2003) Generation of fiber-modified adenovirus vectors containing heterologous peptides in both the HI loop and C terminus of the fiber knob. J Gene Med 5(4):267–276
32. Magnusson MK, Henning P, Myhre S, Wikman M, Uil TG, Friedman M et al (2007) Adenovirus 5 vector genetically re-targeted by an affibody molecule with specificity for tumor antigen HER2/neu. Cancer Gene Ther 14 (5):468–479
33. Dmitriev IP, Kashentseva EA, Curiel DT (2002) Engineering of adenovirus vectors containing heterologous peptide sequences in the C terminus of capsid protein IX. J Virol 76 (14):6893–6899
34. Poulin KL, Lanthier RM, Smith AC, Christou C, Risco Quiroz M, Powell KL et al (2010) Retargeting of adenovirus vectors through genetic fusion of a single-chain or single-domain antibody to capsid protein IX. J Virol 84(19):10074–10086
35. Vigne E, Mahfouz I, Dedieu JF, Brie A, Perricaudet M, Yeh P (1999) RGD inclusion in the hexon monomer provides adenovirus type 5-based vectors with a fiber knob-independent pathway for infection. J Virol 73 (6):5156–5161

36. Campos SK, Barry MA (2004) Rapid construction of capsid-modified adenoviral vectors through bacteriophage lambda red recombination. Hum Gene Ther 15(11):1125–1130
37. Coughlan L, Alba R, Parker AL, Bradshaw AC, McNeish IA, Nicklin SA et al (2010) Tropism-modification strategies for targeted gene delivery using adenoviral vectors. Viruses 2 (10):2290–2355
38. Soboleski MR, Oaks J, Halford WP (2005) Green fluorescent protein is a quantitative reporter of gene expression in individual eukaryotic cells. FASEB J 19(3):440–442
39. Smale ST (2010) Luciferase assay. Cold Spring Harb Protoc. doi:10.1101/pdb.prot5421
40. Sato A, Klaunberg B, Tolwani R (2004) In vivo bioluminescence imaging BLI : an overview. Comp Med 54(6):631–634
41. Potts KG, Favis NA, Pink DB, Vincent KM, Lewis JD, Moore RB, Hitt MM, Evans D (2017) Oncolytic vaccinia virus F4L (ribonucleotide reductase) mutants promote antitumor immunity with superior safety in bladder cancer models. EMBO Mol Med 9 (5):638–654
42. Hoganson DK, Sosnowski BA, Pierce GF, Doukas J (2001) Uptake of adenoviral vectors via fibroblast growth factor receptors involves intracellular pathways that differ from the targeting ligand. Mol Ther 3(1):105–112
43. Meulenbroek RA, Sargent KL, Lunde J, Jasmin BJ, Parks RJ (2004) Use of adenovirus protein IX (pIX) to display large polypeptides on the virion–generation of fluorescent virus through the incorporation of pIX-GFP. Mol Ther 9 (4):617–624
44. Rogée S, Grellier E, Bernard C, Loyens A, Beauvillain J-C, D'halluin J-C et al (2007) Intracellular trafficking of a fiber-modified adenovirus using lipid raft/caveolae endocytosis. Mol Ther 15(11):1963–1972
45. Leopold PL, Ferris B, Grinberg I, Worgall S, Hackett NR, Crystal RG (1998) Fluorescent virions: dynamic tracking of the pathway of adenoviral gene transfer vectors in living cells. Hum Gene Ther 9(3):367–378
46. MacLeod SH (2013) Binding, internalization, and transgene expression of an adenoviral vector retargeted to HER3/4. Dissertation, University of Alberta
47. Evans RK, Nawrocki DK, Isopi LA, Williams DM, Casimiro DR, Chin S et al (2004) Development of stable liquid formulations for adenovirus-based vaccines. J Pharm Sci 93 (10):2458–2475
48. Public Health Agency of Canada (2015) Canadian biosafety standard. Public Health Agency of Canada, Ottawa, p 151
49. Parks RJ, Chen L, Anton M, Sankar U, Rudnicki MA, Graham FL (1996) A helper-dependent adenovirus vector system: removal of helper virus by Cre-mediated excision of the viral packaging signal. Proc Natl Acad Sci U S A 93(24):13565–13570
50. National Research Council (US) Committee on Methods of Producing Monoclonal Antibodies (1999) In vitro production of monoclonal antibody. In: Monoclonal antibody production. National Academies Press (US), Washington (DC)
51. Condit RC (2001) Principles of virology. In: Knipe DK, Howley PM (eds) Fields virology, 4th edn. Lippincott-Raven, Philadelphia, USA, p 3280
52. Bilbao G, Contreras JL, Gomez-Navarro J, Curiel DT (1998) Improving adenoviral vectors for cancer gene therapy. Tumor Target 1 (3):59–79

Chapter 19

Isolation of Human Mesenchymal Stem Cells for Studying ErbB Receptor Signaling

Chao Chen and Hongxing Jiang

Abstract

ErbB receptor signaling plays pivotal roles in tumorigenesis, cancer development, and drug resistance. A better understanding of ErbB signaling is required to achieve better clinical outcomes in cancer treatment. With increasing evidence showing the link between human mesenchymal stem cells (MSCs) and cancer, there is a growing interest in studying the role of ErbB receptor signaling in the regulation of MSCs. For this purpose, obtaining quality primary human MSCs is of great importance. This chapter describes the method of isolating primary human MSCs, aiming to offer researchers of this field a useful tool.

Key words Mesenchymal stem cells, Isolation

1 Introduction

ErbB receptors belong to the subclass I of receptor tyrosine kinase superfamily and include four members: ErbB1 (also known as epidermal growth factor receptor, EGFR), ErbB2, ErbB3, and ErbB4. ErbB receptors comprise an extracellular ligand-binding domain, a single transmembrane domain, and a cytoplasmic tyrosine kinase domain [1]. Upon binding to epidermal growth factor (EGF)-family ligands, ErbB receptors form a homodimer or heterodimer, which triggers the phosphorylation of specific tyrosine residues in the C-terminal domain of ErbB receptors. The phosphorylated tyrosines in turn trigger association of intracellular proteins and activate a complex network of downstream signaling pathways [2]. ErbB receptors play pivotal roles in cell pathophysiology through regulating a complex network of intracellular signaling [1, 2]. In the past two decades, the roles and mechanisms of ErbB signaling in oncogenesis have been well studied. Several ErbB-targeting drugs have successfully advanced to the bedside for breast cancer treatment, and substantial improvement on survival rate has been achieved [3, 4]. However, drug resistance limits

Zhixiang Wang (ed.), *ErbB Receptor Signaling: Methods and Protocols*, Methods in Molecular Biology, vol. 1652, DOI 10.1007/978-1-4939-7219-7_19, © Springer Science+Business Media LLC 2017

the efficacy of these compounds in a number of cancer types and therefore identifying new therapeutic targets is needed [5].

Mesenchymal stem cells (MSCs) are a group of pluripotent progenitor cells that are capable of multilineage differentiation upon appropriate stimulation [6]. Accumulating evidence shows the cross talk between MSCs and cancer cells during cancer development and progression. It has been reported that MSCs regulate properties of breast cancer stem cells (CSCs) through paracrine cytokines and microRNAs networks, and promote breast tumor growth and metastasis through forming an immunosuppressive microenvironment for cancer cells [7–10]. In addition, MSCs are able to promote angiogenesis for tumor growth through secreting various angiogenic factors, such as vascular endothelial growth factor (VEGF) [11–13]. Interestingly, the expression profile of MSCs from cancer patients differs from that of MSCs from healthy donors, including the elevated expression levels of BMP2, 4, 6 and EGFR [14, 15], suggesting MSCs might be reciprocally influenced by the tumorigenic microenvironment. With their unique homing property and plasticity, however, MSCs could serve as an effective tool to deliver anticancer therapeutic drugs or genes to cancer sites [16, 17]. For instance, MSCs expressing ErbB2/neu provoke protective immunity against breast cancer in mice [18].

In the last decade, the interplay between ErbB and MSCs in carcinogenesis and tissue regeneration has gained increasing interest. Human primary MSCs spontaneously express EGFR/ErbB1 and are responsive to ErbB1 inhibitor treatment [19]. EGF stimulates proliferation and mobility of human bone marrow MSCs via the activation of EGFR, ERK, and protein kinase B/Akt, but does not alter their differentiation potential toward adipogenic and osteogenic lineages [20]. Similarly, heparin-binding EGF (HB-EGF)/ErbB1 signaling promotes proliferation, but inhibits adipogenic, osteogenic, and chondrogenic differentiation of MSCs in vitro [21]. MSCs expressing ErbB2 and ErbB4 mediate a favorable protection in breast cancer and myocardial infarction [18, 22]. However, MSCs promote colorectal cancer progression through paracrine ErbB3 signaling [23]. Further studies are required to understand whether ErbB signaling indeed regulates the proliferation and immune modulation of MSCs in cancer patients in vivo.

In this chapter we describe the method of isolating primary human bone marrow MSCs, which could be subsequently used to decipher the roles of ErbB signaling in the complex interaction between MSCs and cancer cells in the tumorigenic microenvironment at various stages of cancer development.

2 Materials

2.1 Reagents

1. Dulbecco's Modified Eagle's Medium—high glucose (DMEM) (Invitrogen).
2. Fetal Bovine Serum (FBS) (Invitrogen).
3. 100× Antibiotic/antimycotic (Invitrogen).
4. 100× GlutaMAX (Invitrogen).
5. FGF-2 (EMD Millipore).
6. 1× Phosphate buffered saline without calcium and magnesium (PBS) (Invitrogen).
7. 0.05% trypsin (Invitrogen).
8. Ficoll-Paque Premium (GE Healthcare).
9. Complete cell culture medium: 10% FBS in DMEM supplemented with 0.29 mg/mL GlutaMAX, 1% antibiotic/antimycotic, and 4 ng/mL FGF-2.

2.2 Equipment

1. 100-mm TC-treated culture dish (Corning).
2. 15- and 50-mL conical polypropylene tubes (BD Falcon).
3. Hemocytometer (American Optical: 1483).
4. Inverted phase contrast microscope (Nikon: TMS-F).
5. Tissue culture incubator, 37 °C, 5% CO_2 (Thermo Scientific: 3110).
6. Recovery cell culture freezing medium (Invitrogen: 12648010).
7. Sterile polypropylene cryogenic tubes (Corning: 430659).
8. Freezing container (Nalgene: 5100-0001).
9. Allegra X-15R benchtop centrifuge with swinging buckets (Beckman Coulter: 392932).

3 Methods

1. The isolation procedure should be completed within 24 h after receiving the bone marrow sample. Write down all the information of each sample.
2. Thoroughly clean the outside of specimen containers containing bone marrow sample with 70% ethanol and open the containers inside the biosafety cabinet.
3. Transfer the bone marrow sample into 50-mL conical tubes, rinse the specimen container with 1× PBS and transfer it to the same 50-mL conical tubes.
4. Centrifuge at 800 × g for 10 min at room temperature.

5. Discard the supernatant, and resuspend the sample.
6. Add 15-mL Ficoll-Paque Premium solution to new 50-mL conical tubes.
7. Slowly load the resuspended sample on top of Ficoll-Paque solution (*see* **Note 1**).
8. Centrifuge at 400 × *g* for 25 min at room temperature in a swinging bucket rotor (*see* **Note 2**).
9. After centrifuge, there will be four layers (*see* Fig. 1), collect mononuclear cells at the interface of upper supernatant and Ficoll-Paque solution.
10. Transfer collected cells into a new 15-mL conical tube.
11. Add 1× PBS to 15 mL, centrifuge at 300 × *g* for 5 min at room temperature.
12. Remove supernatant and repeat **step 11** twice.
13. Remove supernatant and resuspend cells with 9-mL complete cell culture medium. Take 20 μL for cell counting with a hemocytometer and a microscope.
14. Adjust cell concentration to 2×10^6 cells/mL with the appropriate volume of complete cell culture medium.
15. Seed 2×10^7 cells to each 100-mm culture dish with 10 mL of culture medium per dish and culture the cells in an incubator at 37 °C and with humidified 5% CO_2.
16. Replace culture medium every 3 days until attached cells reach 90% confluence.
17. Detach subconfluent cells with 0.05% trypsin for 1 min at 37 °C or until cells fully detached, inactivate trypsin with complete cell culture medium, and transfer cells into 15 mL conical tubes. Take 20 μL cell suspensions for counting.
18. Centrifuge for 5 min at 300 × *g* at room temperature. Discard supernatant.
19. Resuspend cells to 1×10^6 cells/mL with the appropriate volume of recovery cell culture freezing medium and aliquot into sterile cryogenic tubes.
20. Freeze cells in Mr. Frosty Freezing container overnight at −80 °C and transfer to liquid nitrogen for future use (*see* **Note 3**).

4 Notes

1. When loading samples to Ficoll-Paque solution, make sure the Pipet-Aid pipette controller is in slow setting.

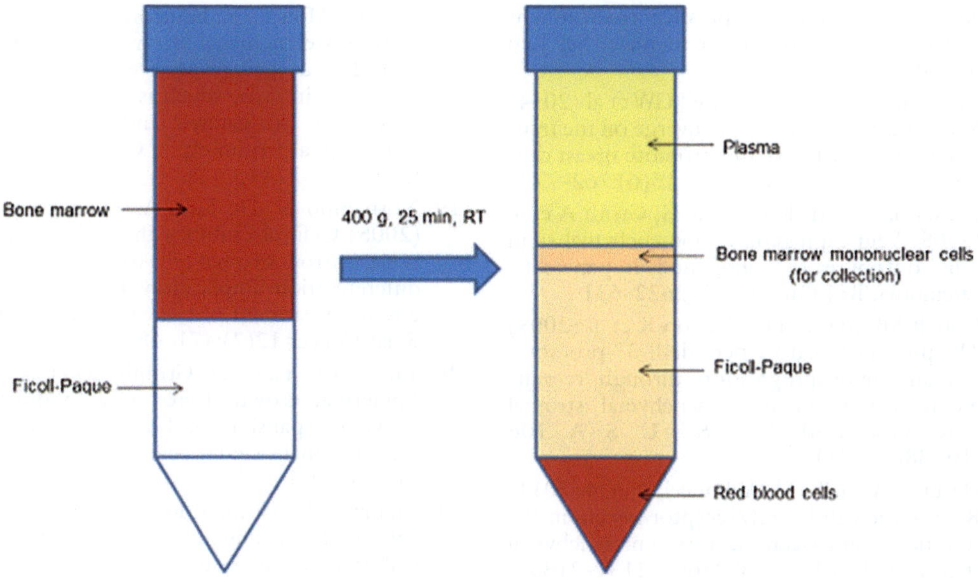

Fig. 1 Distribution and appearance of bone marrow components prior to and after Ficoll-Paque centrifugation

2. When centrifuging samples on top of Ficoll-Paque solution, set the acceleration and deceleration of the centrifuge to "no brake" or "slow" mode.
3. Characterization of isolated human bone marrow MSCs has been reported in our publication [24]. Cultured MSCs exhibit fibroblast-like morphology, and express CD105, CD13, not CD45. They are capable to undergo adipogenic, chondrogenic, and osteogenic differentiation with proper tissue culture conditions.

References

1. Hynes NE, Lane HA (2005) Erbb receptors and cancer: the complexity of targeted inhibitors. Nat Rev Cancer 5(5):341–354
2. Yarden Y, Sliwkowski MX (2001) Untangling the erbb signalling network. Nat Rev Mol Cell Biol 2(2):127–137
3. Hynes NE, MacDonald G (2009) Erbb receptors and signaling pathways in cancer. Curr Opin Cell Biol 21(2):177–184
4. Martinello R, Milani A, Geuna E et al (2016) Investigational erbb-2 tyrosine kinase inhibitors for the treatment of breast cancer. Expert Opin Investig Drugs 25(4):393–403
5. Holohan C, Van Schaeybroeck S, Longley DB et al (2013) Cancer drug resistance: an evolving paradigm. Nat Rev Cancer 13(10):714–726
6. Pittenger MF, Mackay AM, Beck SC et al (1999) Multilineage potential of adult human mesenchymal stem cells. Science 284(5411):143–147
7. Karnoub AE, Dash AB, Vo AP et al (2007) Mesenchymal stem cells within tumour stroma promote breast cancer metastasis. Nature 449(7162):557–563
8. Liu S, Ginestier C, SJ O et al (2011) Breast cancer stem cells are regulated by mesenchymal stem cells through cytokine networks. Cancer Res 71(2):614–624
9. Ljujic B, Milovanovic M, Volarevic V et al (2013) Human mesenchymal stem cells

creating an immunosuppressive environment and promote breast cancer in mice. Sci Rep 3:2298

10. Cuiffo BG, Campagne A, Bell GW et al (2014) Msc-regulated micrornas converge on the transcription factor foxp2 and promote breast cancer metastasis. Cell Stem Cell 15(6):762–774
11. Beckermann BM, Kallifatidis G, Groth A et al (2008) Vegf expression by mesenchymal stem cells contributes to angiogenesis in pancreatic carcinoma. Br J Cancer 99(4):622–631
12. Coffelt SB, Marini FC, Watson K et al (2009) The pro-inflammatory peptide ll-37 promotes ovarian tumor progression through recruitment of multipotent mesenchymal stromal cells. Proc Natl Acad Sci U S A 106 (10):3806–3811
13. De Luca A, Gallo M, Aldinucci D et al (2011) Role of the egfr ligand/receptor system in the secretion of angiogenic factors in mesenchymal stem cells. J Cell Physiol 226(8):2131–2138
14. Hofer EL, La Russa V, Honegger AE et al (2005) Alteration on the expression of il-1, pdgf, tgf-beta, egf, and fgf receptors and c-fos and c-myc proteins in bone marrow mesenchymal stroma cells from advanced untreated lung and breast cancer patients. Stem Cells Dev 14 (5):587–594
15. McLean K, Gong Y, Choi Y et al (2011) Human ovarian carcinoma-associated mesenchymal stem cells regulate cancer stem cells and tumorigenesis via altered bmp production. J Clin Invest 121(8):3206–3219
16. Studeny M, Marini FC, Dembinski JL et al (2004) Mesenchymal stem cells: potential precursors for tumor stroma and targeted-delivery vehicles for anticancer agents. J Natl Cancer Inst 96(21):1593–1603
17. Hong IS, Lee HY, Kang KS (2014) Mesenchymal stem cells and cancer: friends or enemies? Mutat Res 768:98–106
18. Romieu-Mourez R, Francois M, Abate A et al (2010) Mesenchymal stromal cells expressing erbb-2/neu elicit protective antibreast tumor immunity in vivo, which is paradoxically suppressed by ifn-gamma and tumor necrosis factor-alpha priming. Cancer Res 70 (20):7742–7747
19. Normanno N, De Luca A, Aldinucci D et al (2005) Gefitinib inhibits the ability of human bone marrow stromal cells to induce osteoclast differentiation: implications for the pathogenesis and treatment of bone metastasis. Endocr Relat Cancer 12(2):471–482
20. Tamama K, Fan VH, Griffith LG et al (2006) Epidermal growth factor as a candidate for ex vivo expansion of bone marrow-derived mesenchymal stem cells. Stem Cells 24 (3):686–695
21. Krampera M, Pasini A, Rigo A et al (2005) Hb-egf/her-1 signaling in bone marrow mesenchymal stem cells: inducing cell expansion and reversibly preventing multilineage differentiation. Blood 106(1):59–66
22. Liang X, Ding Y, Zhang Y et al (2015) Activation of nrg1-erbb4 signaling potentiates mesenchymal stem cell-mediated myocardial repairs following myocardial infarction. Cell Death Dis 6:e1765
23. De Boeck A, Pauwels P, Hensen K et al (2013) Bone marrow-derived mesenchymal stem cells promote colorectal cancer progression through paracrine neuregulin 1/her3 signalling. Gut 62 (4):550–560
24. Chen C, Uludag H, Wang ZX et al (2012) Macrophages inhibit migration, metabolic activity and osteogenic differentiation of human mesenchymal stem cells in vitro. Cells Tissues Organs 195(6):473–483

INDEX

Zhixiang Wang (ed.), *ErbB Receptor Signaling: Methods and Protocols*, Methods in Molecular Biology, vol. 1652, DOI 10.1007/978-1-4939-7219-7, © Springer Science+Business Media LLC 2017